Cell Cycle Regulation and Differentiation in Cardiovascular and Neural Systems

Antonio Giordano · Umberto Galderisi
Editors

Cell Cycle Regulation and Differentiation in Cardiovascular and Neural Systems

Springer

Editors
Antonio Giordano
College of Science & Technology
Sbarro Institute for Cancer Research
Temple University
N. 12th Street 1900
19122 Philadelphia, Pennsylvania
USA
giordano@temple.edu

Umberto Galderisi
Dipto. Medicina Sperimentale
Università Napoli Federico II
Sezione di Biotecnologia e Biologia Molecolare
Via Costantinopoli 16
80138 Napoli
Italy
umberto.galderisi@unina2.it

ISBN 978-1-4899-8218-6 ISBN 978-1-60327-153-0 (eBook)
DOI 10.1007/978-1-60327-153-0
Springer New York Dordrecht Heidelberg London

Printed on acid-free paper

Humana Press is part of Springer Science+Business Media (www.springer.com)

Preface

The brain and heart have been used as metaphors for thinking and feeling, for cognition and emotion. These ideas have been demonstrated to have a high scientific value. In fact, complex physiopathological relationships exist between these two organs. This book will deal with several aspects of the cardiovascular and nervous systems from a new viewpoint. It will describe the differences and similarities in their differentiation pathways with a peculiar emphasis on the role of cell cycle regulation and cell differentiation.

Cell cycle exit represents the fundamental step to trigger differentiation of cells and induction of a novel program of gene expression leading to the elaboration of a specialized phenotype. Moreover, there is evidence demonstrating that several components of the cell cycle machinery play a major role also in cell specification and differentiation both in neural and cardiovascular systems.

The differentiation process will be evaluated starting from the most early cell precursor, i.e., stem cells. The attention of readers will be focused also on transcription factors with differentiating properties and on their relationship with cell cycle regulators.

In summary, this book will offer an in-depth analysis of the differentiation process in two systems that have profound relationships with one another and, therefore, will help us to better understand their biology by providing the tools to dissect the molecular basis of pathological conditions.

The book will prompt the scientific community to perceive cell cycle regulation and differentiation under a novel and more comprehensive light.

Contents

Contributors

Marco Bacigaluppi, M.D. Neuroimmunology Unit, DIBIT and Institute of Experimental Neurology (InSpe), San Raffaele Scientific Institute, Via Olgettina 58, 20123 Milan, Italy

Elena Brini, Ph.D. Neuroimmunology Unit, DIBIT and Institute of Experimental Neurology (InSpe), San Raffaele Scientific Institute, Via Olgettina 58, 20123 Milan, Italy

Bruce A. Bunnell, Ph.D. Department of Pharmacology, Division of Gene Therapy, Center for Gene Therapy, Tulane National Primate Research Center, Tulane University Health Sciences Center, 18703 Three Rivers Road, Covington, LA 70433, USA; Center for Gene Therapy, School of Medicine, Tulane University, New Orleans, LA, USA; Division of Gene Therapy, Tulane National Primate Research Center, Tulane University Health Sciences Center, Covington, LA, USA

Erica Butti, Ph.D. Neuroimmunology Unit, DIBIT and Institute of Experimental Neurology (InSpe), San Raffaele Scientific Institute, Via Olgettina 58, 20123 Milan, Italy

Gérard Cabello, Ph.D. INRA UMR 866 Différenciation Cellulaire et Croissance, (INRA, Université Montpellier 1, Université Montpellier 2), 2, place Viala, 34060 Montpellier Cedex 1, France

François Casas, Ph.D. INRA UMR 866 Différenciation Cellulaire et Croissance, (INRA, Université Montpellier 1, Université Montpellier 2), 2, place Viala, 34060 Montpellier Cedex 1, France

Antonino Cascino, Ph.D.† Department of Experimental Medicine, Section of Biotechnology and Molecular Biology, Second University of Naples, Via L. De Crecchio, 7, Naples, Italy

Marilena Cipollaro, M.D. Department of Experimental Medicine, Section of Biotechnology and Molecular Biology, Second University of Naples, Via L. De Crecchio, 7, Naples, Italy

†Deceased

Chiara Cossetti, B.Sc. Neuroimmunology Unit, DIBIT and Institute of Experimental Neurology (InSpe), San Raffaele Scientific Institute, Via Olgettina 58, 20123 Milan, Italy

Melania Cusimano, B.Sc. Neuroimmunology Unit, DIBIT and Institute of Experimental Neurology (InSpe), San Raffaele Scientific Institute, Via Olgettina 58, 20123 Milan, Italy

Amalia Forte, Ph.D. Department of Experimental Medicine, Section of Biotechnology and Molecular Biology, Second University of Naples, Via L. De Crecchio, 7, Naples, Italy

Umberto Galderisi, Ph.D. Department Experimental Medicine, Second University of Naples, Via L. De Crecchio 7, 80138 Napoli, Italy

Cristina Giacinti, Ph.D. Sbarro Institute for Cancer Research and Molecular Medicine, College of Science and Technology, Temple University, Philadelphia, PA, USA; Department of Histology and Medical Embryology, University "La Sapienza", Rome, Italy; Department of Human Pathology and Oncology, University of Siena, Siena, Italy

Antonio Giordano, M.D. Sbarro Institute for Cancer Research and Molecular Medicine, College of Science and Technology, Temple University, Philadelphia, PA, USA; Department of Histology and Medical Embryology, University "La Sapienza", Rome, Italy; Department of Human Pathology and Oncology, University of Siena, Siena, Italy

Ryoichiro Kageyama, M.D., Ph.D. Institute for Virus Research, Kyoto University, Shogoin-Kawahara, Sakyo-ku, Kyoto 606-8507, Japan

Shihuan Kuang, Ph.D. Department of Animal Sciences, Purdue University, 901 West State Street, West Lafayette, IN 47906, USA

Ronglih Liao, Ph.D. Cardiac Muscle Research Laboratory, Cardiovascular Division, Department of Medicine, Brigham and Women's Hospital, Harvard Medical School, 77 Avenue Louis Pasteur, New Research Building, NRB 431, Boston, MA 02115, USA

Gianvito Martino, M.D. Neuroimmunology Unit, DIBIT and Institute of Experimental Neurology (InSpe), San Raffaele Scientific Institute, Via Olgettina 58, 20123 Milan, Italy

W. Robb MacLellan, Ph.D. The Cardiovascular Research Laboratory, David Geffen School of Medicine at UCLA, 675 C.E. Young Dr., MRL 3-645, Los Angeles, CA 90095-1760, USA

Rebecca Matsas, Ph.D. Laboratory of Cellular and Molecular Neurobiology, Hellenic Pasteur Institute, 127. Vas. Sofias Avenue, Athens 11521, Greece

Toshiyuki Ohtsuka, M.D., Ph.D. Institute for Virus Research, Kyoto University, Shogoin-Kawahara, Sakyo-ku, Kyoto 606-8507, Japan

Stefano Pluchino, M.D. Neuroimmunology Unit, DIBIT and Institute of Experimental Neurology (InSpe), San Raffaele Scientific Institute, Via Olgettina 58, 20123 Milan, Italy

Panagiotis K. Politis, Section of Histology, Center of Basic Research, Biomedical Research Foundation, Academy of Athens, Soranou Efessiou 4, 11527 Athens, Greece

Cynthia Ripoll, Department of Pharmacology, Division of Gene Therapy, Center for Gene Therapy, Tulane National Primate Research Center, Tulane University Health Sciences Center, 18703 Three Rivers Road, Covington, LA 70433, USA; Center for Gene Therapy, School of Medicine, Tulane University, New Orleans, LA, USA

Michael A. Rudnicki, Ph.D. The Sprott Center for Stem Cell Research, Ottawa Health Research Institute, Regenerative Medicine Program, 501 Smyth Road, Ottawa, ON, Canada, K1H 8L6; Department of Cellular and Molecular Medicine, Faculty of Medicine, University of Ottawa, 451 Smyth Road, Ottawa, ON, Canada, K1H 8M5

Regina L. Sohn, M.D., Ph.D. Cardiac Muscle Research Laboratory, Cardiovascular Division, Department of Medicine, Brigham and Women's Hospital, Harvard Medical School, 77 Avenue Louis Pasteur, New Research Building, NRB 431, Boston, MA 02115, USA

Dimitra Thomaidou, Ph.D. Laboratory of Cellular and Molecular Neurobiology, Hellenic Pasteur Institute, 127. Vas. Sofias Avenue, Athens 11521, Greece

Chantal Wrutniak-Cabello, Ph.D. INRA UMR 866 Différenciation Cellulaire et Croissance, (INRA, Université Montpellier 1, Université Montpellier 2), 2, place Viala, 34060 Montpellier Cedex 1, France

Lucia Zanotti, B.Sc. Neuroimmunology Unit, DIBIT and Institute of Experimental Neurology (InSpe), San Raffaele Scientific Institute, Via Olgettina 58, 20123 Milan, Italy

Part I
Introduction to Cell Cycle and Differentiation

Chapter 1
Short Introduction to the Cell Cycle

Antonio Giordano and Umberto Galderisi

Abstract Molecular pathways that regulate cell growth and differentiation are now beginning to be understood. This is mainly due to the identification of molecules that orchestrate the cell cycle. Cyclins and the cyclin-dependent kinases (CDKs) are the major players in the control of cell cycle progression. Cyclins do not have enzymatic activity and CDKs are inactive without a partner cyclin. The CDKs regulate the function of multiple proteins involved in DNA replication and mitosis by phosphorylating them at specific regulatory sites, activating some and inhibiting others to coordinate their activities. In this way cyclin/CDK complexes coordinate an ordered passage from a cell cycle phase to the next one. Multiple levels of regulation of cyclin/CDK complexes are present in "cell machinery" to obtain a tight control of cell cycle progression. In this chapter we will address these items.

Keywords Cell cycle · Cyclins · Cyclin-dependent kinases (CDKs) · Cyclin kinase inhibitors (CKIs) · Retinoblastoma proteins

1 Introduction

In order for an organism to develop, two kinds of processes are fundamental: cell division and cell differentiation. Both of these formative pathways must be carefully regulated and coordinated for normal growth to occur. The mechanisms that control cell growth and differentiation are now beginning to be understood. This is mainly due to the identification of molecules that orchestrate the cell cycle.

What is a "cell cycle"? It is the series of ordered events in a eukaryotic cell between one cell division and the next. During this time period a cell duplicates its chromosomes and then gives rise to two daughter cells, each with the same DNA

U. Galderisi (✉)
Department Experimental Medicine, Second University of Naples, Via L. De Crecchio 7, 80138 Napoli, Italy
e-mail: umberto.galderisi@unina2.it

A. Giordano, U. Galderisi (eds.), *Cell Cycle Regulation and Differentiation in Cardiovascular and Neural Systems*, DOI 10.1007/978-1-60327-153-0_1,

content. Cell cycle can be described in terms of four distinct phases: G_1 phase, S phase, G_2 phase, and M phase. During G_1 phase a strong increase in RNA and protein synthesis occurs before cell duplicates its DNA (S phase). During G_2 phase DNA replication is completed and cell increases in size in order to give rise to the two daughter cells. Following G_2 phase, cell enters into mitosis (M phase), during which nuclear division occurs. At the end of this process cell splits into two new cells (cytokinesis). Cells can also enter in a quiescent state (G_0 phase) where they arrest cellular division for long periods of times even indefinitely. This exit from cell cycle represents the fundamental step to trigger differentiation of cells and induction of a novel program of gene expression leading to the elaboration of a specialized phenotype [1, 2].

On the other hand, cells can permanently arrest division due to age or accumulated DNA damage and enter the "senescent state." Senescence is the alternative to apoptosis, the programmed cell death, that eliminates "damaged" cells [3, 4].

Key molecules controlling cell cycle progression are the cyclins and the cyclin-dependent kinases (CDKs). Cyclins are the regulatory subunits and CDKs the catalytic subunits of an activated heterodimer. Cyclins do not have enzymatic activity and CDKs are inactive without a partner cyclin. The CDKs regulate the function of multiple proteins involved in DNA replication and mitosis by phosphorylating them at specific regulatory sites, activating some and inhibiting others to coordinate their activities. In this way cyclin/CDK complexes coordinate an ordered passage from a cell cycle phase to the next one [5].

Multiple levels of regulation of cyclin/CDK complexes are present in "cell machinery" to obtain a tight control of cell cycle progression. Such regulation relies on (1) controlled expression and degradation of cyclins, (2) activating and inhibitory phosphorylation and dephosphorylation of the CDKs, and (3) expression and destruction of inhibitory proteins, the CDK inhibitors (CKIs), that associate with CDK/cyclin complexes [5–11].

Cell concentration of cyclins changes during cell cycle since they are produced and degraded as needed. For example, growth factors can promote G_1 cyclin production, then to pass to the next phase of cycle the cyclins have to be degraded by proteases [12–14].

CDKs are regulated by phosphorylation and dephosphorylation of critical residues. Important targets of this regulation are some threonine and tyrosine residues present in the ATP-binding domain. The cyclin kinase inhibitors (CKI), also called CDKIs, negatively regulate CDK activity. There are two CKI classes. The first one includes $p21^{Cip1}$, $p27^{Kip1}$, and $p57^{Kip2}$. These bind to G_1/S CDKs; second class is the INK4 family (p15/INK4B, p16/INK4A, p18/INK4C, and p19/INK4D). These act on cyclin D complexed to CDK4 or CDK6 [15–18].

2 The G_1 Phase

In a cell, following extrinsic or intrinsic cell growth signaling, the cyclin/CDK complexes of G_1 phase are activated to drive the cell in S phase. This is accomplished by promoting the expression of transcription factors that in turn induce the expression of S cyclins and of enzymes required for DNA replication. The G_1 cyclin/CDK complexes also promote the degradation of molecules that function as S phase inhibitors by targeting them for ubiquitination, that is, a proteolytic process that degrades proteins. The G_1/S transition is the key step for cell cycle progression and is controlled by D-type cyclins/CDK4, D-type cyclins/CDK6, which act in mid-G_1, and by cyclin E/CDK2, which operates in late G_1 [1, 19–23].

Unlike many other cyclins, the level of D-type cyclins (cyclin D1, D2, and D3) does not change during the cell cycle, but rather their levels are controlled mainly by mitogens. Several evidences have proved that D-type cyclins are functionally redundant. Nevertheless, each of them has unique tissue-specific functions [24, 25].

One key substrate of cyclin D/CDK complexes is the nuclear tumor suppressor pRb (and its related proteins pRb2/p130 and p107), which is phosphorylated on serine and threonine residues during G_1 phase. pRb phosphorylation results in the liberation of E2F transcription factors, whose activity is required for entry into S phase [1, 19–23].

There are at least eight members of E2F transcription factor family [26]. In addition, several E2F isoforms are generated by alternative splicing. E2F1–E2F5 are the most studied members of this family and a have well-recognized role in cell cycle regulation through interaction with RB family proteins. These E2F proteins can recognize specific DNA *cis*-elements forming heterodimers with partially related proteins called DP. In this way they activate transcription from genes responsible for cell cycle control, initiation of replication, and DNA synthesis (DHFR, thymidine kinase, HsOrc1, and DNA polymerase alpha, PCNA, cyclin E, cyclin A, cdc2), as well as several proto-oncogenes such as c-myb, B-myb, and c-myc [26, 27].

Cell cycle progression is strictly associated with Rb activity. The cell responds to mitogenic stimuli and progresses through the various phases of the cell cycle only during a limited phase of its cycle. In fact, the cell needs stimulation only during the first two-thirds of its G_1 phase where it may decide to continue its advance and complete its cell cycle. This point is termed "restriction point" (R point) or "checkpoint"; it is a central event in normal cellular proliferation control. It has been demonstrated that pRb is the molecular device that serves as the R point switch [28–30].

pRb is hypophosphorylated in resting G_0 cells, is increasingly phosphorylated during progression through G_1, and is maintained in a hyperphosphorylated state until late mitosis [31, 32]. pRb phosphorylation seems to be related to mitogenic signals, which converge on the cell cycle machinery, represented by the cyclin D1/CDK4 (CDK6) complex in the early and mid-G_1, and composed of cyclin E/CDK2 in late G_1 [1, 20, 22].

The phosphorylation Rb protein releases E2F transcription factors and allows the expression of proteins required for S phase progression. Among these the expression of cyclin E determines a positive feedback of Rb phosphorylation, since cyclin E in complex with CDK2 will continue to phosphorylate Rb, contributing to an irreversible transition into the S phase [1, 20, 22].

The G_1/S progression is regulated also by members of CKI family. In fact, formation of active complexes among D-type cyclins and CDK4 and CDK6 is influenced by the INK4 proteins, namely $p15^{INK4b}$, $p16^{INK4a}$, $p18^{INK4c}$, and $p19^{INK4d}$. These proteins bind to the catalytic subunits and inhibit the association of D-type cyclins with CDKs [5, 15, 16].

Proteins of the second class of CKI ($p21^{CIP1}$, $p27^{KIP1}$, and $p57^{KIP2}$) are broad-spectrum inhibitors of different cyclin–CDK complexes [10, 15, 16, 33].

Delineating the mode of action for this protein family is more complicated, since these proteins can have a role both in promoting cyclin/CDK complex formation and in inhibiting their activity. It has been observed that $p21^{CIP1}$ levels increase immediately following mitogenic stimulation of serum-starved human fibroblasts, before declining at the G_1/S boundary [34]. Moreover, the assembly of cyclin D/CDK4 correlates concomitantly with the binding of CIP–KIP proteins and is impaired in mouse embryonic fibroblasts lacking either $p21^{CIP1}$ or $p27^{KIP1}$. In order to explain this double role of CIP–KIP family member, a titration model has been proposed, where cyclin D/CDK4/6 complexes act as activators of cyclin E/CDK2 complexes by titrating CIP/KIP proteins away from, and thus releasing the inhibition of, cyclin E/CDK2 complexes [2, 15, 16, 33, 35].

Cyclin D/CDK4/6 and cyclin E/CDK2 complexes are regulated by several proteins, including Wee1/Myt1/Mik1 kinases, CDK-activating kinase (CAK), and Cdc25 phosphatase. CAK is a complex that is composed of a catalytic subunit, p40MO15, also called CDK7, a regulatory subunit, cyclin H, and an assembly factor MAT1. CAK adds an activating phosphate to the CDK proteins, while Wee1 adds an inhibitory phosphate. The presence of both activating and inhibitory phosphates inactivates cyclins/CDKs complexes. To activate the complexes, component of the Cdc25 phosphatase family activates CDK/cyclin complexes deleting the Wee1 phosphorylation [16, 36–39].

In summary, biochemical events are protein synthesis and degradation, primarily phosphorylation, dephosphorylation, promoting the complex formation and/or inhibiting its activity, with the overall mission to either prevent or induce a new cell cycle through the Rb pathway.

3 The S and G_2 Phases

In S phase of cell cycle DNA replication occurs. Main regulator of this and related events is the cyclin A. In fact, cyclin A is thought to contribute to the G_1/S transition, S phase progression, and G_2/M transition. This protein is expressed at low levels in G_1, then its levels increase from S phase through G_2, finally during M phase it declines again [40, 41].

Support for a key role of cyclin A in S phase comes from the observations that microinjection of neutralizing antibodies against cyclin A resulted in a failure to replicate DNA in fibroblasts. Moreover, cyclin A knockout drosophila embryos cannot enter mitosis. Cyclin A associates with two CDKs: CDK2 and Cdc2 (now called CDK1). Cyclin A/CDK2 complex is present in both S and G_2 phase, cyclin A/Cdc2 activity is present only in G_2 [42–44].

Several studies suggest that cyclin A/CDK2 complex is required for S phase progression, and cyclin A/Cdc2 activity is required for G_2/M progression. In fact, in cell-free extracts CDK2 is essential for DNA synthesis, while mouse cells expressing temperature-sensitive mutated Cdc2 arrest specifically in G_2 phase. A few targets for CDK2 and Cdc2 kinases have been identified; among these are histone proteins. This is in agreement with the observation that cyclin A/CDKs phosphorylate proteins that make up the pre-replication complexes assembled during G_1 phase on DNA replication origins. These phosphorylation events activate already assembled pre-replication complex and, at the same time, prevent formation of new complexes to ensure that every region of DNA will be replicated once and only once [2, 42–44].

The RB protein is also a target of cyclin A/CDK complexes. It has been demonstrated that RB, with a mutation in its phosphorylation sites, was capable of blocking progression through S phase, suggesting that the continued hyperphosphorylation of RB could be a key event for cell cycle progression. It is interesting to note that RB represses both cyclin A and Cdc2 expression; this could create a fine "feedback regulation loop" [2, 20, 22].

4 The M Phase

During S and G_2 phases are synthesized mitotic cyclin/CDK complexes that are kept inactive till entry into M phase when they promote the initiation of mitosis by stimulating downstream proteins involved in chromosome condensation and mitotic spindle assembly [45, 46].

Cyclin B/Cdc2 complex is the main component of the mitotic promoting factor (MPF), whose activity triggers mitotic entry. The expression of cyclin B changes through the cell cycle: cyclin B is initially synthesized during S phase, increases in G_2, and declines by proteasome degradation during the anaphase of mitotic division. Cyclin B1, B2, and B3 are the three cyclin B isoforms so far described [47–50].

The activity of MPF complex is determined through phosphorylation of Cdc2 as well as subcellular localization of the cyclin B/Cdc2 complex. Cyclins B1 and B2 are cytoplasmic till mitotic division, whereas cyclin B3 appears to be nuclear. At the beginning of mitosis, cyclin B1 and B2 translocate to the nucleus prior to nuclear-envelope breakdown, this nuclear localization is necessary for normal cyclin B activity [48, 51, 52].

During G_2 phase, the Wee1 and Myt1 kinases phosphorylate and inactivate Cdc2. At mitotic onset, these inhibitory phosphorylations are removed by Cdc25 phosphatases, such as Cdc25A, Cdc25B, and Cdc25C. Also Polo-like kinase-1 protein has a role in mitotic entry through activation of Cdc25C and nuclear translocation of Cdc25C and cyclin B. Moreover, as for other CDKs, CAK phosphorylation is required for complete activation of Cdc2 [40, 53–57].

As described, progression from G_2 to M phase is driven by activation of the cyclin B/Cdc2. However, this complex must be active from the prophase to the metaphase of mitotic division. Subsequent entry into the anaphase critically relies on the sudden destruction of the cyclin B/Cdc2 activity. At the end of the metaphase, the anaphase-promoting complex (APC) destroys cyclin B to allow mitosis to proceed. The APC, a polyprotein complex with ubiquitin ligase activity, recruits cyclin B, causes its ubiquitination, and thus targets it for degradation by the 26S proteasome. To complete mitosis, APC promotes also degradation of structural proteins associated with the chromosomal kinetochore [58–60].

5 Cell Cycle Checkpoints

In normal cells, there are several "checkpoints" that allow the cell to determine if conditions are favorable for the cell cycle to continue. If a cell fails to meet the requirements of a phase it will not be allowed to proceed to the next phase until the requirements have been met. Alternatively, cells may undergo apoptosis with or without growth arrest or enter an irreversible G_0 state, called senescence [61, 62].

Several checkpoints are designed to ensure that damaged or incomplete DNA is not inherited by daughter cells. During a life cycle, continuous DNA damages occur into cells. These are caused either by endogenous phenomena, such as the activity of reactive oxygen species (ROS), DNA replication errors, stalling of DNA replication forks, or by exogenous factors, such as ionizing radiation, UV lights, chemicals. Check on DNA status occurs at the end of the G_1 phase, G_2 phase, and after replication (S phase). Moreover, at the end of the M phase a checkpoint is present to stop cytokinesis in case the chromosomes are not properly aligned on the mitotic spindle (spindle checkpoint). A delay in activation or inactivation of a particular set of cyclin/CDK complex is the way through which checkpoints arrest cell cycle progression [61, 62].

6 The G_1 Checkpoint

The strict regulation of CDK phosphorylation on tyrosine and residues is a key event of G_1 checkpoint. For example, UV lights induce a phosphorylation on Tyr 17 of CDK4, and this blocks the activity of cyclin/CDK complex and a G_0/G_1 arrest will occur. During cell cycle, CDK2 is inhibited by phosphorylation on Tyr 15, and Cdc25 phosphatases relieve this block. DNA damaging agents interfere with CDC25 enzymes and maintain CDK2 in its phosphorylated form to arrest cell cycle progression [63–67].

G_1 arrest associated with DNA damages is heavily dependent on p53-related pathways. This is a transcription factor that is mutated in a high percentage of human tumors. Several carcinogens that cause DNA damages (such as ionizing radiations) trigger a strong p53 activation and G_1 arrest. Cells lacking a functional p53 enter S phase regardless of radiations [68–73].

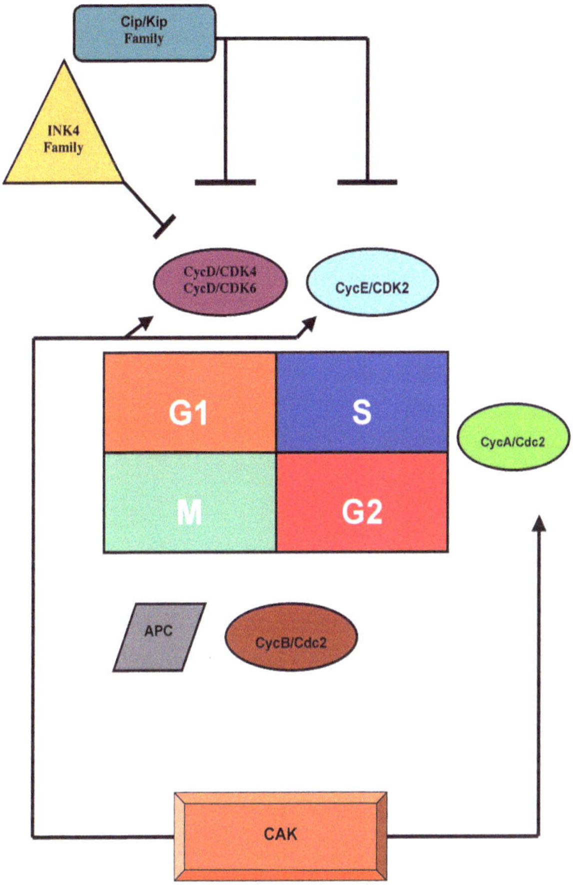

Fig. 1.1 Schematic diagram of cell cycle. Cyclins/CDKs and some of the proteins involved in their regulation are depicted in the picture. Cyclin/CDK complexes promote cell cycle progression through phosphorylation of several targets. Cyclin/CDK complexes are inactivated by cyclin kinase inhibitors, such as those belonging to INK4 family and to Cip/Kip family. Cyclin kinase activators (CAK) act on cyclin/CDKs and promote cell cycle progression. APC: anaphase-promoting complex

The promoter of $p21^{CIP1}$ genes has binding sites for p53 transcription factor. For this reason, p53 activation can induce $p21^{CIP1}$ gene expression, besides several other genes. $p21^{CIP1}$ can inhibit the activity of cyclin/CDK complexes acting in G_1 phase and arrest cell cycle. In agreement, the overexpression of $p21^{CIP1}$ in cycling cells determines G_1 arrest, while cells lacking this protein exhibit a reduced arrest in G_1 phase, following treatment with ionizing radiations [33, 74–76].

Also the proteins of INK4 family play a role in regulation of cell cycle progression. However, it is not clear if they contribute to cell cycle checkpoint.

7 The S Checkpoint

Few minutes following exposure to DNA damaging agents, eukaryotic cells in cultures show a significant reduction in DNA synthesis. In detail, there is a quick suppression of formation of new DNA replicons and, subsequently, the arrest of initiated replicons. The block of DNA synthesis is accomplished through the inhibition of cyclin A/CDK2 complex. ATM gene has a key role in S checkpoint. This gene is mutated in patients suffering from ataxia-telangiectasia. Mutation of ATM causes defective cell cycle checkpoint along with reduced capacity for repair of DNA double-strand breaks and abnormal apoptosis, all of which contribute to the major features of ataxia-telangiectasia including genome instability, increased cancer risk, and neurodegeneration [29, 77–80].

8 The G_2 Checkpoint

As for the others, also in G_2 checkpoint, regulation of CDK phosphorylation on tyrosine residues is a key event. For example, treatment of cell cultures with DNA damaging agents induces phosphorylation on Thr 14 and Tyr 15 of cdc2 protein, resulting in the inhibition of cyclin B/cdc2 activity. In cell lines harboring mutated cdc2 that cannot be phosphorylated, cell cycle arrest in G_2 phase is partially eliminated. The phosphorylation/dephosphorylation of cdc2 relies on a complex pathway including the Cdc25 phosphatases. These proteins have to be phosphorylated to be active on cdc2. In cells treated with DNA damaging agents the Cdc25 enzymes are not phosphorylated, which in turn maintain cdc2 in inactive state and block cell cycle progression [81–85].

Inactivation of cdc2 can be reached also by regulation of cyclin B/cdc2 cellular distribution. During S/G_2 phases, this complex is present in cytoplasm, then it is transferred to nucleus as cell traverses G_2/M phases. Ionizing radiation treatment of cell cultures can induce an arrest of cell cycle progression and accumulation of cyclin B/cdc2 into cytoplasms [86, 87].

9 The Spindle Checkpoint

This checkpoint ensures proper chromosome segregation, avoiding aneuploidia. The spindle checkpoint delays anaphase onset until all chromosomes are correctly attached in a bipolar fashion to the mitotic spindle. The core spindle checkpoint proteins are Mad and Bub proteins that were identified in budding yeast by genetic screens for mutants that failed to arrest in mitosis when the spindle was destroyed [88–91].

Complex molecular interactions regulate both chromosome attachment and formation of microtubules. Checkpoint proteins can monitor these interactions. In fact, lack of microtubule attachment elicits the checkpoint response. This is due to a careful screening of tension forces. Tension is established across the sister kinetochores by the pulling forces of the spindle when a chromosome is attached to microtubules from opposite poles. Experiment aiming to the laser ablation of the last unattached kinetochore induced the elimination of the checkpoint-dependent arrest and the cell reached anaphase without a proper completion of mitosis [92–94].

Spindle checkpoint has a major role also in controlling duplication of centrosomes. These organelles duplicate before cells enter M phase. Failure in duplication event can determine polyploidy because most cells return to interphase without division. On the other site, if centrosomes duplicate more than once in a cell cycle, chromosomes are unequally distributed to daughter cells because of multipolar spindle assembly [92–94].

References

1. Pucci B, Kasten M, Giordano A (2000) Cell cycle and apoptosis. Neoplasia 2(4):291–299
2. Harper JV, Brooks G (2005) The mammalian cell cycle: an overview. Methods Mol Biol 296:113–153
3. Collado M, Blasco MA, Serrano M (2007) Cellular senescence in cancer and aging. Cell 130(2):223–233
4. Crews DE (2007) Senescence, aging, and disease. J Physiol Anthropol 26(3):365–372
5. Donjerkovic D, Scott DW (2000) Regulation of the G1 phase of the mammalian cell cycle. Cell Res 10(1):1–16
6. Chiariello M, Gomez E, Gutkind JS (2000) Regulation of cyclin-dependent kinase (Cdk) 2 Thr-160 phosphorylation and activity by mitogen-activated protein kinase in late G1 phase. Biochem J 349 Pt 3:869–876
7. Lolli G, Johnson LN (2005) CAK-cyclin-dependent activating kinase: a key kinase in cell cycle control and a target for drugs? Cell Cycle 4(4):572–577
8. Hu X et al (2002) Ubiquitin/proteasome-dependent degradation of D-type cyclins is linked to tumor necrosis factor-induced cell cycle arrest. J Biol Chem 277(19): 16528–16537
9. Perkins ND (2002) Not just a CDK inhibitor: regulation of transcription by p21(WAF1/CIP1/SDI1). Cell Cycle 1(1):39–41
10. Slingerland J, Pagano M (2000) Regulation of the cdk inhibitor p27 and its deregulation in cancer. J Cell Physiol 183(1):10–17
11. Yang W et al (2007) Myostatin induces cyclin D1 degradation to cause cell cycle arrest through a phosphatidylinositol 3-kinase/AKT/GSK-3 beta pathway and is antagonized by insulin-like growth factor 1. J Biol Chem 282(6):3799–3808
12. Minshull J (1993) Cyclin synthesis: who needs it? Bioessays 15(3):149–155
13. Hershko A (1999) Mechanisms and regulation of the degradation of cyclin B. Philos Trans R Soc Lond B Biol Sci 354(1389):1571–1575; discussion 1575–1576
14. Alao JP (2007) The regulation of cyclin D1 degradation: roles in cancer development and the potential for therapeutic invention. Mol Cancer 6:24
15. De Clercq A, Inze D (2006) Cyclin-dependent kinase inhibitors in yeast, animals, and plants: a functional comparison. Crit Rev Biochem Mol Biol 41(5):293–313
16. Fischer PM, Endicott J, Meijer L (2003), Cyclin-dependent kinase inhibitors. Prog Cell Cycle Res 5:235–248
17. Santamaria D, Ortega S (2006) Cyclins and CDKS in development and cancer: lessons from genetically modified mice. Front Biosci 11:1164–1188

18. Sridhar J, Akula N, Pattabiraman N (2006) Selectivity and potency of cyclin-dependent kinase inhibitors. Aaps J 8(1):E204–221
19. Harbour JW et al (1999) Cdk phosphorylation triggers sequential intramolecular interactions that progressively block Rb functions as cells move through G1. Cell 98(6): 859–869
20. Kasten MM, Giordano A (1998) pRb and the cdks in apoptosis and the cell cycle. Cell Death Differ 5(2):132–140
21. Sherr CJ (1994) G1 phase progression: cycling on cue. Cell 79(4):551–555
22. Stiegler P, Kasten M, Giordano A (1998) The RB family of cell cycle regulatory factors. J Cell Biochem Suppl 30–31:30–36
23. Watanabe Y et al (1999) pRb phosphorylation is regulated differentially by cyclin-dependent kinase (Cdk) 2 and Cdk4 in retinoic acid-induced neuronal differentiation of P19 cells. Brain Res 842(2):342–350
24. Ortega S, Malumbres M, Barbacid M (2002) Cyclin D-dependent kinases, INK4 inhibitors and cancer. Biochim Biophys Acta 1602(1):73–87
25. Motokura T, Arnold A (1993) Cyclin D and oncogenesis. Curr Opin Genet Dev 3(1):5–10
26. Bracken AP et al (2004) E2F target genes: unraveling the biology. Trends Biochem Sci 29(8):409–417
27. Lavia P, Jansen-Durr P (1999) E2F target genes and cell-cycle checkpoint control. Bioessays 21(3):221–230
28. Lundberg AS, Weinberg RA (1999) Control of the cell cycle and apoptosis. Eur J Cancer 35(14):1886–1894
29. Blagosklonny MV, Pardee AB (2002) The restriction point of the cell cycle. Cell Cycle 1(2):103–110
30. DelSal G, Loda M, Pagano M (1996) Cell cycle and cancer: critical events at the G1 restriction point. Crit Rev Oncog 7(1–2):127–142
31. Claudio PP et al (1996) Functional analysis of pRb2/p130 interaction with cyclins. Cancer Res 56(9):2003–2008
32. Weinberg RA (1995) The retinoblastoma protein and cell cycle control. Cell 81(3):323–330
33. Harper JW et al (1995) Inhibition of cyclin-dependent kinases by p21. Mol Biol Cell 6(4):387–400
34. Li Y et al (1994) Cell cycle expression and p53 regulation of the cyclin-dependent kinase inhibitor p21. Oncogene 9(8):2261–2268
35. Pavletich NP (1999) Mechanisms of cyclin-dependent kinase regulation: structures of Cdks, their cyclin activators, and Cip and INK4 inhibitors. J Mol Biol 287(5):821–828
36. Kellogg DR (2003) Wee1-dependent mechanisms required for coordination of cell growth and cell division. J Cell Sci 116(Pt 24):4883–4890
37. Perry JA, Kornbluth S (2007) Cdc25 and Wee1: analogous opposites? Cell Div 2:12
38. Rudolph J (2007) Cdc25 phosphatases: structure, specificity, and mechanism. Biochemistry 46(12):3595–3604
39. Wells NJ et al (1999) The C-terminal domain of the Cdc2 inhibitory kinase Myt1 interacts with Cdc2 complexes and is required for inhibition of G(2)/M progression. J Cell Sci 112(Pt 19):3361–3371
40. Strausfeld UP et al (1996) Both cyclin A and cyclin E have S-phase promoting (SPF) activity in Xenopus egg extracts. J Cell Sci 109(Pt 6):1555–1563
41. Woo RA, Poon RY (2003) Cyclin-dependent kinases and S phase control in mammalian cells. Cell Cycle 2(4):316–324
42. Hayashi S, Yamaguchi M (1999) Kinase-independent activity of Cdc2/cyclin A prevents the S phase in the Drosophila cell cycle. Genes Cells 4(2):111–122
43. Jacobs HW, Keidel E, Lehner CF (2001) A complex degradation signal in Cyclin A required for G1 arrest, and a C-terminal region for mitosis. Embo J 20(10):2376–2386
44. Girard F et al (1991) Cyclin A is required for the onset of DNA replication in mammalian fibroblasts. Cell 67(6):1169–1179

45. Kishimoto T, Okumura E (1997) In vivo regulation of the entry into M-phase: initial activation and nuclear translocation of cyclin B/Cdc2. Prog Cell Cycle Res 3:241–249
46. Nurse P (1994) Ordering S phase and M phase in the cell cycle. Cell 79(4):547–550
47. Tokumoto T et al (1997) Initiation of cyclin B degradation by the 26S proteasome upon egg activation. J Cell Biol 138(6):1313–1322
48. Tschop K et al (2006) Human cyclin B3. mRNA expression during the cell cycle and identification of three novel nonclassical nuclear localization signals. FEBS J 273(8): 1681–1695
49. Bellanger S, de Gramont A, Sobczak-Thepot J (2007) Cyclin B2 suppresses mitotic failure and DNA re-replication in human somatic cells knocked down for both cyclins B1 and B2. Oncogene 26(51):7175–7184
50. Chesnel F et al (2006) Cyclin B dissociation from CDK1 precedes its degradation upon MPF inactivation in mitotic extracts of Xenopus laevis embryos. Cell Cycle 5(15):1687–1698
51. Gong D et al (2007) Cyclin A2 regulates nuclear-envelope breakdown and the nuclear accumulation of cyclin B1. Curr Biol 17(1):85–91
52. Walsh S, Margolis SS, Kornbluth S (2003) Phosphorylation of the cyclin b1 cytoplasmic retention sequence by mitogen-activated protein kinase and Plx. Mol Cancer Res 1(4):280–289
53. Booher RN, Holman PS, Fattaey A (1997) Human Myt1 is a cell cycle-regulated kinase that inhibits Cdc2 but not Cdk2 activity. J Biol Chem 272(35):22300–22306
54. Berry LD, Gould KL (1996) Regulation of Cdc2 activity by phosphorylation at T14/Y15. Prog Cell Cycle Res 2:99–105
55. Hoffmann I, Karsenti E (1994) The role of cdc25 in checkpoints and feedback controls in the eukaryotic cell cycle. J Cell Sci Suppl 18:75–79
56. Jessus C, Ozon R (1995) Function and regulation of cdc25 protein phosphate through mitosis and meiosis. Prog Cell Cycle Res 1:215–228
57. Mueller PR et al (1995) Myt1: a membrane-associated inhibitory kinase that phosphorylates Cdc2 on both threonine-14 and tyrosine-15. Science 270(5233):86–90
58. Baker DJ et al (2007) Mitotic regulation of the anaphase-promoting complex. Cell Mol Life Sci 64(5):589–600
59. Fang G, Yu H, Kirschner MW (1999) Control of mitotic transitions by the anaphase-promoting complex. Philos Trans R Soc Lond B Biol Sci 354(1389):1583–1590
60. Page AM, Hieter P (1999) The anaphase-promoting complex: new subunits and regulators. Annu Rev Biochem 68:583–609
61. Pietenpol JA, Stewart ZA (2002) Cell cycle checkpoint signaling: cell cycle arrest versus apoptosis. Toxicology 181–182:475–481
62. van Vugt MA, Bras A, Medema RH (2005) Restarting the cell cycle when the checkpoint comes to a halt. Cancer Res 65(16):7037–7040
63. Peeper DS, van der Eb AJ, Zantema A (1994) The G1/S cell-cycle checkpoint in eukaryotic cells. Biochim Biophys Acta 1198(2–3):215–230
64. Little JB (1968) Delayed initiation of DNA synthesis in irradiated human diploid cells. Nature 218(5146):1064–1065
65. Gu Y, Rosenblatt J, Morgan DO (1992) Cell cycle regulation of CDK2 activity by phosphorylation of Thr160 and Tyr15. EMBO J 11(11):3995–4005
66. Sebastian B, Kakizuka A, Hunter T (1993) Cdc25M2 activation of cyclin-dependent kinases by dephosphorylation of threonine-14 and tyrosine-15. Proc Natl Acad Sci U S A 90(8): 3521–3524
67. Terada Y et al (1995) Requirement for tyrosine phosphorylation of Cdk4 in G1 arrest induced by ultraviolet irradiation. Nature 376(6538):358–362
68. Levine AJ, Momand J, Finlay CA (1991) The p53 tumour suppressor gene. Nature 351(6326):453–456
69. Kastan MB et al (1991) Participation of p53 protein in the cellular response to DNA damage. Cancer Res 51(23 Pt 1):6304–6311

70. Maltzman W, Czyzyk L (1984) UV irradiation stimulates levels of p53 cellular tumor antigen in nontransformed mouse cells. Mol Cell Biol 4(9):1689–1694
71. Campbell C et al (1993) Wavelength specific patterns of p53 induction in human skin following exposure to UV radiation. Cancer Res 53(12):2697–2679
72. Di Leonardo A et al (1994) DNA damage triggers a prolonged p53-dependent G1 arrest and long-term induction of Cip1 in normal human fibroblasts. Genes Dev 8(21):2540–2551
73. Nelson WG, Kastan MB (1994) DNA strand breaks: the DNA template alterations that trigger p53-dependent DNA damage response pathways. Mol Cell Biol 14(3):1815–1823
74. Xiong Y et al (1993) p21 is a universal inhibitor of cyclin kinases. Nature 366(6456):701–704
75. Harper JW et al (1993) The p21 Cdk-interacting protein Cip1 is a potent inhibitor of G1 cyclin-dependent kinases. Cell 75(4):805–816
76. Deng C et al (1995) Mice lacking p21CIP1/WAF1 undergo normal development, but are defective in G1 checkpoint control. Cell 82(4):675–684
77. Gottifredi V, Prives C (2005) The S phase checkpoint: when the crowd meets at the fork. Semin Cell Dev Biol 16(3):355–368
78. Painter RB, Young BR (1976) Formation of nascent DNA molecules during inhibition of replicon initiation in mammalian cells. Biochim Biophys Acta 418(2):146–153
79. Painter RB, Young BR (1980) Radiosensitivity in ataxia-telangiectasia: a new explanation. Proc Natl Acad Sci U S A 77(12):7315–7317
80. Beamish H et al (1996) Defect in multiple cell cycle checkpoints in ataxia-telangiectasia postirradiation. J Biol Chem 271(34):20486–20493
81. O'Connell MJ, Walworth NC, Carr AM (2000) The G2-phase DNA-damage checkpoint. Trends Cell Biol 10(7):296–303
82. Lock RB, Ross WE (1990) Inhibition of p34cdc2 kinase activity by etoposide or irradiation as a mechanism of G2 arrest in Chinese hamster ovary cells. Cancer Res 50(12):3761–3766
83. Lock RB, Ross WE (1990) Possible role for p34cdc2 kinase in etoposide-induced cell death of Chinese hamster ovary cells. Cancer Res 50(12):3767–3771
84. Paules RS et al (1995) Defective G2 checkpoint function in cells from individuals with familial cancer syndromes. Cancer Res 55(8):1763–1773
85. Herzinger T et al (1995) Ultraviolet B irradiation-induced G2 cell cycle arrest in human keratinocytes by inhibitory phosphorylation of the cdc2 cell cycle kinase. Oncogene 11(10):2151–2156
86. Gallant P, Nigg EA (1992) Cyclin B2 undergoes cell cycle-dependent nuclear translocation and, when expressed as a non-destructible mutant, causes mitotic arrest in HeLa cells. J Cell Biol 117(1):213–224
87. Pines J, Hunter T (1989) Isolation of a human cyclin cDNA: evidence for cyclin mRNA and protein regulation in the cell cycle and for interaction with p34cdc2. Cell 58(5):833–846
88. Rudner AD, Murray AW (1996) The spindle assembly checkpoint. Curr Opin Cell Biol 8(6):773–780
89. Murray AW (1995) The genetics of cell cycle checkpoints. Curr Opin Genet Dev 5(1):5–11
90. Lenart P, Peters JM (2006) Checkpoint activation: don't get mad too much. Curr Biol 16(11):R412–414
91. Vanoosthuyse V et al (2004) Kinetochore targeting of fission yeast Mad and Bub proteins is essential for spindle checkpoint function but not for all chromosome segregation roles of Bub1p. Mol Cell Biol 24(22):9786–9801
92. Pinsky BA, Biggins S (2005) The spindle checkpoint: tension versus attachment. Trends Cell Biol 15(9):486–493
93. Lens SM, Medema RH (2003) The survivin/Aurora B complex: its role in coordinating tension and attachment. Cell Cycle 2(6):507–510
94. Goulding SE, Earnshaw WC (2005) Shugoshin: a centromeric guardian senses tension. Bioessays 27(6):588–591

Chapter 2
The Basic Helix-Loop-Helix Transcription Factors in Neural Differentiation

Toshiyuki Ohtsuka and Ryoichiro Kageyama

Abstract During the development of the central nervous system, neural stem cells initially expand their own population by symmetric cell divisions, in which both progeny re-enter the cell cycle. By mid-gestation, the cells initiate neurogenesis by adopting a mode of asymmetric cell division, in which one daughter cell differentiates into a neuron while the other continues to cycle in the ventricular zone. Neural stem cells gradually alter their characteristics during development and thus give rise to different types of neurons over time, and finally switch to gliogenesis. The basic helix-loop-helix (bHLH) genes coordinately govern these processes and play a key role in the fate choice and the cell diversity. The repressor-type bHLH gene *Hes* is essential for maintenance of neural stem cells. *Hes* genes antagonize the activator-type bHLH genes such as *Mash1*, *Math*, and *Neurogenin (Ngn)*, which induce neuronal differentiation by activating the neuronal-specific genes. The activator-type bHLH genes not only promote the neuronal fate determination but also regulate the neuronal subtype specification. They also induce expression of Notch ligands such as Delta, which activate Notch signaling and upregulate *Hes1* and *Hes5* expression in neighboring cells, thereby maintaining these cells undifferentiated. Thus, the activator-type and repressor-type bHLH genes regulate each other, allowing only subsets of cells to undergo differentiation while keeping others to stay neural stem cells. This regulation is essential for generation of complex brain structures of appropriate size, shape, and cell arrangement.

R. Kageyama (✉)
Institute for Virus Research, Kyoto University, Shogoin-Kawahara, Sakyo-ku, Kyoto 606-8507, Japan
e-mail: rkageyam@virus.kyoto-u.ac.jp

A. Giordano, U. Galderisi (eds.), *Cell Cycle Regulation and Differentiation in Cardiovascular and Neural Systems*, DOI 10.1007/978-1-60327-153-0_2,

Keywords Neurogenesis · Gliogenesis · bHLH transcription factors · Neural stem cells · Notch signaling · Cell differentiation

1 Introduction

During the development of the central nervous system (CNS), multipotent neural progenitor cells in the ventricular zone (VZ) initially expand their own population by symmetric cell divisions, in which both progeny re-enter the cell cycle. By mid-gestation, the cells initiate neurogenesis by adopting a mode of asymmetric cell division, in which one daughter cell differentiates into a neuron while the other continues to cycle [1, 2]. Many of the progenitor cells in the VZ exhibit characteristics of neural stem cells, which include the capacity to generate many cell types, including neurons, astrocytes, and oligodendrocytes (multipotentiality), and the ability to divide symmetrically to duplicate their own population or divide asymmetrically to continually replenish the progenitor pool (self-renewal) [3, 4].

Neuroepithelial cells are first generated from the ectoderm, forming the neural plate. These cells are neural stem cells and undergo a symmetric cell division to produce more neural stem cells [5, 6]. After neural tube is formed from the neural plate, neuroepithelial cells become radial glia by acquiring some glial features. Radial glia has a cell body in the ventricular zone and long radial fibers extending from the internal surface to the pial (outer) surface. This cell type was long thought as specialized glia that guides neuronal migration along the radial fibers, but recent studies revealed that it is an embryonic neural stem cell [5–9]. Radial glia undergoes many rounds of asymmetric cell divisions, forming one radial glial cell and one neuron (or a neuronal precursor) from each cell division. After production of neurons, radial glial cells finally give rise to glial cells such as oligodendrocytes and astrocytes. Thus, neural stem cells change their characteristics of both morphology and competency over time during development. It takes a certain period of time for neural stem cells to change their characteristics, and maintenance of these cells until late stages is essential to generate all cell types. Premature differentiation allows differentiation of early-born cell types only and disorganizes the shape and cytoarchitecture of the brain. It has been shown that multiple bHLH genes play a critical role in regulation of neural stem cell differentiation [10, 11].

There are two types of bHLH genes, the repressor type and the activator type. The repressor-type bHLH genes include *Hes* genes, mammalian homologs of *Drosophila hairy* and *Enhancer of split* [*E*(*spl*)], while the activator-type bHLH genes include *Mash1*, *Math*, *Neurogenin* (*Ngn*), and *NeuroD*, homologs of *Drosophila* proneural genes *achaete-scute* complex and *atonal*. *Hes* genes negatively regulate neuronal differentiation and maintain neural stem cells while *Mash1*, *Math*, *Ngn*, and *NeuroD* promote neurogenesis. This chapter describes an overview of the roles of bHLH genes in neural differentiation and their significance in brain morphogenesis.

2 Maintenance of Neural Stem Cells by the Repressor-Type bHLH Genes: *Hes*

2.1 Structure and Transcriptional Activities of Hes Factors

There are seven members in the *Hes* family. Among them, *Hes1*, *Hes3*, and *Hes5* are highly expressed by neural stem cells in the developing nervous system [12–14]. Hes factors have three characteristic conserved domains: the bHLH domain, the Orange domain (the helix 3-helix 4 domain), and the WRPW (Trp-Arg-Pro-Trp) domain, which are essential for transcriptional activities (Fig. 2.1A).

The bHLH domain in the amino-terminal region is important for dimer formation and DNA binding [12]. bHLH factors form homodimers or heterodimers through the HLH domain and bind to DNA targets via the basic regions. Hes factors have a conserved proline residue in the middle of the basic region (Fig. 2.1A, asterisk), suggesting that this may be involved in the specificity of the target DNA sequences, although the exact significance of this proline residue remains to be determined. Hes1 exhibits a higher binding affinity to the N box (CACNAG) and the class C site (CACGCG) than to the E box (CANNTG) sequence and represses the expression of neurogenic bHLH factors such as Mash1 (active repression, Fig. 2.1B), while other activator-type bHLH factors bind to the E box with a higher affinity (Fig. 2.1D).

Hes factors also have a less conserved region, called the Orange domain [15], located just downstream of the bHLH domain (Fig. 2.1A). This domain is suggested to consist of two amphipathic helices and is known to confer specificity for protein–protein interaction between the bHLH factors [15, 16]. For example, the Hes-related bHLH factor Hairy interacts with the bHLH factor Scute efficiently, while another Hes-related bHLH factor E(spl)m8 does not, and this difference in the interaction specificity is attributed to the Orange domain [15]. This domain is also shown to mediate transcriptional repression [17], although a co-repressor interacting with this domain is not known yet.

Another structural feature is a repression domain called the WRPW domain (Trp-Arg-Pro-Trp sequence) located at or near the carboxyl terminus (Fig. 2.1A). The co-repressor TLE/Grg, a homolog of *Drosophila* Groucho, interacts with the WRPW domain and modifies the chromatin structure by recruiting the histone deacetylase Rpd3, thereby inactivating the chromatin and transcription [18, 19]. It is suggested that Groucho mediates long-range transcriptional repression that can affect over distances of several kilobases in *Drosophila* embryos [20]. *Hes* genes thereby actively repress transcription and thus are classified into the repressor-type bHLH genes.

The target genes for Hes factors include the activator-type bHLH genes such as *Mash1*. Hes1 represses *Mash1* expression by directly binding to the promoter (Fig. 2.1B) [21]. Hes1 can bind to these sites not only as a homodimer (Fig. 2.1Ba) but also as a heterodimer with Hes-related bHLH factors such as Hesr [22] and exerts strong repressor activity (Fig. 2.1Bb). The activator-type bHLH factors form a heterodimer with another bHLH activator E47 (e.g., Mash1–E47 heterodimer)

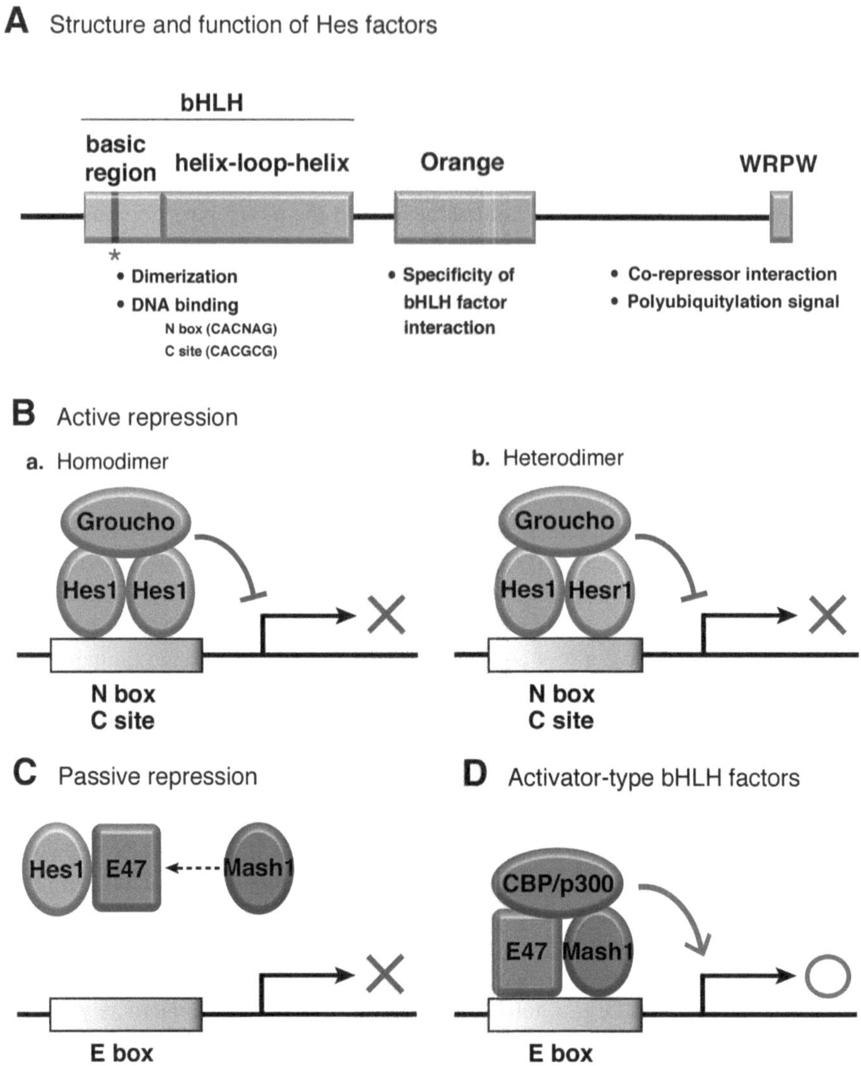

Fig. 2.1 Structure and transcriptional activities of Hes factors. (**A**) The conserved domains of Hes factors. The basic, HLH, Orange, and WRPW domains and their functions are indicated. Asterisk indicates a conserved proline residue in the basic region. (**B**) Active repression: Hes factors bind to the N box or class C site by forming homodimers (**a**) or heterodimers with Hesr (**b**) and actively repress transcription by interacting with co-repressors, such as Groucho homologs. (**C**) Passive repression: Hes factors form non-DNA-binding heterodimers with bHLH activators such as E47 and inhibit transcriptional activation. (**D**) Activation: activator-type bHLH factors such as Mash1 and E47 form heterodimers that bind to the E box and activate transcription by recruiting co-activators such as CBP/p300

and promote neuronal differentiation from neural stem cells (Fig. 2.1D). Hes1 forms heterodimers with other bHLH factors such as Mash1 and its partner E47, which do not bind to DNA (non-functional heterodimers), and thus can passively inhibit neurogenesis by sequestering them (Fig. 2.1C) [12]. Thus, Hes1 antagonizes Mash1 and inhibits neuronal differentiation via two different mechanisms: repressing the expression at the transcriptional level and inhibiting the activity at the protein–protein interaction level.

Likewise, the HLH factor Id, which lacks the basic region [23] and thus lacks the DNA-binding activity, can inhibit neurogenesis by forming a non-functional heterodimer with neurogenic bHLH factors. It was reported that Id also forms a heterodimer with Hes1, but this heterodimer does not bind to DNA [24]. There are also several Hes-related bHLH genes such as Hesr/Hey/HRT/Herp/CHF/Gridlock [22] and Heslike [25], which compose distinct subfamilies of repressor-type bHLH factors.

2.2 Regulation of Hes Genes by Notch Signaling

It has been well characterized that expression of *Hes1* and *Hes5* is regulated by Notch signaling [26, 27]. Notch, a transmembrane protein, is activated by the ligands Delta and Jagged, which are also transmembrane proteins expressed by neighboring cells (Fig. 2.2). Upon activation, Notch is processed to release the intracellular domain (ICD), which is transferred into the nucleus and forms a complex with the DNA-binding protein RBP-J [28, 29]. In the *Hes1* promoter, there are two tandem repeats of the RBP-J binding sites (the core sequence: TGGGAA) at nucleotide positions -70 and -84 (relative to the transcription initiation site). The *Hes5* promoter also has two RBP-J binding sites at nucleotide positions -77 and -293. RBP-J itself is a transcriptional repressor and represses *Hes1* and *Hes5* expression by binding to their promoters (Fig. 2.2). However, when RBP-J forms a complex with Notch ICD, this complex becomes a transcriptional activator and induces *Hes1* and *Hes5* expression (Fig. 2.2). Thus, Notch activation leads to upregulation of *Hes1* and *Hes5* expression. Notch is known to inhibit neuronal differentiation and maintain neural stem cells [30]. In the absence of *Hes1* and *Hes5*, however, Notch fails to inhibit neuronal differentiation, indicating that *Hes1* and *Hes5* are essential effectors of Notch signaling [27]. *Hes1* and *Hes5* are expressed within the VZ throughout the developing CNS, where they function downstream of Notch signaling as negative regulators of neuronal differentiation [27, 31–34] and promotes a neural stem cell identity [35].

In contrast to *Hes1* and *Hes5*, there is no evidence that *Hes3* expression is controlled by Notch signaling [36]. In addition, initial *Hes1* expression occurs at early stages before *Notch* and *Delta* are expressed (see below), indicating that Notch signaling is not the sole regulator of *Hes* expression. BMP (bone morphogenic protein), Shh (sonic hedgehog), and Wnt signaling pathways have also been shown to induce *Hes* expression [37–39]. Thus, *Hes* expression is controlled by multiple signaling pathways, in addition to Notch signaling.

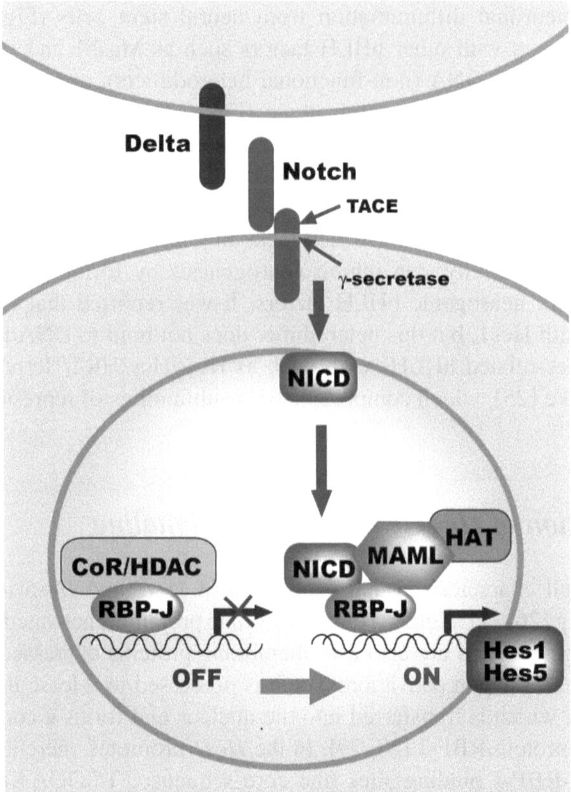

Fig. 2.2 Notch–Hes signaling. Notch signaling is triggered when the ligands (DSL family) expressed on the surface of neighboring cells interact with Notch receptors. Upon ligand binding, Notch receptors undergo successive proteolytic cleavage. The first cleavage by an extracellular protease of the ADAM (a disintegrin and metalloprotease) family, TNF-α converting enzyme (TACE), generates an active membrane-tethered form of Notch. The truncated product is further processed by the γ-secretase activity of Presenilins 1 and 2, which release the intracellular domain of Notch (NICD) from the plasma membrane. The NICD translocates to the nucleus and associates with the DNA-binding transcription factor RBP-J. As a result, RBP-J is converted from a transcriptional repressor to an activator. In this process, NICD, RBP-J, and mastermind-like proteins (MAML family, MAML1-3) assemble on target DNA and form a RBP-J–NICD–MAML ternary complex. This transcriptional activation complex is formed through displacement of the co-repressor complex (CoR/HDAC: histone deacetylase) and recruitment of the co-activators (HAT: histone acetyltransferase). Thus, Notch signaling activates the transcription of target genes, such as *Hes1* and *Hes5*. Hes factors then subsequently repress the transcription of proneural genes such as *Mash1*

2.3 Maintenance of Neural Stem Cells by Hes Genes

Hes genes are expressed by neural stem cells. At the initial stage, *Hes1* and *Hes3* are widely expressed by neuroepithelial cells (Fig. 2.3) [14, 40]. However, *Hes3* expression is gradually downregulated in the ventral part of the neural tube and

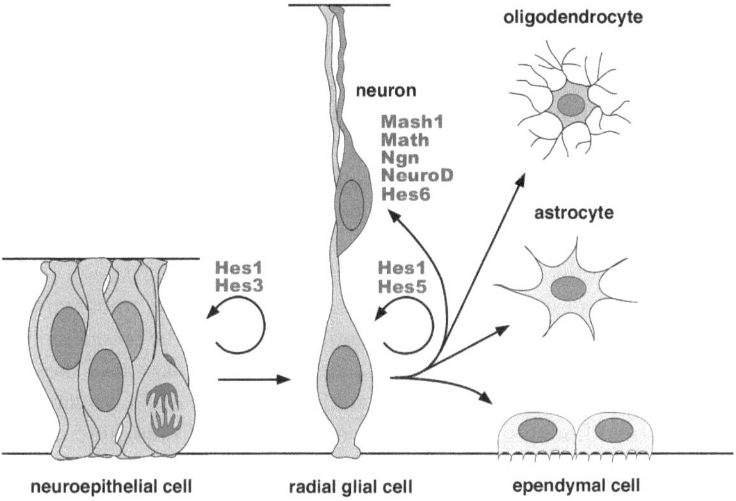

Fig. 2.3 Maintenance of neural stem cells by *Hes* genes. Neuroepithelial cells initially form the neural plate and thin wall of the neural tube. These cells gradually develop into radial glial cells, which have a cell body in the ventricular zone (VZ) and a radial fiber reaching the pial surface. Radial glial cells give rise to neurons by asymmetric cell divisions. After the production of distinct types of neurons, radial glial cells give rise to or differentiate into non-neuronal cells such as oligodendrocytes, astrocytes, and epedymal cells. *Hes* genes maintain neuroepithelial cells and radial glial cells during early development and promote astrocyte formation during late development. Based on their expression patterns, *Hes1* and *Hes3* are important for the maintenance of neuroepithelial cells, while *Hes1* and *Hes5* are required for most radial glial cells. Activator-type bHLH genes such as *Mash1*, *Math*, *Ngn*, and *NeuroD* promote neurogenesis. Unlike other *Hes* genes, *Hes6* also promotes neurogenesis

disappears from most regions at later stage except for the isthmus, the boundary between the midbrain and hindbrain [14, 41]. As *Hes3* expression is downregulated, *Hes5* expression is upregulated (Fig. 2.3). Upregulation of *Hes5* expression synchronizes with that of Delta and Notch expression, suggesting that *Hes5* expression is controlled by Notch signaling while initial *Hes1* and *Hes3* expression is not. *Hes1* expression is maintained even after *Hes3* expression is repressed, but it is likely that *Hes1* expression at later stages may depend on Notch signaling.

Roles of *Hes* genes in neural stem cells have been investigated by gain-of-function and loss-of-function experiments. Misexpression of *Hes1*, *Hes3*, or *Hes5* in the embryonic brain inhibits neuronal differentiation and maintains radial glial cells [31, 35, 42]. Conversely, in *Hes1;Hes5* double knockout mice, many radial glial cells are not maintained and prematurely differentiate into neurons [27, 40, 43, 44]. Furthermore, *Hes1*(-/-);*Hes5*(-/-) neurospheres do not expand properly even in the presence of bFGF and EGF, in contrast to the wild-type neurospheres, which proliferate extensively [27]. Thus, *Hes1* and *Hes5* are essential for maintenance and proliferation of neural stem cells by preventing premature onset of the neurogenic bHLH gene expression in the embryonic brain.

Although many radial glial cells prematurely differentiate into neurons in *Hes1;Hes5* double knockout mice, neuroepithelial cells and some radial glial cells are still maintained, suggesting that *Hes3* may compensate for *Hes1* and *Hes5* deficiency. Agreeing with this notion, in *Hes1;Hes3;Hes5* triple knockout mice, many neuroepithelial cells prematurely differentiate into neurons as early as E8.5, in contrast to the wild type, in which neuroepithelial cells do not differentiate into neurons [40]. Furthermore, in the triple-mutant mice, virtually all radial glial cells prematurely differentiate into neurons by E10.0 at the expense of the later born cell types: later born neurons, oligodendrocytes, astrocytes, and ependymal cells [40]. Thus, *Hes1*, *Hes3*, and *Hes5* are essential to generate cells in correct numbers and full diversity by maintaining neural stem cells until later stages. The premature neuronal differentiation in *Hes*-mutant mice is associated with upregulation of the activator-type bHLH genes such as *Mash1* and *Math3* [40]. Thus, it is likely that *Hes* genes regulate the normal timing of differentiation by repressing premature onset of the activator-type bHLH genes.

Even in *Hes1;Hes3;Hes5* triple-mutant mice, still neuroepithelial cells are initially formed, suggesting that formation of neuroepithelial cells is independent of *Hes* gene activities (Fig. 2.4). It remains to be determined which genes are responsible for the initial formation of neural stem cells. However, in the absence of *Hes* genes, neuroepithelial cells and radial glial cells prematurely differentiate, indicating that their maintenance depends on *Hes* gene activities [40]. Neural stem cells thus change their characteristics over time as follows: *Hes*-independent neuroepithelial cells, *Hes*-dependent neuroepithelial cells (transitory neuroepithelial cells), and *Delta/Notch/Hes*-dependent radial glial cells (Fig. 2.4). Based on their expression patterns, *Hes1* and *Hes3* are important for the maintenance of neuroepithelial cells, while *Hes1* and *Hes5* are required for most radial glial cells.

Notch and Delta are not expressed at the neuroepithelial stage, during which the initial *Hes1* and *Hes3* expression occurs, indicating that *Hes1* and *Hes3* expression in neuroepithelial cells is not controlled by Notch signaling [40]. After E8.5 onward, *Hes5* expression starts and *Hes3* expression is downregulated while *Hes1* expression is maintained. The onset of *Hes5* expression coincides with that of *Notch* and *Delta* expression, indicating that expression of *Hes5*, and possibly *Hes1* at this stage also, is controlled by Notch signaling. It remains to be determined which upstream factors are responsible for the initial expression of *Hes1* and *Hes3*. Recent studies revealed that there are at least two types of neural stem cells depending on the developmental stages: primitive and definitive neural stem cells [45]. Definitive neural stem cells are derived from later stages and depend on Notch signaling, while primitive neural stem cells are derived from earlier stages and do not depend on Notch signaling but on LIF signaling [45]. Because initial *Hes1* and *Hes3* expression in neuroepithelial cells is not controlled by Notch signaling, this expression could be controlled by LIF signaling or by a related signaling pathway. Cytokine signaling is known to regulate neural stem cells. In response to the activation of cytokine receptors, JAK2 phosphorylates tyrosine residues of STAT3, and this phosphorylated STAT3 can promote maintenance of neural stem cells. Interestingly, JAK2–STAT3 signaling depends on Notch signaling [46]. The Notch effectors Hes1 and Hes5 physically

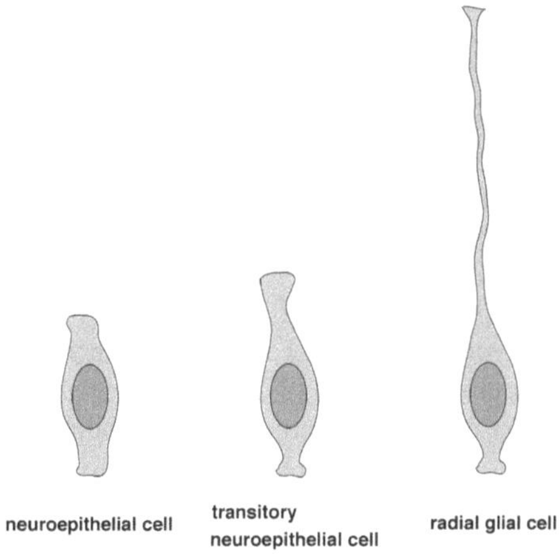

	neuroepithelial cell	transitory neuroepithelial cell	radial glial cell
Hes dependency	No	Yes	Yes
Delta/Notch	No	No	Yes
Neurogenesis	No	No	Yes

Fig. 2.4 Change of characteristics of neural stem cells during development. Neuroepithelial cells are formed independently of *Hes* genes, but their maintenance critically depends on *Hes* genes and not on Delta/Notch signaling. Radial glial cells depend on Delta, Notch, and Hes activities. Thus, neural stem cells change their characteristics over time as follows: *Hes*-independent neuroepithelial cells, *Hes*-dependent neuroepithelial cells (transitory neuroepithelial cells), and *Delta/Notch/Hes*-dependent radial glial cells, which produce neurons by asymmetric cell divisions

interact with both JAK2 and STAT3, and this complex facilitates the phosphorylation and activation of STAT3 by JAK2 [46], thus highlighting the significance of the cross talk between the Notch–Hes and JAK–STAT pathways in neural stem cells. In *Drosophila*, *E(spl)* expression is controlled by Notch while *hairy* expression is not. Based on the expression profiles, *Hes1* and *Hes3* expressed at early stages are similar to *hairy*, while *Hes1* and *Hes5* expressed at later stages are similar to *E(spl)*. Thus, it is likely that *Hes1* has mixed features of *Drosophila hairy* and *E(spl)*. Primitive neural stem cells thus could be Hes-dependent or Hes-independent neuroepithelial cells, while definitive neural stem cells could be Hes-dependent radial glial cells.

Neural stem cells are also important for maintenance of the structures of the developing nervous system. Both neuroepithelial and radial glial cells have epithelial features such as the apical junctional complex and the basal lamina, which form the inner and outer barriers of the neural tube, respectively. In the absence of *Hes* genes, both the apical junctional complex and the basal lamina are disrupted due to the premature loss of neural stem cells, leading to spilling of neurons into the lumen as well as into the surrounding tissues [40]. Thus, neural stem cells are essential for the structural integrity of the nervous system. In wild-type embryos, by the time neural stem cells disappear, the ependymal cells are differentiated at the apical side and form the apical junctional complex, while astrocytes are differentiated and contribute to the basal lamina formation at the basal side. These results indicate that the normal timing of cell differentiation and maintenance of neural stem cells are important for the structural integrity of the nervous system.

Hes-related bHLH genes, *Hesr1* and *Hesr2*, are also expressed by neural stem cells in the embryonic brain, and misexpression of *Hesr1* and *Hesr2* promotes maintenance of neural stem cells [47]. *Hesr* expression is also controlled by Notch signaling, and Hesr and Hes proteins form heterodimers and act as repressors [22]. Thus, it is possible that Hesr and Hes cooperatively regulate maintenance of neural stem cells.

3 Neurogenesis by the Activator-Type bHLH Genes: *Mash1*, *Math*, *Ngn*, and *NeuroD*

3.1 Neuronal Determination and Subtype Specification

In mammals, bHLH genes homologous to *Drosophila achaete–scute* complex (*Mash*) and atonal (*Math*, *Ngn*, *NeuroD*) have been identified. These activator-type bHLH genes are expressed by neural precursor cells and/or differentiating neurons (Fig. 2.5). These factors form a heterodimer with a ubiquitously expressed bHLH factor, E47, and activate neuronal-specific gene expression by binding to the E box (Fig. 2.1D). Misexpression of these genes in neural stem cells promotes the determination of neuronal fate and accelerates the neurogenesis. Some cyclin-dependent kinase (CDK) inhibitors are involved in the cell cycle exit which accompanies the neuronal differentiation. For example, $p27^{Xic1}$ is shown to play a role in the coordination of cell cycle exit and differentiation and enhance neurogenesis by stabilizing the neurogenin protein during early neurogenesis in *Xenopus* embryos [48]. It was reported in neuroblastoma cells that NeuroD and E47 activate the expression of *TrkB* and $p21^{Cip1}$ by binding to the E box sequences in their promoters [49]. This establishes a direct transcriptional link between the regulation of cell cycle and neuronal differentiation.

These bHLH genes promote the neuronal subtype specification as well. For example, in the forebrain, *Mash1* is expressed in the ventral region and specifies the GABAergic neurons while *Ngn2* is expressed in the dorsal region and specifies the glutamatergic neurons [50, 51]. In the absence of *Ngn2*, *Mash1* is ectopically expressed in the dorsal telencephalon, leading to ectopic GABAergic neurogenesis in the dorsal telencephalon. Likewise in the peripheral nervous system, the activator-type bHLH genes regulate the neuronal subtype specification. *Ngn1* and *Ngn2* promote generation of sensory neurons while *Mash1* promotes autonomic

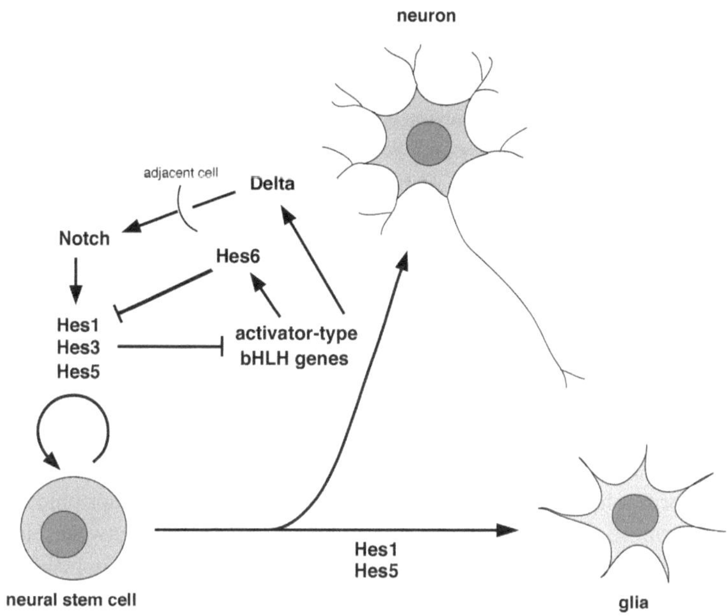

Fig. 2.5 The bHLH gene network in neural development. *Hes1*, *Hes3*, and *Hes5* repress activator-type bHLH gene expression and maintain neural stem cells. In contrast, activator-type bHLH genes induce *Hes6*, which inhibits Hes1 and promotes neuronal differentiation. Thus, these bHLH genes regulate each other in a mutually antagonistic manner. Activator-type bHLH genes also play an important role in keeping neighboring cells as neural stem cells via Delta–Notch signaling by activating the expression of Notch ligands such as *Delta* in differentiating cells. Neural stem cells that retain *Hes1*/*Hes5* expression finally become glial cells

neurogenesis from neural crest stem cells [52]. Furthermore, *Ngn1* and *Ngn2* are required for the specification of distinct sensory precursor populations. *Ngn2* is essential for development of the epibranchial placode-derived cranial sensory ganglia [53], while *Ngn1* is essential for the determination of neuronal precursors for proximal cranial sensory ganglia [54]. In dorsal root ganglia (DRG), *Ngn2* is required only for the development of early-generated TrkC+ and TrkB+ neurons, whereas *Ngn1* is required for most or all later generated TrkA+ neurons [55]. For other examples, it was reported that the bHLH genes *Nscl-1* and *Nscl-2* synergistically determine the GnRH (gonadotropin releasing hormone)-1 neurons [56], and *Hand2* determines the noradrenergic phenotype in the sympathetic nervous system [57]. Thus, the activator-type bHLH genes can regulate the neuronal subtype identity, in addition to the pan-neuronal phenotypes. Similarly in *Drosophila* peripheral nervous system, *achaete-scute* complex bHLH genes promote generation of sensory organs while *atonal* bHLH gene promotes generation of chordotonal organs. It has been shown that the basic regions of the bHLH factors are important for the neuronal subtype-specific activities [58].

The activator-type bHLH factors not only induce the neuronal-specific gene expression but also inhibit the glia-specific gene expression. *Glial fibrillary acidic protein (GFAP)*, an astroglia-specific gene, is upregulated by Stat1/3 and Smad1, which are bridged by the co-activator p300 [59]. Ngn1 sequesters the p300/Smad complex from the glial promoters and thereby inhibits glial-specific gene expression [60]. Thus, the activator-type bHLH factors seem to reinforce the process of neuronal fate specification by inhibiting the alternative fate. Agreeing with this notion, in *Mash1;Math3* double knockout or *Mash1;Ngn2* double knockout mice, the cells that should normally become neurons adopt the glial fate instead [61, 62]. Thus, there is a fate switch from neurons to glia in the absence of the activator-type bHLH genes, indicating that *Mash1*, *Math*, and *Ngn* regulate neuronal versus glial cell fate determination.

Proneural bHLH genes such as *Mash1* override the inhibitory activities of *Hes* genes and promote neuronal differentiation. This process involves another member of the *Hes* family, *Hes6*, which has a unique function in neuronal differentiation. Hes6 can form a heterodimer with Hes1, but this complex does not bind to DNA [63]. Furthermore, Hes6 was shown to inhibit the interaction between Hes1 and Groucho/TLE/Grg and induce degradation of Hes1 protein [64]. As a result, Mash1 is relieved from Hes-induced inhibition. Thus, Hes6 inhibits Hes1 activity but supports Mash1 and promotes neuronal differentiation in the developing brain and retina (Fig. 2.5) [63–65]. *Hes6* expression is induced by the activator-type bHLH genes such as *Neurogenin* (Fig. 2.5) [65]. Thus, the proneural bHLH genes inhibit *Hes1/3/5* genes by inducing *Hes6*, while *Hes1/3/5* genes inhibit the proneural bHLH genes, indicating that these bHLH genes regulate each other in a mutually antagonistic manner (Fig. 2.5). *Hes6* also has an activity to inhibit astrocyte differentiation independently of its ability to promote neurogenesis [66].

While the activator-type bHLH genes promote neuronal specification, they are also important for maintenance of neural stem cells. These genes are known to activate expression of Notch ligands such as *Delta* (Fig. 2.5). In *Mash1;Math3* double knockout mice, *Delta* expression is lost in the hindbrain, leading to inactivation of Notch signaling and premature loss of neural stem cells [67]. Thus, the activator-type bHLH genes also play an important role in keeping neighboring cells as neural stem cells via Delta–Notch signaling (Fig. 2.5).

3.2 Combinatorial Functions of bHLH Genes for the Cell-Type Diversity

Although the activator-type bHLH genes regulate the neuronal subtype specificity, they alone are not sufficient to generate diverse types of neurons. For example, in the retina, stem cells give rise to six types of neurons, which are aligned in specific layers. The activator-type bHLH genes alone are not sufficient but coexpression of homeodomain genes is required for specification of six types of neurons [68, 69]. It is likely that homeodomain genes regulate the layer identity while activator-type bHLH genes determine the neuronal fate specific for each layer.

In some cases, combinations of two distinct bHLH genes can induce different types of neurons from those induced by each bHLH gene alone. The bHLH factor Heslike is structurally related to Hes factors, but unlike Hes, it has neither a proline residue in the basic region nor the WRPW domain at the carboxy terminus [25], indicating that *Heslike* does not belong to the *Hes* family. Heslike is coexpressed with Mash1 in the ventricular zone that gives rise to GABAergic neurons. The expression occurs mainly in the midbrain and diencephalon, but not in the hindbrain or spinal cord. Heslike or Mash1 alone does not give rise to GABAergic neurons whereas they together can give rise to GABAergic neurons in the midbrain and diencephalon [25]. Thus, combinations of bHLH genes can specify different types of neurons from those induced by each bHLH gene alone. The bHLH genes *Olig1* and *Olig2* are structurally related to *Drosophila* proneural gene *atonal* but form a distinct subfamily. Interestingly, these bHLH genes alone promote oligodendrocyte development but not neuronal development [70, 71]. However, when *Ngn2* is coexpressed, *Olig2* can promote motor neuron development [72, 73]. Thus, *Olig2* regulates development of both neurons and oligodendrocytes in a context-dependent manner (whether or not *Ngn2* is coexpressed). These results indicate that different combinations of bHLH genes can specify different types of cells, which could be one of the strategies for limited numbers of transcription factors to regulate specification of a diversity of cell types.

4 Gliogenesis by the bHLH Genes: *Hes*, *Olig*, and *Scl*

At later stages, when gliogenesis occurs, *Hes1* and *Hes5* are transiently expressed by astrocytes in the developing brain [38, 74] and by Müller glial cells in the developing retina [75, 76]. Misexpression of *Hes1* and *Hes5* at later stages increases generation of astrocytes in the brain and Müller glial cells in the retina (Fig. 2.5) [35, 75, 76]. Conversely, in the absence of *Hes1* and *Hes5*, production of Müller glial cells is decreased [75–77]. Thus, *Hes1* and *Hes5* are involved in gliogenesis at later stages, indicating that *Hes* genes exhibit different activities depending on their developmental stages: maintenance of neural stem cells at early stages and promotion of gliogenesis at later stages. However, it remains to be determined whether *Hes1* and *Hes5* instructively promote the glial fate or just maintain neural stem cells until the gliogenic phase. It has been shown that the proneural bHLH gene *Neurogenin1 (Ngn1)* has two activities: promotion of neurogenesis and inhibition of gliogenesis [60]. Ngn1 sequesters the CBP–Smad1 transcriptional complex away from the glial-specific promoters and recruits the complex to the neuronal-specific promoters, thereby promoting neurogenesis while inhibiting alternative fates. Conversely, inactivation of the proneural genes such as *Mash1*, *Ngn2*, and *Math3* blocks neurogenesis while enhancing gliogenesis [61, 62]. Thus, suppression of the proneural genes could be one of the major mechanisms for *Hes1*- and *Hes5*-induced gliogenesis. Recent studies show that, at least, subsets of GFAP-positive glia-like cells serve as neural stem cells in the adult brain, and *Hes1* and *Hes5* may be involved in formation/maintenance of such adult neural stem cells.

Numerous studies have revealed that *Olig1* and *Olig2* genes are essential for oligodendrocyte development [70, 71]. *Olig* genes act downstream of Shh and regulate the establishment of the pMN domain and the sequential generation of motor neurons and oligodendrocytes in developing neural tube. Ectopic expression of *Olig1* promotes oligodendrocyte formation in the developing cortex [70], and forced expression of *Olig2* and *Nkx2.2* in the spinal cord promotes ectopic induction of oligodendrocyte precursors and precocious oligodendrocyte differentiation [71]. Olig proteins seem to promote oligodendrocyte cell fate while inhibiting astrocyte development. Conversely, the bHLH gene *Scl* regulates astrocyte versus oligodendrocyte cell fate acquisition and V2b versus V2a interneuron cell fate acquisition in the p2 domain of the developing spinal cord [78]. *Scl* function is necessary and sufficient for p2-associated astrocyte development, and it represses the production of oligodendrocyte precursors through cross-antagonistic interactions between *Scl* and *Olig2* [78].

5 Cell Cycle Regulation and Boundary Formation by Hes1

The developing nervous system is partitioned into many compartments by boundaries, such as the zona limitans intrathalamica (Zli) and the isthmus (Fig. 2.6A). The Zli is the boundary between the thalamus and the prethalamus, whereas the isthmus is the boundary between the midbrain and the hindbrain. Boundaries are formed by specialized neuroepithelial or radial glial cells, which have unique features, including slow proliferation, delayed or no neurogenesis, and organizer activities that regulate specificity of neighboring compartments. For example, the Zli and the isthmus function as organizing centers by secreting morphogens such as Shh and Wnt1/Fgf8, respectively, and by regulating the regional specification of neighboring compartments [79] (Fig. 2.6A). Cells migrate within each compartment but do not usually cross boundaries; thus, each compartment forms a unit that consists of a distinct set of cell types [79].

The isthmic cells express *Hes1* and *Hes3* and do not give rise to any neurons. In the absence of *Hes* genes, however, proneural genes are ectopically expressed in boundaries, leading to ectopic neurogenesis and the loss of organizer activity [41, 80]. The isthmic cells prematurely lose *Wnt1* and *Fgf8* expression, and as a result, the midbrain and hindbrain neurons are not properly specified. For example, oculomotor and trochlear nuclei and dopaminergic neurons of the midbrain and locus ceruleus neurons of the hindbrain are missing in *Hes1;Hes3* double-mutant embryos [41]. Thus, *Hes1* and *Hes3* are essential for maintenance of the isthmic organizer and development of the midbrain and hindbrain. In zebrafish, Hes-related bHLH genes *her3* and *her5* have similar activities, inhibiting neurogenesis and contributing to the formation of the midbrain–hindbrain boundary [81, 82]. It has been reported that the expression of *her3* and *her5* does not depend on Notch signaling, suggesting that *Hes* expression in boundaries of the mouse nervous system could also be independent of this pathway.

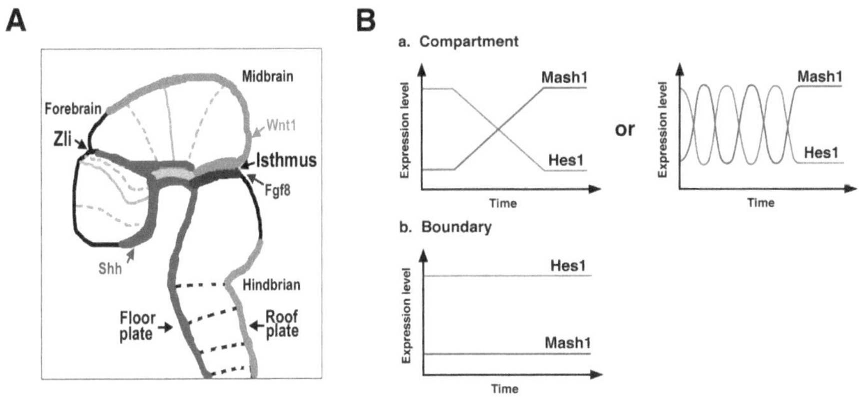

Fig. 2.6 Differential *Hes1* expression between compartments and boundaries. (**A**) Lateral view of the developing central nervous system (CNS) in the mouse at E10.5. The CNS is partitioned into many compartments by boundaries, such as the zona limitans intrathalamica (Zli) and the isthmus. These boundaries and the roof and floor plates function as organizing centers by expressing *Shh*, *Wnt1*, or *Fgf8*. Cells in these compartments undergo active proliferation and neurogenesis, whereas those in boundaries undergo slower proliferation and no neurogenesis. (**B**) The *Hes1* expression mode is different between compartments and boundaries: (**a**) variable expression (could be oscillatory) in compartments and (**b**) persistently high expression in boundaries. In compartments, when Hes1 levels are low, Mash1 levels are high, and vice versa. These cells finally lose *Hes1* expression and differentiate into neurons. By contrast, in boundaries, *Hes1* is persistently expressed at high levels, and neurogenesis is inhibited. This difference in the *Hes1* expression modes may confer compartment versus boundary characteristics

Although *Hes1* is expressed in both compartments and boundaries, the mode of expression is different in the two structures [80]. In compartments, Hes1 levels are variable: high levels occur in some cells, whereas, in others, levels are lower (Fig. 2.6Ba) [80]. It was found that *Hes1* expression is oscillating in these cells (Fig. 2.6Ba) [83]. By contrast, Hes1 is persistently expressed at high levels by many boundary cells (Fig. 2.6Bb) [80]. There is an inverse correlation between Hes1 and Mash1 levels: cells that express high levels of Hes1 express low levels of Mash1 and vice versa (Fig. 2.6Ba). Within boundaries, persistent and high levels of *Hes1* expression constitutively repress the expression of proneural genes, such as *Mash1*, thereby inhibiting neurogenesis (Fig. 2.6Bb) [80].

It has been shown that Hes1 regulates cell cycle progression. During the G1 phase, cyclin-dependent kinase (CDK) promotes cell cycle progression by forming complexes with cyclins, whereas the CDK inhibitors p21 and p27 antagonize this process. Low levels of Hes1 promote cell proliferation by downregulating p21 and p27 [84]. However, persistent and high levels of *Hes1* expression have been shown to inhibit the cell cycle, probably because Hes1 also represses the expression of some cell cycle regulators such as E2F-1, which promotes the G1–S phase transition [17, 80, 85, 86]. Thus, within boundaries, persistent and high levels of *Hes1* expression may contribute to slowing cell proliferation as well as to the inhibition of differentiation, raising the possibility that persistent versus variable *Hes1* expression differentially regulates cell cycle progression and boundary versus compartment characteristics (Fig. 2.6B).

6 Perspective

Neural stem cells change their competency over time during development, giving rise to different types of cells sequentially. Thus, the timing of differentiation and the number of cell cycle by the onset of differentiation should be strictly regulated to ensure normal brain size and cellular components. Previous studies have revealed that the balance between the repressor-type and the activator-type bHLH genes governs the stem cell versus neuronal fate choice, but it remains to be determined how their exact temporal expression is controlled. For example, it is not known which factors induce the initial *Hes* expression independently of Notch signaling, how the onset of expression of Notch and its ligand is controlled, and which factors upregulate the expression of activator-type bHLH genes to sufficient level for initiating the mode of neurogenesis. Changes of competency of neural stem cells during development are likely due to changes of sets of transcription factors expressed in neural stem cells, but it is totally unknown how these changes occur over time. Elucidation of the mechanism which controls the timing of such gene expression and neuronal/glial differentiation will be required for further understanding of neural development.

References

1. Takahashi T, Nowakowski RS, Caviness VS Jr (1994) Mode of cell proliferation in the developing mouse neocortex. Proc Natl Acad Sci U S A 91:375–379
2. Chenn A, McConnell SK (1995) Cleavage orientation and the asymmetric inheritance of Notch1 immunoreactivity in mammalian neurogenesis. Cell 82:631–641
3. Weiss S, Reynolds BA, Vescovi AL, Morshead C, Craig CG, van der Kooy D (1996) Is there a neural stem cell in the mammalian forebrain? Trends Neurosci 19:387–393
4. Weissman IL, Anderson DJ, Gage F (2001) Stem and progenitor cells: origins, phenotypes, lineage commitments, and transdifferentiations. Annu Rev Cell Dev Biol 17:387–403
5. Fujita S (2003) The discovery of the matrix cell, the identification of the multipotent neural stem cell and the development of the central nervous system. Cell Struct Funct 28:205–228
6. Alvarez-Buylla A, GarciaVerdugo JM, Tramontin AD (2001) A unified hypothesis on the lineage of neural stem cells. Nat Rev Neurosci 2:287–293
7. Malatesta P, Hartfuss E, Götz M (2000) Isolation of radial glial cells by fluorescent-activated cell sorting reveals a neuronal lineage. Development 127:5253–5263
8. Noctor SC, Flint AC, Weissmann TA, Dammerman RS, Kriegstein AR (2001) Neurons derived from radial glial cells establish radial units in neocortex. Nature 109:714–720
9. Anthony TE, Klein C, Fishell G, Heintz N (2004) Radial glia serve as neuronal progenitors in all regions of the central nervous system. Neuron 41:881–890
10. Bertrand N, Castro DS, Guillemot F (2002) Proneural genes and the specification of neural cell types. Nat Rev Neurosci 3:517–530
11. Ross SE, Greenberg ME, Stiles CD (2003) Basic helix-loop-helix factors in cortical development. Neuron 39:13–25

12. Sasai Y, Kageyama R, Tagawa Y, Shigemoto R, Nakanishi S (1992) Two mammalian helix-loop-helix factors structurally related to Drosophila *hairy* and *Enhancer of split*. Genes Dev 6:2620–2634
13. Akazawa C, Sasai Y, Nakanishi S, Kageyama R (1992) Molecular characterization of a rat negative regulator with a basic helix-loop-helix structure predominantly expressed in the developing nervous system. J Biol Chem 267:21879–21885
14. Allen T, Lobe CG (1999) A comparison of *Notch*, *Hes* and *Grg* expression during murine embryonic and post-natal development. Cell Mol Biol 45:687–708
15. Dawson SR, Turner DL, Weintraub H, Parkhurst SM (1995) Specificity for the hairy/enhancer of split basic helix-loop-helix (bHLH) proteins maps outside the bHLH domain and suggests two separable modes of transcriptional repression. Mol Cell Biol 15:6923–6931
16. Taelman V, Van Wayenbergh R, Solter M, Pichon B, Pieler T, Christophe D, Bellefroid EJ (2004) Sequences downstream of the bHLH domain of the Xenopus hairy-related transcription factor-1 act as an extended dimerization domain that contributes to the selection of the partners. Dev Biol 276:47–63
17. Castella P, Sawai S, Nakao K, Wagner JA, Caudy M (2000) HES-1 repression of differentiation and proliferation in PC12 cells: role for the helix 3-helix 4 domain in transcription repression. Mol Cell Biol 20 6170–6183
18. Paroush Z, Finley LR Jr, Kidd T, Wainwright SM, Ingham PW, Brent R, Ish-Horowictz D (1994) Groucho is required for Drosophila neurogenesis, segmentation, and sex determination and interacts directly with hairy-related bHLH proteins. Cell 79:805–815
19. Grbavec D, Stifani S (1996) Molecular interaction between TLE1 and the carboxyl-terminal domain of HES-1 containing the WRPW motif. Biochem Biophys Res Commun 223:701–705
20. Zhang H, Levine M (1999) Groucho and dCtBP mediate separate pathways of transcriptional repression in the Drosophila embryo. Proc Natl Acad Sci U S A 96:535–540
21. Chen H, Thiagalingam A, Chopra H, Borges MW, Feder JN, Nelkin BD, Baylin SB, Ball DW (1997) Conservation of the *Drosophila* lateral inhibition pathway in human lung cancer: a hairy-related protein (HES-1) directly represses achaete-scute homolog-1 expression. Proc Natl Acad Sci USA 94:5355–5360
22. Iso T, Sartorelli V, Chung G, Shichinohe T, Kedes L, Hamamori Y (2001) HERP, a new primary target of Notch regulated by ligand binding. Mol Cell Biol 21:6071–6079
23. Benezra R, Davis RL, Lockshon D, Turner DL, Weintraub H (1990) The protein Id: a negative regulator of helix-loop-helix DNA binding proteins. Cell 61 49–59
24. Jögi A, Persson P, Grynfeld A, Pahlman S, Axelson H (2002) Modulation of basic helix-loop-helix transcription complex formation by Id proteins during neuronal differentiation. J Biol Chem 277:9118–9126
25. Miyoshi G, Bessho Y, Yamada S, Kageyama R (2004) Identification of a novel basic helix-loop-helix gene, *Heslike*, and its role in GABAergic neurogenesis. J Neurosci 24: 3672–3682
26. Jarriault S, Brou C, Logeat F, Schroeter EH, Kopan R, Israel A (1995) Signalling downstream of activated mammalian Notch. Nature 377:355–358
27. Ohtsuka T, Ishibashi M, Gradwohl G, Nakanishi S, Guillemot F, Kageyama R (1999) Hes1 and Hes5 as Notch effectors in mammalian neuronal differentiation. EMBO J 18: 2196–2207
28. Honjo T (1996) The shortest path from the surface to the nucleus: RBP-J κ/Su(H) transcription factor. Genes Cells 1:1–9
29. Selkoe D, Kopan R (2003) Notch and presenilin: regulated intramembrane proteolysis links development and degeneration. Annu Rev Neurosci 26:565–597
30. Gaiano N, Fishell G (2002) The role of Notch in promoting glial and neural stem cell fates. Annu Rev Neurosci 25:471–490
31. Ishibashi M, Moriyoshi K, Sasai Y, Shiota K, Nakanishi S, Kageyama R (1994) Persistent expression of helix-loop-helix factor HES-1 prevents mammalian neural differentiation in the central nervous system. EMBO J 13:1799–1805

32. Tomita K, Ishibashi M, Nakahara K, Ang SL, Nakanishi S, Guillemot F, Kageyama R (1996) Mammalian hairy and Enhancer of split homolog 1 regulates differentiation of retinal neurons and is essential for eye morphogenesis. Neuron 16:723–734
33. Kageyama R, Ohtsuka T (1999) The Notch-Hes pathway in mammalian neural development. Cell Res 9:179–188
34. Nakamura Y, Sakakibara S, Miyata T, Ogawa M, Shimazaki T, Weiss S, Kageyama R, Okano H (2000) The bHLH gene hes1 as a repressor of the neuronal commitment of CNS stem cells. J Neurosci 20:283–293
35. Ohtsuka T, Sakamoto M, Guillemot F, Kageyama R (2001) Roles of the basic helix-loop-helix genes *Hes1* and *Hes5* in expansion of neural stem cells of the developing brain. J Biol Chem 276:30467–30474
36. Nishimura M, Isaka F, Ishibashi M, Tomita K, Tsuda H, Nakanishi S, Kageyama R (1998) Structure, chromosomal locus, and promoter of mouse *Hes2* gene, a homologue of *Drosophila hairy* and *Enhancer of split*. Genomics 49:69–75
37. Issac PS, Ziff EB (1998) Genetic elements regulating HES-1 induction in Wnt-1-transformed PC12 cells. Cell Growth Differ 9:827–836
38. Nakashima K, Takizawa T, Ochiai W, Yanagisawa M, Hisatsune T, Nakafuku M, Miyazono K, Kishimoto T, Kageyama R, Taga T (2001) BMP2-mediated alteration in the developmental pathway of fetal mouse brain cells from neurogenesis to astrocytogenesis. Proc Natl Acad Sci U S A 98:5868–5873
39. Solecki DJ, Liu XL, Tomoda T, Fang Y, Hatten ME (2001) Activated Notch2 signaling inhibits differentiation of cerebellar granule neuron precursors by maintaining proliferation. Neuron 31:557–568
40. Hatakeyama J, Bessho Y, Katoh K, Ookawara S, Fujioka M, Guillemot F, Kageyama R (2004) *Hes* genes regulate size, shape and histogenesis of the nervous system by control of the timing of neural stem cell differentiation. Development 131:5539–5550
41. Hirata H, Tomita K, Bessho Y, Kageyama R (2001) Hes1 and Hes3 regulate maintenance of the isthmic organizer and development of the mid/hindbrain. EMBO J 20:4454–4466
42. Hirata H, Ohtsuka T, Bessho Y, Kageyama R (2000) Generation of structurally and functionally distinct factors from the basic helix-loop-helix gene *Hes3* by alternative first exons. J Biol Chem 275:19083–19089
43. Ishibashi M, Ang SL, Shiota K, Nakanishi S, Kageyama R, Guillemot F (1995) Targeted disruption of mammalian *hairy* and *Enhancer of split* homolog-1 (*HES-1*) leads to up-regulation of neural helix-loop-helix factors, premature neurogenesis, and severe neural tube defects. Genes Dev 9:3136–3148
44. Cau E, Gradwohl G, Casarosa S, Kageyama R, Guillemot F (2000) *Hes* genes regulate sequential stages of neurogenesis in the olfactory epithelium. Development 127:2323–2332
45. Hitoshi S, Seaberg RM, Koscik C, Alexson T, Kusunoki S, Kanazawa I, Tsuji S, van der Kooy D (2004) Primitive neural stem cells from the mammalian epiblast differentiate to definitive neural stem cells under the control of Notch signaling. Genes Dev 18:1806–1811
46. Kamakura S, Oishi K, Yoshimatsu T, Nakafuku M, Masuyama N, Gotoh Y (2004) Hes binding to STAT3 mediates crosstalk between Notch and JAK-STAT signaling. Nat Cell Biol 6: 547–554
47. Sakamoto M, Hirata H, Ohtsuka T, Bessho Y, Kageyama R (2003) The basic helix-loop-helix genes Hesr1/Hey1 and Hesr2/Hey2 regulate maintenance of neural precursor cells in the brain. J Biol Chem 278:44808–44815
48. Vernon AE, Devine C, Philpott A (2003) The cdk inhibitor p27Xic1 is required for differentiation of primary neurones in Xenopus. Development 130:85–92
49. Liu Y, Encinas M, Comella JX, Aldea M, Gallego C (2004) Basic helix-loop-helix proteins bind to TrkB and p21(Cip1) promoters linking differentiation and cell cycle arrest in neuroblastoma cells. Mol Cell Biol 24:2662–2672
50. Fode C, Ma Q, Casarosa S, Ang SL, Anderson DJ, Guillemot F (2000) A role for neural determination genes in specifying the dorsoventral identity of telencephalic neurons. Genes Dev 14:67–80

51. Parras CM, Schuurmans C, Scardigli R, Kim J, Anderson DJ, Guillemot F (2002) Divergent functions of the proneural genes Mash1 and Ngn2 in the specification of neuronal subtype identity. Genes Dev 16:324–338
52. Lo L, Dormand E, Greenwood A, Anderson DJ (2002) Comparison of the generic neuronal differentiation and neuron subtype specification functions of mammalian *achaete-scute* and *atonal* homologs in cultured neural progenitor cells. Development 129:1553–1567
53. Fode C, Gradwohl G, Morin X, Dierich A, LeMeur M, Goridis C, Guillemot F (1998) The bHLH protein NEUROGENIN 2 is a determination factor for epibranchial placode-derived sensory neurons. Neuron 20:483–494
54. Ma Q, Chen Z, del Barco Barrantes I, de la Pompa JL, Anderson DJ (1998) Neurogenin1 is essential for the determination of neuronal precursors for proximal cranial sensory ganglia. Neuron 20:469–482
55. Ma Q, Fode C, Guillemot F, Anderson DJ (1999) Neurogenin1 and neurogenin2 control two distinct waves of neurogenesis in developing dorsal root ganglia. Genes Dev 13:1717–1728
56. Kruger M, Ruschke K, Braun T (2004) NSCL-1 and NSCL-2 synergistically determine the fate of GnRH-1 neurons and control necdin gene expression. EMBO J 23:4353–4364
57. Morikawa Y, D′Autreaux F, Gershon MD, Cserjesi P (2007) Hand2 determines the noradrenergic phenotype in the mouse sympathetic nervous system. Dev Biol 307:114–126
58. Chien C, Hsiao CD, Jan LY, Jan YN (1996) Neuronal type information encoded in the basic-helix-loop-helix domain of proneural genes. Proc Natl Acad Sci U S A 93:13239–13244
59. Nakashima K, Yanagisawa M, Arakawa H, Kimura N, Hisatsune T, Kawabata M, Miyazono K, Taga T (1999) Synergistic signaling in fetal brain by STAT3-Smad1 complex bridged by p300. Science 284:479–482
60. Sun Y, NadalVicens M, Misono S, Lin MZ, Zubiaga A, Hua X, Fan G, Greenberg ME (2001) Neurogenin promotes neurogenesis and inhibits glial differentiation by independent mechanisms. Cell 104:365–376
61. Tomita K, Moriyoshi K, Nakanishi S, Guillemot F, Kageyama R (2000) Mammalian *achaete-scute* and *atonal* homologs regulate neuronal versus glial fate determination in the central nervous system. EMBO J 19:5460–5472
62. Nieto M, Schuurmans S, Britz O, Guillemot F (2001) Neural bHLH genes control the neuronal versus glial fate decision in cortical progenitors. Neuron 29:401–413
63. Bae S, Bessho Y, Hojo M, Kageyama R (2000) The bHLH gene *Hes6*, an inhibitor of *Hes1*, promotes neuronal differentiation. Development 127:2933–2943
64. Gratton MO, Torban E, Jasmin SB, Theriault FM, German MS, Stifani S (2003) Hes6 promotes cortical neurogenesis and inhibits Hes1 transcription repression activity by multiple mechanisms. Mol Cell Biol 23:6922–6935
65. Koyano-Nakagawa N, Kim J, Anderson D, Kintner C (2000) Hes6 acts in a positive feedback loop with the neurogenins to promote neuronal differentiation. Development 127:4203–4216
66. Jhas S, Ciura S, Belanger-Jasmin S, Dong Z, Llamosas E, Theriault FM, Joachim K, Tang Y, Liu L, Liu J, Stifani S (2006) Hes6 inhibits astrocyte differentiation and promotes neurogenesis through different mechanisms. J Neurosci 26:11061–11071
67. Ohsawa R, Ohtsuka T, Kageyama R (2005) Mash1 and Math3 are required for development of branchiomotor neurons and maintenance of neural progenitors. J Neurosci 25:5857–5865
68. Hatakeyama J, Tomita K, Inoue T, Kageyama R (2001) Roles of homeobox and bHLH genes in specification of a retinal cell type. Development 128:1313–1322
69. Inoue T, Hojo M, Bessho Y, Tano Y, Lee JE, Kageyama R (2002) *Math3* and *NeuroD* regulate amacrine cell fate specification in the retina. Development 129:831–842
70. Lu QR, Cai L, Rowitch D, Cepko CL, Stiles CD (2001) Ectopic expression of *Olig1* promotes oligodendrocyte formation and reduces neuronal survival in developing mouse cortex. Nat Neurosci 4:973–974
71. Zhou Q, Choi G, Anderson DJ (2001) The bHLH transcription factor Olig2 promotes oligodendrocyte differentiation in collaboration with Nkx2.2. Neuron 31:791–807
72. Mizuguchi R, Sugimori M, Takebayashi H, Kosako H, Nagao M, Yoshida S, Nabeshima Y, Shimamura K, Nakafuku M (2001) Combinatorial roles of Olig2 and Neurogenin2 in the

coordinated induction of pan-neuronal and subtype-specific properties of motoneurons. Neuron 31:757–771

73. Novitch BG, Chen AI, Jessell TM (2001) Coordinate regulation of motor neuron subtype identity and pan-neuronal properties by the bHLH repressor Olig2. Neuron 31:773–789
74. Wu Y, Liu Y, Levine EM, Rao MS (2003) Hes1 but not Hes5 regulates an astrocyte versus oligodendrocyte fate choice in glial restricted precursors. Dev Dyn 226:675–689
75. Hojo M, Ohtsuka T, Hashimoto N, Gradwohl G, Guillemot F, Kageyama R (2000) Glial cell fate specification modulated by the bHLH gene *Hes5* in mouse retina. Development 127:2515–2522
76. Furukawa T, Mukherjee S, Bao ZZ, Morrow EM, Cepko CL (2000) rax, Hes1, and notch1 promote the formation of Müller glia by postnatal retinal progenitor cells. Neuron 26:383–394
77. Takatsuka K, Hatakeyama J, Bessho Y, Kageyama R (2004) Roles of the bHLH gene Hes1 in retinal morphogenesis. Brain Res 1004:148–155
78. Muroyama Y, Fujiwara Y, Orkin SH, Rowitch DH (2005) Specification of astrocytes by bHLH protein SCL in a restricted region of the neural tube. Nature 438:360–363
79. Kiecker C, Lumsden A (2005) Compartments and their boundaries in vertebrate brain development. Nat Rev Neurosci 6:553–564
80. Baek JH, Hatakeyama J, Sakamoto S, Ohtsuka T, Kageyama R (2006) Persistent and high levels of Hes1 expression regulate boundary formation in the developing central nervous system. Development 133:2467–2476
81. Hans S, Scheer N, Riedl I, v Weizsäcker E, Blader P, Campos-Ortega JA (2004) her3, a zebrafish member of the hairy-E(spl) family, is repressed by Notch signaling. Development 131:2957–2969
82. Ninkovic J, Tallafuss A, Leucht C, Topczewski J, Tannhauser B, Solnica-Krezel L, Bally-Cuif L (2005) Inhibition of neurogenesis at the zebrafish midbrain-hindbrain boundary by the combined and dose-dependent activity of a new hairy/E(spl) gene pair. Development 132: 75–88
83. Shimojo H, Ohtsuka T, Kageyama R (2008) Oscillations in Notch signaling regulate maintenance of neural progenitors. Neuron 58:52–64
84. Murata K, Hattori M, Hirai N, Shinozuka Y, Hirata H, Kageyama R, Sakai T, Minato N (2005) Hes1 directly controls cell proliferation through the transcriptional repression of p27Kip1. Mol Cell Biol 25:4262–4271
85. Ström A, Arai N, Leers J, Gustafsson JÅ (2000) The *Hairy* and *Enhancer of Split* homologue-1 (HES-1) mediates the proliferative effect of 17 – estradiol on breast cancer cell lines. Oncogene 19:5951–5953
86. Hartman J, Müller P, Foster JS, Wimalasena J, Gustafsson JÅ, . Ström A (2004) HES-1 inhibits 17 -estradiol and heregulin- 1-mediated upregulation of E2F-1. Oncogene 23 8826–8833

Chapter 3
Transcription Factors and Muscle Differentiation

Gérard Cabello, François Casas, and Chantal Wrutniak-Cabello

Abstract Through their ability to repress irreversible cell cycle arrest, some transcription factors such as cellular oncogenes are considered as potent repressors of myoblast differentiation. Interestingly, their expression and/or activity are regulated by ligand-dependent transcription factors. Indeed, functional interactions between these proteins are deeply involved in the shift from proliferation to differentiation. Whereas c-Myc efficiently represses myoblast withdrawal from the cell cycle and terminal differentiation, c-Jun exerts a dual myogenic influence. In proliferating cells, it inhibits cell cycle arrest through sequestration of MyoD in a TR/c-Jun/MyoD complex; in parallel, high levels of Jun/Fos AP-1 complexes inhibit MyoD and myogenin expression. At the onset of myoblast differentiation, RXR dissociates the TR/c-Jun/MyoD complex, thus restoring a full MyoD transcriptional activity; in addition, it allows the liganded T3 nuclear receptor to inhibit TPA-inducible AP-1 activity, an event favouring c-Jun/ATF2 formation, a complex stimulating myogenin expression. In addition, inhibition of TPA-inducible AP-1 activity induces the expression of BTG1, a potent coactivator of myogenic factors transcriptional activity. In parallel, RXR expression, through the synthesis of its truncated form addressed into mitochondria, could potentiate the activity of p43, a mitochondrial T3-dependent transcription factor. This results in a stimulation of mitochondrial activity inducing a strong downregulation of c-Myc expression leading to cell cycle arrest and terminal differentiation. Overall, targeting of cellular oncogenes by ligand-dependent transcription factors regulates myoblast withdrawal from the cell cycle. In addition, it appears that RXR has to be considered as a master gene able to reverse the myogenic influence of c-Jun and TR, also involved in a downregulation of c-Myc expression.

G. Cabello (✉)
INRA UMR 866 Différenciation Cellulaire et Croissance (INRA, Université Montpellier 1, Université Montpellier 2), 2, place Viala, 34060 Montpellier Cedex 1, France
e-mail: cabello@supagro.inra.fr

A. Giordano, U. Galderisi (eds.), *Cell Cycle Regulation and Differentiation in Cardiovascular and Neural Systems*, DOI 10.1007/978-1-60327-153-0_3,

Keywords bHLH transcription factors · Cellular oncogenes · Triiodothyronine (T3) nuclear receptor · Myoblast differentiation · T3 mitochondrial receptors · Retinoic acid receptors

1 Introduction

Skeletal muscle development involves complex events, occurring during embryonic life and initiated in the somitic area. Acquisition of the muscle lineage results from signals secreted by the neural tubule and the notochord inducing the expression of Pax3, one of the earlier markers of the myogenic precursors, in the dorso-median part of the dermomyotome [1]. Expression of Myf5 and MyoD is detected thereafter in the highly proliferating cells myocytes and myoblasts, which undergo terminal differentiation. This last process leads to the formation of myotubes, plurinucleated structures considered as precursors of muscle fibres, after a phase of maturation. Along all these processes the crucial role of myogenic, muscle-specific bHLH transcription factors (Myf5, MyoD, myogenin, MRF4) is well established. In vitro experiments have established the myogenic potency of each factor in demonstrating that independent overexpression of Myf5, MyoD, myogenin or MRF4 induces a skeletal muscle phenotype in non-myogenic cells [2–6]. In vivo experiments using gene disruption in mice have provided a more clear-cut knowledge of their respective involvement in the different steps of muscle development. While Myf5 and MyoD appear to be involved in acquisition of the muscle phenotype [7], myogenin plays a major role in the induction of terminal differentiation [8, 9]. For a long time, the influence of MRF4 has been restricted to myotube maturation; however, MRF4 mRNA levels in mice display a particular biphasic pattern, with a transient expression detected between days 9 and 11 post-coitum (pc) and a permanent expression occurring after day 16 pc [10]. In a more recent study, Kassar-Duchossoy et al. [11] demonstrated that, in parallel to Myf5 and MyoD, the transient earlier expression of MRF4 was also involved in the induction of muscle lineage.

The importance of processes following the determination of the myogenic lineage is well illustrated by the absence of a significant increase observed in the number of muscle fibres following the perinatal period in most vertebrates. Consequently, this number partly defines the potency of postnatal muscle growth, which essentially occurs through myofibre hypertrophy. Therefore, all processes of muscle tissue development during the foetal life display particular importance. Submitted to tight regulation, myoblast terminal differentiation represents a major step in muscle development. As only postmitotic cells are able to undergo this process, irreversible myoblast withdrawal from the cell cycle is an absolute prerequisite for myotube formation and expression of major muscle-specific proteins. This event is of crucial importance, with an occurrence too early leading to a reduced duration of myoblast proliferation decreasing both the number of cells able to fuse into myotubes and consequently the definitive number of muscle fibres. Conversely, an exit failure from the cell cycle would lead to impaired myotube formation and muscle mass development.

Several studies have focused on the influence of MyoD on the irreversible cell cycle arrest allowing terminal differentiation through an interaction with the retinoblastoma protein (Rb). Rb phosphorylation is strongly involved in myoblast withdrawal from the cell cycle by regulating the formation of Rb/E2F complexes. Their formation leads to inactivation of the E2F/DP1 transcriptional complex normally upregulating the expression of genes needed for cell cycle completion [12, 13]. After Rb hyperphosphorylation by cyclinD/cdk4 and cyclinD/cdk6 complexes, E2F is released allowing cell proliferation [14]. Interestingly, by interacting with Rb, and probably inhibiting its phosphorylation, MyoD restores the ability of Rb to sequestrate E2F and stop progression of the cell cycle [15]. Therefore, MyoD appears to be involved in irreversible myoblast withdrawal from the cell cycle. In addition, through its transcriptional activity, this myogenic factor also upregulates the expression of numerous muscle-specific genes involved in differentiation, like myogenin, and in muscle functionality, such as genes encoding contractile proteins. However, several studies have suggested an influence of cell cycle control on MyoD transcriptional activity. For instance, cyclin D1 expression in proliferating myoblasts increases cdk4 nuclear localization, thus favouring the formation of MyoD/cdk4 complexes and inhibiting the transcriptional activity of the myogenic factor [16]. In agreement with this possibility, other studies have shown that cyclin D1/cdk inhibits MyoD and myogenin activity [17, 18]. Consequently, inhibition of the activity of these complexes by p21 and p16 CKI leads to a stimulation of MyoD transcriptional activity [18]. As MyoD induces p21 expression, such a mechanism provides a positive amplification loop able to enhance MyoD activity.

Intriguingly, however, in vitro experiments show that although highly expressed during the proliferation phase in practically all kinds of myoblast cultures, MyoD fails to induce myoblast withdrawal from the cell cycle and in turn terminal differentiation. In most myoblast cell lines, terminal differentiation is only inducible after decreasing the concentration of serum in the culture medium. In other words, repressive MyoD-dependent or independent mechanisms occurring in proliferating myoblasts must be overcome to allow the induction of differentiation by myogenic factors.

This aspect has been particularly addressed while studying MyoD interacting partners. MyoD transcriptional activity is exerted through dimeric complexes with E12/E47 proteins [19]. Several factors have been characterized for their ability to interfere with this transcriptional activity. Among them, Id proteins displaying a helix-loop-helix (HLH) domain do not bind to DNA but either compete with MyoD for dimerization with E12/E47 proteins or directly interact with myogenic factors to produce transcriptional inactive complexes [20]. Strikingly, their high expression level recorded during proliferation strongly decreases at the induction of terminal differentiation [20]. A similar expression pattern has also been described for Mist 1 and MyoR, two other bHLH proteins repressing MyoD transcriptional activity by either competitive binding to E boxes (MyoD responsive elements) or sequestration of MyoD in inactive complexes [21, 22]. Other negative regulators such as Twist and ZEB have also been identified [23, 24]. Theoretically, the influence of these negative regulators acting by competition with myogenic factors should be overcome by MyoD overexpression. In contrast, however, several experimental data indicate that MyoD or myogenin overexpression fails to induce terminal differentiation in myoblasts either expressing high levels of proto-oncogenes [25, 26] or displaying a depressed mitochondrial activity [27]. These results clearly suggest that, besides the expression and/or functionality of myogenic factors, important inhibiting processes non-related to MyoD competitors need turning off to allow myoblast differentiation.

The possibility of turning on mechanisms to potentiate the activity of MRFs at the onset of terminal differentiation has also been investigated, alongside more recent underlining of the role of transcriptional coactivators. The ubiquitous acetyl-transferase coactivator CBP/p300 interacts with numerous transcription factors including myogenic factors, increasing their activity [28, 29]. Although CBP or p300 expression does not clearly differ during myoblast proliferation and differentiation, changes in p300 ability to target PCAF for interaction with MyoD and acetylation of this MRF could play an important role in MyoD activity [30, 31]. More recently, it has also been shown that the expression of BTG1 (B-cell translocated gene 1), another MyoD and myogenin coactivator [32], is induced at the onset of differentiation [33, 34]. Therefore, changes in the activity (CBP/p300) or the expression (BTG1) of MRF coactivators are probably involved in the induction of complete MyoD and myogenin activity, needed to undergo the differentiation processes.

Through their ability to repress irreversible cell cycle arrest, some transcription factors such as cellular oncogenes are considered as potent repressors of myoblast differentiation. This influence occurs at least partly independently of MRF and needs inhibiting to allow terminal differentiation. Interestingly, their expression and/or activity are regulated by ligand-dependent transcription factors. Indeed, functional interactions between these proteins are deeply involved in the shift from proliferation to differentiation. This review is focused on such mechanisms, and more particularly on the interplay with cellular oncogenes such as c-Myc and c-Jun, and ligand-dependent transcription factors like triiodothyronine nuclear receptors (TR) and 9-*cis*-retinoic acid receptors (RXR). Moreover, the unexpected influence of mitochondrial transcription factors targeting cellular oncogenes will also be reviewed. Our aim is to illustrate with some examples how these mechanisms could allow irreversible myoblast withdrawal from the cell cycle, in some cases by reversing the myogenic potency of the transcription factor.

2 Cellular Oncogenes

2.1 c-Jun and the AP-1 Complex

The AP-1 (activator protein 1) transcription factor is involved in multiple cellular regulations including proliferation and differentiation. Initially known as different heterodimers of members of the Jun (c-Jun, JunB and JunD) and Fos (c-Fos, FosB, Fra1 and Fra2) families [35–40], they also include homo- or heterodimerical complexes of Jun proteins displaying reduced stability. However, AP-1 complexes have progressively revealed an important diversity with members of the ATF/CREB family (ATF2, ATF3, ATF4) [41] shown to interact with members of the Jun family. Lastly, Maf proteins were also identified as physical partners of Jun and Fos proteins in these complexes [42, 43].

All these proteins share a charged DNA binding domain rich in basic amino acids (b), adjacent to a common amphipathic leucine zipper dimerization sequence (zip) [44, 45]. A more divergent transactivation domain is located in their N-terminal part (Fig. 3.1). Dimerization seems to be required to induce high-affinity binding of AP-1 to specific response elements. However, each complex displays a particular affinity for different responsive elements occurring in the promoter of AP-1 target genes. Jun/Fos heterodimers preferentially bind to TPARE (tetradecanoyl phorbol acetate-responsive element) [36], whereas Jun/ATF bind to CRE (cAMP-responsive element) [41] and Jun/Maf to MARE (Maf-responsive elements) [42]. These different patterns of sequence recognition suggest that c-Jun could activate the expression of different sets of genes according to the identity of its heterodimerization partner, with different consequences at the cellular level and in particular on terminal differentiation.

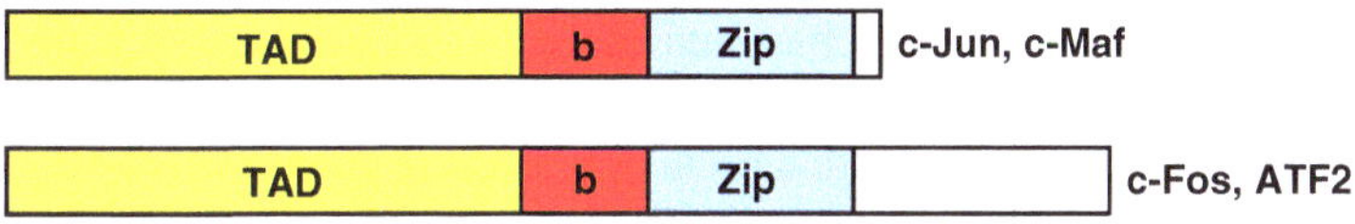

Fig. 3.1 Functional domains of c-Jun, c-Maf, c-Fos and ATF2 proteins. TAD: transactivation domain; b: basic domain; Zip: leucine zipper dimerization domain

The c-Jun promoter is highly conserved between species and displays binding sites for different transcription factors (for review [46]). In particular, one AP-1 binding site which preferentially binds c-Jun/ATF2 heterodimers has been characterized in the proximal region of the murine promoter [47]. A second AP-1 binding site identified in the distal part of this regulatory region mediates the influence of TPA or growth factor on the expression of this cellular oncogene [48, 49]. Consequently, c-Jun can positively autoregulate its own expression through the binding of Jun/ATF complexes in particular to these responsive elements. In addition, MEF-2 (myocyte enhancer factor-2) responsive elements have also been identified [50] in agreement with the observation that MEF-2A increases c-Jun expression during myoblast differentiation [51].

Activation of preformed AP-1 complexes is induced by several kinases mediating the influence of either TPA and growth factors (ERK: extracellular-signal regulated kinases; FRK: Fos-regulating kinase) or cytokines and genotoxic factors (JNK, p38). Whereas ERK and FRK phosphorylate c-Fos [52, 53], p38 phosphorylates ATF2 [54], leading to an activation of each protein. In addition, JNK (Jun N-terminal kinase), belonging to the MAPK family, phosphorylates several serines and threonines in the N-terminal part of c-Jun. This event increases the affinity for DNA [55] and the transcriptional activity of c-Jun through recruitment of CBP, a histone acetylase coactivator [56, 57]. In addition, it increases the stability of the protein by reducing its ubiquitination level [58].

2.1.1 The c-Jun Myogenic Influence

Different experimental approaches decreasing c-Jun expression and/or activity in fibroblasts have clearly established the important role of c-Jun in processes of cell growth. Microinjections of c-Jun antibodies or of AP-1 binding sites in order to sequestrate AP1 complexes induce marked alterations in cell proliferation and inhibit cell entry into S phase [59, 60]. Conversely, according to Pfarr et al. [61] c-Jun overexpression increases the proportion of cells in S, G2 and M relative to the G1 phase. Studies surrounding the mechanisms supporting this influence have led to the identification of AP-1 binding sequences in the promoter of genes involved in the cell cycle, with a positive regulation of cyclin D1 expression by AP1 activity [62]. In addition, the strong alteration in the proliferation rate of c-Jun-depleted fibroblasts is related to an increased expression of p53 and its target p21 [63, 64]. Altogether, these data underlining the c-Jun regulation of cell cycle genes, in particular p53 and p21 involved in myoblast withdrawal from the cell cycle [64], suggest the need to inhibit c-Jun expression and/or activity to allow irreversible cell cycle arrest in muscle cells. However, other studies have provided a different theory suggesting more subtle influences. For instance, Riabowol et al. [60] reported that the progressive alteration of proliferation occurring during fibroblast senescence is associated with changes in the nature of predominant AP-1 complexes, thus suggesting that not all complexes promote an efficient cell growth. Although not yet performed in myoblasts, these studies indicate that some AP-1 transcription factors could exert a strong repression of myoblast withdrawal from the cell cycle.

The influence of AP-1 on cell differentiation depends on the cell type. For instance, c-Fos and c-Jun expression is induced during myeloid cell differentiation [65], whereas c-Jun overexpression induces macrophage differentiation in U-937 cells [66]. Similarly, stimulation of c-Jun/c-Fos activity by TPA induces human keratinocyte differentiation [67]. In contrast, however, TPA-inducible AP-1 activity (Jun/Fos) is considered a potent repressor of adipocyte [68] or chondrocyte differentiation [69].

In myoblasts, stimulation of AP-1 activity by TPA or okadaic acid abrogates terminal differentiation [70, 71], supporting the initial conclusion that c-Jun should be considered a major myogenic repressor. Although this influence remained more or less understood, it was generally assumed that this repression involved undefined AP-1 complexes. As with overexpression of c-Jun, stimulation of Jun/Fos AP-1 activity by TPA or coexpression of c-Jun/c-Fos or c-Jun/Fra2 efficiently represses myoblast withdrawal from the cell cycle [72]. This suggests that c-Jun exerts essentially its inhibitory myogenic influence through TPA-inducible AP-1 complexes. This possibility agrees with the observation that c-Jun/c-Fos binds to the MyoD promoter and decreases MyoD expression [73]. Similarly, coexpression of c-Jun and c-Fos also reduces myogenin expression at the onset of differentiation [72]. Another explanation of this repressive c-Jun myogenic influence resides in the ability of TPA-inducible AP-1 complexes to directly inhibit the expression of BTG1, a transcriptional coactivator of Myf5, MyoD and myogenin [32]. Furthermore, a TPARE sequence occurring in the sheep myostatin promoter [74] indicates the ability of Jun/Fos complexes to increase the expression of this potent repressor of myoblast differentiation [75]. Lastly, Bengal et al. [76] reported the occurrence of physical interactions of c-Jun with MyoD and myogenin, leading to a potent repression of their transcriptional activity. Overall, it is reasonable to conclude from these studies that c-Jun is essentially a myogenic repressor, targeting either by itself or through AP-1 complexes the expression and activity of myogenic factors and inducing the expression of other myogenic repressors such as myostatin.

However, other studies have provided another theory on the myogenic activity of c-Jun. Thinakaran and Bag [77] reported no impairment of myogenin expression or terminal differentiation in L6 myoblasts with high c-Jun levels. In the same way, the expression of this cellular oncogene transiently increases at the onset of avian myoblast differentiation in contradiction with a potential role as myogenic repressor [72]. This is probably due to the stimulation of c-Jun transcription by MEF-2A [51]. Similarly, Maf expression is directly increased by MyoD during differentiation, indicating that the c-Jun/Maf AP-1 complex does not repress myoblast differentiation [78]. Recent studies have argued against c-Jun being considered only as a myogenic repressor. Indeed it regulates several important differentiation processes through complex mechanisms involving ligand-dependent transcription factors and transcriptional coactivators. Although stable c-Jun overexpression in avian myoblasts significantly delays myoblast withdrawal from the cell cycle through Jun/Fos complexes and physical interaction with MRFs, it subsequently induces a significant stimulation of terminal differentiation through Jun/ATF2 complexes [72]. These data agree with results concerning the myogenic influence of kinases stimulating different AP-1 complexes. Whereas Jun/Fos activity is regulated by the ERK pathway (extracellular signal-regulated kinases) [52, 53], Jun/ATF2 activity is specifically stimulated by the p38 pathway [79, 80]. Interestingly, p38 activation induces L8 or C2C12 myoblast differentiation [81–83] whereas activation of the Ras/MEK/ERK pathway exerts exactly the reverse influence [84].

2.2 The c-Myc Cellular Oncoprotein

In humans, the c-Myc cellular oncogene encodes a major 64 kDa transcription factor, considered, like c-Jun and c-Fos, as an immediate early gene product. It belongs to the Myc family comprising L-Myc and N-Myc, all members displaying a deregulated expression in numerous tumours [85]. The c-Myc protein includes a bHLH and a leucine zipper motif, all located in its terminal part (Fig. 3.2). A transactivation domain with two well-conserved regions, MB1 (Myc Box 1) and MB2 (Myc box 2) located in the N-terminal part of the protein, mediates its transcriptional activity.

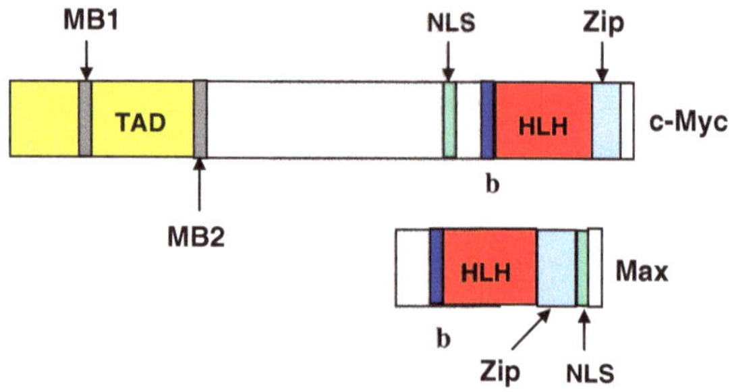

Fig. 3.2 Functional domains of c-Myc and Max proteins. TAD: transactivation domain; b: basic domain; HLH: helix-loop-helix sequence; Zip: leucine zipper dimerization domain; MB1: Myc 1 box; MB2: Myc 2 box; NLS: nuclear localization signal

c-Myc binds to DNA essentially as a heterodimerical complex with Max (Myc-associated factor X), a bHLH/bzip protein devoid of a transactivation domain. The responsive elements recognized are particularly E boxes such as CACGTG, CACATG or CAyGTG. After DNA binding, interaction of c-Myc with proteins involved in the transcriptional machinery such as the TATA box binding protein (TBP), or in chromatin remodelling such as histone acetyltransferase, helicase or CBP/p300, stimulates transcription of its target genes. With a modest transcriptional efficiency it has been proposed that c-Myc weakly activates the expression of a broad spectrum of important regulator genes for cell proliferation [86]. In addition, c-Myc also represses the expression of some of its target genes, in particular, directly reducing the transcription of its own gene according to a negative regulatory loop [87, 88]. According to this repressive transcriptional activity, the inhibition of genes involved in cell adhesion probably contributes to the loss of contact inhibition associated with c-Myc-induced cell transformation. As the c-Myc variant devoid of a transactivation domain conserves this repressive activity and induces a wild-type c-Myc cell transformation and apoptosis [89], this transrepressive activity appears to play an important role in the c-Myc influence on oncogenesis, proliferation and apoptosis [90].

The transcriptional activity of c-Myc is regulated by a large number of proteins able to inhibit Myc activity according to competition for dimerization with Max and/or binding to Myc responsive elements. For instance, proteins belonging from the Mad family interact with Max and repress c-Myc target genes by recruiting a complex including NCor, Sin3a, Sin3b and histone deacetylases 1 and 2 [91].

In addition to the regulation of its activity, c-Myc expression is tightly regulated at the transcriptional and post-transcriptional level. Among numerous regulators of its expression, the pathway c-Src, through the tyrosine kinase abl, has been identified [92]. In addition, the influence of the cAMP pathway, calcium signalling or reactive oxygen species production through the NFκB transcription factor have been particularly studied [93–97]. Stability of c-Myc transcripts is also a major way of controlling c-Myc amounts, with a half-life lower than 30 min. At the protein level, dysregulation of c-Myc translation has been reported in several cancers [98, 99], and an increase in c-Myc protein stability is observed in Burkitt's lymphoma [100].

Indeed, protein half-life seems to be also an important target as c-Myc displays three sequences involved in protein stability: a C-terminal stabilization signal (stabilon) [101], a degradation signal (degron) and an additional destabilizator signal (the D-element) [102]. Therefore, the amount of the c-Myc protein in a particular cell is the result of complex mechanisms influencing transcription, translation and/or stability of the transcript and the protein.

2.2.1 c-Myc Is a Potent Inhibitor of Myoblast Differentiation

Besides a clear activation of cell proliferation, c-Myc is known to abrogate differentiation in numerous cell types. Indeed, a decrease in c-Myc expression is considered an important event required to allow differentiation in most cell types tested. In addition, inhibition of c-Myc expression by itself is often sufficient to induce differentiation, an event strongly inhibited by c-Myc overexpression. The repressive transcriptional activity of c-Myc mentioned previously could play a significant role in the ability of this cellular oncogene to maintain cell proliferation. For instance, c-Myc represses the expression of genes involved in cell differentiation such as c/EBPα in preadipocytes [103, 104] or those inhibiting the cell cycle such as $p27^{KIP1}$ or p21 [105–108].

In myoblasts, as in other cell types, c-Myc expression decreases at the onset of terminal differentiation [109–111], and its overexpression abrogates myoblast differentiation [26, 112]. Strikingly, the influence of some myogenic regulators involves changes in c-Myc expression and/or activity. Expressed in myotome, myoblasts and myofibres, the paired-related homeobox gene Pitx 2C plays a central role during development, possibly by balancing proliferation versus differentiation. Interestingly, its ability to enhance the proliferative potential of myoblasts and decrease their terminal differentiation is clearly associated with a potent rise in c-Myc expression [113]. Similarly, Stat 3, besides its ability to interact with MyoD and inhibit the transcriptional activity of this MRF, also upregulates c-Myc expression thereby promoting cell cycle progression and repressing myoblast withdrawal from the cell cycle [114]. Conversely, in L6E9 myoblasts, TGFβ1 quickly decreases c-Myc expression in a mitogen-rich medium, delays progression through the G1 phase of the cell cycle and stimulates cell commitment to terminal differentiation [115]. Similarly, the murine protein p202a, expressed during the fusion of myoblasts to myotubes [116], can be induced by interferon. Its overexpression in C2C12 myoblasts inhibits proliferation, at least partly by a direct interaction with c-Myc disrupting Myc/Max complexes. This event leads to inhibition of c-Myc transcriptional activity and a consecutive decrease in the expression of its target genes [117]. In the same way, the activity and expression of double-stranded RNA-activated protein kinase PKR increases from the onset of myoblast differentiation, and it has been shown to inhibit proliferation and potentiate differentiation, in association with a downregulation of c-Myc [118].

Undoubtedly, the c-Myc antimyogenic activity is partly exerted at the level of the cell cycle. Earlier studies clearly suggested that c-Myc affects the expression and activity of factors regulating the cell cycle [119–121]. In hematopoietic cells, c-Myc abrogates differentiation via an inhibition of p21 expression, mediated by

Miz-1 [107, 122]. In CV1 cells, Kitaura et al. [123] reported a direct interaction of c-Myc with p21, leading to inactivation of this kinase and completion of the cell cycle. Reciprocally, p21 represses the transcriptional activity of c-Myc, by inhibiting formation of Myc-Max heterodimers. Therefore, the balance of the reciprocal inactivation between c-Myc and p21 may determine the course of cellular proliferation and differentiation processes. Another mechanism mediating the influence of c-Myc on proliferation has been reported by Lasorella et al. [124]. According to this study, c-Myc induces the expression of Id2, which besides inhibiting MyoD activity also interacts with pocket proteins Rb, p107 or p130 which regulate the cell cycle, thus disrupting their antiproliferative activity. Overall, c-Myc inhibits myoblast differentiation by impairing expression and activity of proteins involved in regulating the cell cycle, thereby inhibiting irreversible proliferation arrest.

However, besides its negative influence on myoblast withdrawal from the cell cycle, c-Myc also appears to interfere with the expression and/or activity of myogenic factors. In quail myoblasts, La Rocca et al. reported that v-Myc, the viral counterpart of c-Myc, represses myoblast terminal differentiation and depresses Myf5, MyoD and myogenin expression [125, 126]. According to these authors, v-Myc inhibits MyoD expression by reducing its promoter activity, whereas MyoD overexpression restores terminal differentiation. This suggests that MyoD transcriptional repression could be a major and rate-limiting step in the negative influence of v-Myc or c-Myc on myoblast differentiation. In contrast, Crescenzy et al. [112] had already reported data suggesting that the differentiation block induced by v-Myc in quail myoblasts did not clearly depend on the transcriptional silencing of MyoD. A similar conclusion could be drawn from a study performed in NIH-3T3 cells where even though c-Myc overexpression could suppress the MRF-induced myogenic commitment of these cells, neither additional MyoD nor myogenin overexpression could bypass the myogenic block induced by c-Myc. In agreement with these last results, more recent studies have established that in quail myoblasts (Seyer et al., unpublished data) or murine C2C12 myoblasts [26], c-Myc overexpression only inhibits myogenin expression, without any influence on Myf5 or MyoD expression. Moreover, this cellular oncogene fully abrogates the ability of MyoD or myogenin overexpression to potentiate myoblast differentiation [26].

Overall, in myoblasts, the c-Myc cellular oncogene appears to play a major role in the balance between proliferation and differentiation. Clearly, a high expression level seems incompatible with the induction of a differentiation programme, in agreement with downregulated expression levels in all myoblast lines tested at the onset of differentiation. One major target of c-Myc is probably the irreversible cell cycle arrest, mediated by regulating the expression and/or activity of proteins involved in cell cycle progression, such as p21, Rb, p107 or p130. In addition, despite some discrepancies, myogenin expression seems a major target of c-Myc. Furthermore, several reports also suggest that c-Myc abrogates the ability of MyoD or myogenin to induce myogenic differentiation, but the mechanisms involved remain unclear. One attractive possibility is that the strong repression of myoblast withdrawal from the cell cycle cannot be overcome by MRFs, due to the inefficiency of MyoD/Rb interactions caused either by the c-Myc-induced impairment of Rb antiproliferative activity [124] or the downregulation of p21 [107, 122].

3 Ligand-Dependent Transcription Factors: Triiodothyronine Receptors

3.1 Triiodothyronine (T3) Nuclear Receptors

Nuclear receptors belong to a superfamily of ligand-dependent transcription factors including more than 50 members. A number of them are orphan receptors for which no specific ligands are known. They share a structural similarity with the occurrence of four functional domains: (i) the N-terminal domain, involved in ligand-independent transcriptional activity; (ii) the DNA binding domain, including two zinc fingers largely involved in DNA binding, conferring the recognition of specific hormone responsive elements (HREs), and including a dimerization sequence; (iii) the hinge domain, containing interacting sequences with transcriptional corepressors; (iv) the ligand binding domain including a succession of amphipathic helices allowing dimerization, and a C-terminal sequence allowing interaction with transcriptional coactivators (Fig. 3.3).

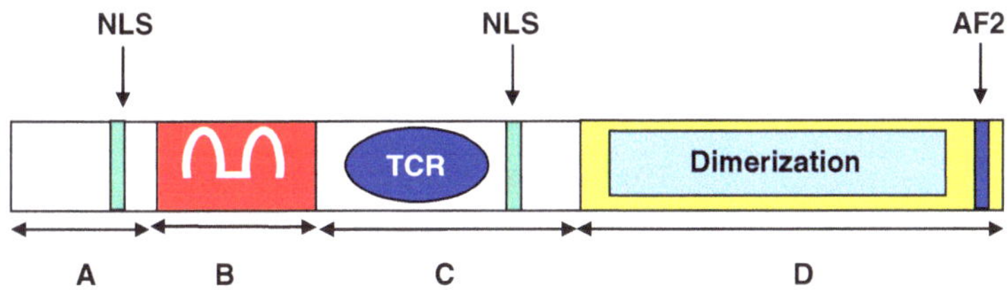

Fig. 3.3 Functional domains of the TRα receptor. (**A**) Ligand-independent transactivation domain; (**B**) DNA binding domain including two zinc fingers; (C) hinge region including interacting sequences with transcriptional corepressors (TCR) and a nuclear localization signal (NLS); (**D**) ligand binding domain including a dimerization sequence and an interacting sequence with transcriptional coactivators (AF2)

In the absence of ligand, members belonging to the glucocorticoid receptor subgroup sequestrate in a cytosolic complex needing hormone binding to translocate into the nucleus and bind to their specific HREs, leading to transcription of their target genes [127]. In contrast, members belonging to the thyroid hormone receptor (TR) subgroup (vitamin D3 receptor, VDR; retinoic acid receptors, RAR; retinoid X receptor, in fact 9-*cis*-retinoic acid receptors, RXR; peroxysome proliferator activated receptors, PPAR) display a constitutive nuclear localization and bind to their

HREs independently of T3 presence [128]. These features confer a dual transcriptional activity to TR. In the absence of their ligands they repress transcription, whereas ligand binding induces a potent stimulation of transcription of their target genes [128]. However, negative HREs have been described for TR, inducing, respectively, a decrease and an increase in gene expression in the presence and absence of T3 [129–131].

Triiodothyronine nuclear receptors are encoded by two genes c-erbAα and c-erbAβ located on chromosomes 17 and 3 in humans [132, 133], respectively. Each gene gives rise to several mRNAs via distinct promoters or alternative splicing processes. In addition to the nuclear receptor TRα, the c-erbAα1 mRNA also encodes three other proteins using internal AUG codons that allow differential translation (Fig. 3.4). Overall, a multitude of proteins are encoded by these two genes (Fig. 3.5) [134], but only in a restricted number do we really know the physiological role.

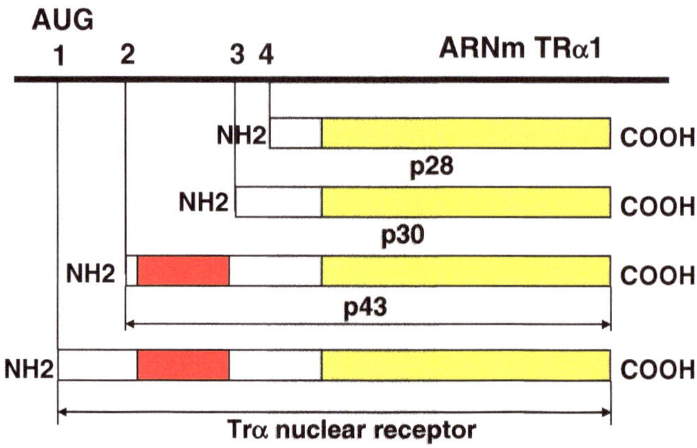

Fig. 3.4 The c-erbAα1 transcript encodes four different proteins by the use of different AUGs. Whereas p30 and TRα display a nuclear localization, p28 and p43 are specifically imported into mitochondria. P30 is believed to act as a negative regulator of TRα and β activity. P43 is a mitochondrial T3 receptor. The function of p28 in mitochondria remains unknown

TRs generally bind in heterodimeric complexes with other receptors belonging to the same subgroup II. TR heterodimers have been described with VDR [135], PPAR [136] or RAR [137, 138], but the TR/RXR heterodimer is considered the major transcriptional complex, possibly representing more than 80% of all TR transcriptional complexes [139]. In the absence of T3, TR interacts with transcriptional corepressors such as NCoR or SMRT, large proteins interacting with HDACs (histone deacetylases) [128]. This induces histone deacetylation and chromatin compaction, resulting in a repression of gene expression. By contrast, in the presence of T3, TR interacts with transcriptional coactivators which are either HACs (Histone acetyltransferases) or proteins recruiting HACs, thereby inducing histone

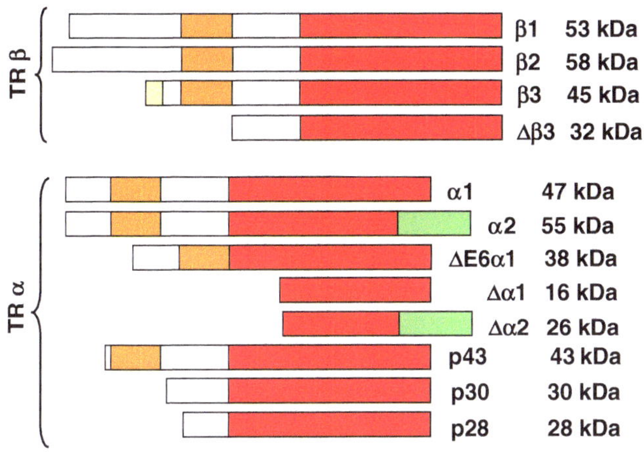

Fig. 3.5 Genes TRβ and TRα encode a multiplicity of proteins. TRβ receptors only differ by their N-terminal sequence. All proteins bind T3. TRα proteins are more or less truncated in their N-terminal part, relatively to the classical TRα nuclear receptor (α1). Due to alternative splicing α2 and Δα2 proteins display a different C-terminal part from other α proteins. α2, Δα2 and Δα1 proteins do not bind T3

acetylation, DNA decompaction and potent stimulation of target gene expression [128].

In addition to this direct transcriptional activity, TRs interact with other transcription factors such as AP-1. In RXR expressing cells, a functional interaction of TR with AP-1 strongly inhibits the transcriptional activity of each transcription factor [140, 141]. Through such mechanisms, TRs also indirectly regulate the expression of genes targeted by different transcription factors. Interestingly, direct and indirect TR transcriptional activities are involved in mechanisms controlling the balance between proliferation and differentiation in myoblasts.

3.1.1 The TR Myogenic Influence

Although the in vivo myogenic influence of T3 has been widely accepted over the last 50 years, it is only recently that its molecular basis has been elucidated. In vitro studies using cultured avian myoblasts demonstrated that physiological amounts of this hormone strongly stimulate myoblast differentiation essentially by potentiating irreversible myoblast withdrawal from the cell cycle [33, 142]. In addition, this influence was increased by TR overexpression and strongly potentiated by RXR coexpression [143]. These data initially led a search for TR/RXR target genes able to induce myoblast differentiation.

Using C2C12 murine myoblasts, Downes et al. [144] first demonstrated that in the presence of T3, TRs directly increase MyoD and myogenin expression, after binding to specific TREs (thyroid hormone responsive elements) occurring in the promoter of these two genes. Increasing the relative amounts MRFs compared to their antagonists could improve their ability to induce terminal differentiation. However, due to the absence of TREs on MyoD and myogenin, this mechanism does not function in avian myoblasts [33]. This suggests that a more general fundamental mechanism supports the TR myogenic influence. In agreement with this hypothesis, it appears that TR exerts its strong myogenic potency by way of different, sometimes contradictory, pathways including interferences with MyoD and AP-1 activities, as outlined in this review.

3.2 T3 Mitochondrial Receptors

The physiological importance of mitochondrial activity has, for a long time, been restricted to its involvement in ATP synthesis and cell fuel metabolism. However, in the last decade, numerous data have underlined its involvement in cell proliferation [145–147] and differentiation [27, 148–151]. Moreover, the discovery of their involvement in the induction of apoptosis (for review [152]) has led to the concept that these organelles influence all developmental processes. In addition to in vitro experiments, studies using transgenic mice have indicated that inhibition of mitochondrial activity induced by Tfam gene disruption (Tfam is a nuclear gene encoding a constitutive transcription factor of the mitochondrial genome) is associated with embryonic lethality around the 11th day *post-coitum* [153]. Furthermore, it is currently well accepted that all genetic abnormalities leading to deficiencies in mitochondrial activity are associated with degenerative diseases affecting a wide variety of tissues, including muscle [154]. All these data clearly suggest that these organelles not only influence in utero developmental processes but also the maintenance and regeneration of adult tissues.

In particular, several studies have reported a significant influence of mitochondrial activity on myoblast differentiation. Whereas inhibition of the organelle activity by different drugs reducing the mitochondrial membrane potential (FCCP) or the mitochondrial protein synthesis (chloramphenicol) strongly depresses terminal differentiation, stimulation of mitochondrial activity enhances myoblast differentiation [26, 27]. In line with these observations, a stimulation of mitochondrial activity occurs just before differentiation and persists throughout the differentiation period [155].

One crucial element involved in the regulation of mitochondrial activity is the control of mitochondrial genome transcription by specific mitochondrial transcription factors. Among them, Tfam is probably the best known, binding to specific sequences of the D-loop, a regulatory region of mitochondrial DNA transcription [156]. Interaction of Tfam with the transcriptional machinery including mt-F1B, mt-F2B and mt-RNA pol leads to a constitutive stimulation of organelle transcription [157, 158], through the synthesis of a polycistronic RNA subsequently cleaved into mature transcripts. More recently, other ligand-dependent transcription factors have been discovered.

3.2.1 P43: A Mitochondrial T3 Receptor Acting as a Ligand-Dependent Transcription Factor

In addition to the T3 nuclear receptor TRα1, the c-erbAα1 transcript also encodes a 43 kDa truncated form of this receptor, synthesized through the use of an internal AUG during the translational process (Fig. 3.3). This protein (p43), devoid of the N-terminal part of the nuclear receptor, is imported into the mitochondrial matrix [159]. P43 is detected in practically all tissues of all studied species (human, rat, mice, rabbit, chicken, bovine, pig, etc.), with the exception of brain tissue [159, 160]. In the mitochondrial matrix, p43 specifically binds to five sequences (mt-TREs) [161] of the mitochondrial genome in dimeric complexes and in the presence of T3 stimulates mitochondrial DNA transcription, protein synthesis, respiratory chain activity and mitochondriogenesis [159, 162].

Other truncated forms of nuclear receptors are also imported into mitochondria, such as mt-PPAR (devoid of the C-terminal part of PPARγ2) or mt-RXR (devoid of the N-terminal part of RXRα) [160, 163]. Although the mitochondrial influence of mt-PPAR remains unknown, similar to p43, mt-RXR is a ligand-dependent transcription factor of the mitochondrial genome. In addition, as its nuclear counterpart, it forms heterodimerical complexes with p43 which bind to mt-TRE, potentiating the transcriptional activity of the mitochondrial T3 receptor [163]. Therefore, the hormonal regulation of mitochondrial transcription displays striking analogies with that occurring for nuclear genes.

3.2.2 The p43 Pathway Induces Irreversible Cell Cycle Arrest and Myoblast Differentiation

In order to test the influence of this mitochondrial transcription factor on myoblast differentiation, p43 has been stably overexpressed in the C2C12 or QM7 cell lines. Strikingly, p43 induces terminal differentiation, even in the presence of high serum concentration in a culture medium considered inadequate to allow myoblast withdrawal from the cell cycle [27]. In addition to this surprising observation, studying the cell cycle by FACS (fluorescent analyzer cell sorter) clearly indicates that stimulation of mitochondrial activity by this pathway involves the induction of irreversible cell cycle arrest needed for terminal differentiation [26]. Moreover, after serum removal, p43 strongly stimulates myoblast differentiation [26, 27]. Conversely, chloramphenicol, a drug specifically inhibiting mitochondrial protein synthesis and consequently the p43 pathway downstream of mt-DNA transcription, abrogates myoblast differentiation, in particular by reducing myoblast withdrawal from the cell cycle [26, 27].

Another interesting finding was that p43 expression not only stimulates myoblast differentiation but also induces a preferential synthesis of slow myosin isoform (MHC I) through upregulation of calcineurin expression, thus leading to the formation of oxidative slow-type myotubes [164]. This last finding has recently been validated in transgenic mice specifically overexpressing p43 in skeletal muscle

[165]. Clearly, this mitochondrial transcription factor plays a major role in the proliferation/differentiation balance and is probably involved in the acquisition of the contractile and metabolic status of muscle fibres.

4 Targeting of Cellular Oncogenes by Ligand-Dependent Transcription Factors Regulates Myoblast Withdrawal from the Cell Cycle

4.1 TR Represses MyoD Transcriptional Activity in Proliferating Myoblasts

Downes et al. [166] initially reported an unexpected finding that could possibly alter TR activity in C2C12 myoblasts. They observed a lack of expression of RXR isoforms in proliferating cells, with expression induced simultaneously with the initiation of terminal differentiation. This surprising result confirmed later on in the avian QM7 myoblast line [143] raised the question of the identity of the transcriptional complex allowing the regulation of T3 target gene expression during myoblast proliferation. This led to the discovery of a complex including TR, c-Jun and MyoD mediating the transcriptional activity of the T3 receptor in the absence of RXR [143, 167]. Indeed, several findings demonstrate the simultaneous occurrence of the oncoprotein and the MRF in this complex: (i) c-Jun induces no TR transcriptional activity in cells not expressing MRFs; (ii) stable or transient MyoD expression in these cells restores this capacity; and (iii) in non-myogenic cells, an antibody raised against c-Jun coimmunoprecipitates TR only after MyoD transfection [167]. Furthermore, it was shown that RXR expression efficiently disrupts this transcriptional complex by inducing formation of the typical TR/RXR heterodimer [167]. These data led to the conclusion that different complexes mediate TR transcriptional activity during myoblast proliferation (TR/c-Jun/MyoD) and terminal differentiation (TR/RXR) with several consequences for gene expression.

The first consequence relates to the differential recognition of TREs by the two complexes. Whereas TR/c-Jun/MyoD transactivates reporter genes driven by DR4, TREpal or Ipal-TREs, TR/RXR only transactivates reporter genes under the control of DR4-TREs, thus indicating possible differences in the set of TR target genes during proliferation and terminal differentiation. In agreement with this possibility, a microarray study performed in our lab clearly indicated that, in addition to a common set of genes, different T3 target genes are recorded during these two phases of the culture. Furthermore, this analysis clearly suggested that RXR expression occurring at the onset of myoblast differentiation allows TR to change the expression pattern of genes clearly involved in the regulation of this process [168].

Another important consequence is that formation of the TR/c-Jun/MyoD complex recruits the myogenic factor into a complex able to bind to different TREs but unable to recognize E boxes, the specific responsive elements of MRFs. Consequently, in proliferating myoblasts, TR efficiently represses MyoD transcriptional activity [169]. The relevance of this mechanism was underlined later on discovering that disruption of the TR/MyoD physical interaction with a novel TR

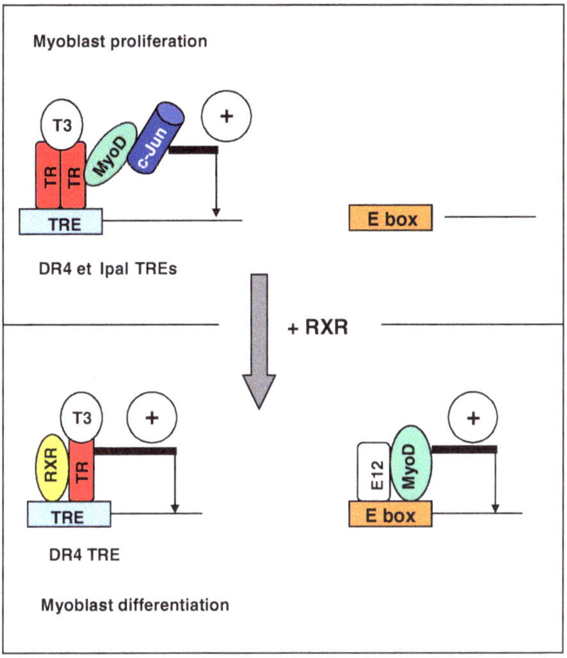

Fig. 3.6 In absence of RXR expression, the transcriptional activity of TR is induced through a TR/c-Jun/MyoD complex, resulting in MyoD sequestration and inhibition of its transcriptional activity on E boxes. Dissociation of this complex by RXR expression occurring at the onset of myoblast differentiation restores MyoD transcriptional activity on its target genes

variant (TR-ΔE6) stimulates myoblast differentiation [170]. Therefore, despite the fact that the T3 pathway displays a strong myogenic influence, it appears that TR must be considered as a MyoD antagonist during myoblast proliferation. However, RXR expression abrogates this repressive activity by disrupting the TR/c-Jun/MyoD complex, thus allowing the MRF to stimulate transcription of muscle-specific genes (Fig. 3.6) [167].

4.2 TR Only Represses AP-1 Activity at the Induction of Myoblast Differentiation

As previously mentioned, in the presence of T3, TR has been shown to inhibit the TPA (tetra decanoyl phorbol acetate)-inducible AP-1 activity mediated by the c-Jun/c-Fos or c-Jun/Fra2 complexes (Jun/Fos complexes) in RXR expressing cells. As these AP-1 complexes are considered major repressors of myoblast withdrawal from the cell cycle and terminal differentiation, such a regulation could play an important role in the transition occurring between proliferation and differentiation. Similar studies to those previously performed in CV1 or HeLa cells [140, 141] led to the conclusion that this mechanism is non-functional in proliferating myoblasts. Furthermore, transient or stable RXR expression in these cells induced the ability of the liganded TR to inhibit basal or TPA-stimulated AP-1 activity in these cells [143, 171]. Therefore, these data clearly suggest that RXR is a major TR partner involved not only in its direct transcriptional activity but also in its inhibition of AP-1 activity.

In terms of myogenesis, this indicates that liganded TR may only facilitate the induction of myoblast differentiation at a particular stage of progression in the differentiation programme, characterized by RXR expression.

4.3 The Dual Myogenic Influence of TR Is Governed by RXR Expression

Overall these data highlight a dual influence of TR in the induction of terminal differentiation. The inhibition of MyoD transcriptional activity by TR in proliferating cells probably delays the induction of terminal differentiation, in a way similar to that occurring with Id, Mist 1 or MyoR. In contrast, with respect to the onset of terminal differentiation, the liganded receptor inhibits the transcriptional activity of AP-1, a major repressor of myoblast differentiation, thus allowing the progression of this process.

One interesting feature of this regulation is the observation that the expression of only one ligand-dependent transcription factor, RXR, plays a key role in the reversion of TR myogenic activity from repression to stimulation of myoblast differentiation. This makes RXR particularly important in the process inducing the shift from proliferation to differentiation. Indeed, RXR expression abolishes the repressive myogenic TR activity by disrupting the TR/c-Jun/MyoD complex and restoring a full MyoD transcriptional activity through E boxes. Simultaneously, RXR expression also allows the liganded TR to inhibit the repressive AP-1 activity. These data underline that the induction of myoblast differentiation involves reversing the influence of one transcription factor, TR, from a mechanism initially preventing an anticipated differentiation to another stimulating terminal differentiation. As TR and T3 are present at very early stages of embryonic development [172], the requirement of RXR expression for the functionality of this mechanism prevents an anticipated myoblast differentiation induced by T3 and TR. Such a process is probably needed to optimize muscle tissue development.

4.4 RXR Expression Also Reverses the Myogenic Influence of c-Jun Through TR-Dependent Mechanisms

Two independent studies each reported that c-Jun physically interacts with MyoD and Myogenin, repressing their transcriptional activity [76, 173]. Interestingly, some indications clearly suggest that this interaction strongly resembles that previously described in this review concerning TR, c-Jun and MyoD: (i) similar to TR, c-Jun represses MyoD activity; (ii) identical c-Jun functional domains show involvement in the repression of MyoD transcriptional activity and the stimulation of TR activity [76, 167]; (iii) TR physically interacts with MyoD [167], and MyoD with c-Jun [76]. Therefore, TR interaction with MyoD induces the recruitment of c-Jun in a complex inducing TR activity in the absence of RXR but repressing MyoD activity through an E box. Similar interactions have been shown with Myogenin, resulting in inhibition of the transcriptional activity of this MRF [76, 169]. As discussed earlier, RXR expression occurring at the onset of differentiation disrupts these complexes, thereby abolishing this inhibiting activity of MRF via TR and c-Jun. These results indicate that c-Jun is involved in mechanisms repressing MRF transcriptional activity in proliferating myoblasts that are relieved by RXR expression during terminal differentiation.

Furthermore, as previously mentioned, RXR expression occurring at the onset of myoblast differentiation also allows the liganded TR to repress TPA-inducible AP-1 activity. In agreement with this event, target genes of c-Jun/c-Fos are down-regulated at the onset of myoblast differentiation, as shown for c-Fos and Fra2. Consequently, their expression level decreases after the induction of differentiation. In parallel, as the level of another c-Jun partner, ATF2, remains unchanged, this event favours preferential formation of c-Jun/ATF2 heterodimers relative to c-Jun/c-Fos or c-Jun/Fra2 complexes [72]. In contrast to the Jun/Fos AP-1 complex which efficiently represses myogenic differentiation, c-Jun/Fra2 stimulates myogenin promoter activity and myogenin mRNA and protein expression and potentiates terminal differentiation. In accordance with this, stimulation of myogenin expression by c-Jun is observed during differentiation but is abolished after cotransfection of a dominant negative form of ATF2, indicating a major role of the c-Jun/ATF2 complex [72]. Such a mechanism could explain the unexpected observation that c-Jun overexpression delays avian myoblast withdrawal from the cell cycle, but stimulates terminal differentiation when induced by decreasing serum concentration in the culture medium [72].

All these data demonstrate that, as shown for TR myogenic activity, c-Jun exerts a dual myogenic influence either alone or through different AP-1 complexes. The oncoprotein first represses myoblast withdrawal from the cell cycle and thereafter stimulates terminal differentiation. RXR expression occurring at the onset of myogenic differentiation seems again to play a major role in this reversal.

4.4.1 BTG1, A Downstream Target of AP-1 Activity

The *btg1* (B-cell translocation gene 1) gene coding sequence was isolated from a translocation break point *t*(8; 12) (q24; q22) in a case of B-cell chronic lymphocytic leukaemia [174]. The encoded protein belongs to the BTG family, all members displaying a high degree of conservation in two regions of 22 and 20 amino acids named boxes A and B [175]. BTG1 and related proteins inhibit proliferation of several cell lines and are therefore believed to be members of an antiproliferative protein family [176, 177].

BTG1 expression is induced at the onset of myoblast differentiation [33, 34]. At this step, the protein displays a nuclear and cytosolic localization, due to the presence of an atypical but functional nuclear localization signal and a nuclear export signal. In addition, domains involved in cytosolic or nuclear retention have been characterized [178]. The importance of this gene for myoblast differentiation has been concluded from studies demonstrating that BTG1 nuclear localization is enhanced by positive myogenic regulators such as T3 or by transient cAMP release [33]. Studies concerning the induction of BTG1 expression at the onset of differentiation established that TPA-inducible AP-1 activity (c-Jun/c-Fos) efficiently represses BTG1 expression at the transcriptional level. This led to the concept that inhibition of this activity by TR occurring at the induction of differentiation is an important event allowing BTG1 expression [34].

Studies of the influence of BTG1 in myoblasts established that its overexpression potentiates myoblast irreversible cell cycle arrest and stimulates terminal differentiation [34]. In addition, use of BTG1 mutants displaying a specific nuclear or cytosolic localization led to the conclusion that the myogenic activity of this protein is induced at the nuclear level. The search for the molecular basis of this myogenic influence led to the demonstration that BTG1 physically interacts with nuclear receptors known to stimulate myoblast differentiation, such as TR or retinoic acid receptors [32]. Even more interesting was the discovery that BTG1 also interacts with Myf5, MyoD and myogenin [32]. Furthermore, a similar interaction also occurs with c-Jun; as BTG1 is expressed at a step where c-Jun/ATF2 complexes are prevalent AP-1 heterodimers, it also concerns a transcription factor displaying a positive myogenic influence. As a result, these physical interactions lead to a potent stimulation of the transcriptional activity of each BTG1 partner. As deletion of BTG1 sequences abrogating these physical interactions fully suppresses this stimulation and the myogenic influence of BTG1, these data clearly indicate that BTG1 stimulates myoblast differentiation through its transcriptional coactivator activity [32].

Overall, these data provide evidence for the inhibition of AP-1 activity occurring at the onset of myoblast differentiation inducing the expression of a transcriptional coactivator which in turn stimulates the activities of TR, RAR, c-Jun/ATF2 and myogenic factors. This thereby implies an important role of AP-1 activity regulation in the ability of these factors to induce and/or to stimulate terminal differentiation.

4.5 c-Myc, a Target Involved in the Control of Myoblast Differentiation by Mitochondrial TR Isoforms

Several studies have been performed in order to understand the unexpected myogenic influence of the T3-dependent mitochondrial transcription factor p43, a TR-related protein encoded by the same gene that codes for the T3 nuclear receptor TRα. Experiments conducted in human fibroblasts indicated that overexpression of p43 induces a defective myogenic phenotype in these cells, which express only the Myf5 MRF and some muscle-specific proteins such as desmin, slow MHC and connectin, and displays a strong ability to fuse into myotube-like structures [179]. This observation raises the question of the possible involvement of this pathway in early processes of muscle development, in particular in determining muscle lineage.

Moreover, in experiments dedicated to better understanding the molecular mechanisms driving the influence of p43 on myoblast terminal differentiation, it appeared that overexpression of this receptor induces myogenin expression in the presence of high serum levels in culture medium and an increased expression of this MRF in differentiating myoblasts. In contrast, by blocking the p43 pathway, chloramphenicol abrogates myogenin expression even after serum removal [27]. However, c-Myc is probably a major target of this pathway. While p43 overexpression severely depresses the expression of c-Myc, blocking this pathway with chloramphenicol significantly increases c-Myc expression [26]. The direct implication of c-Myc on the myogenic influence of p43 is demonstrated following the restoration of c-Myc expression levels in p43 expressing cells which abrogates the myogenic influence of this mitochondrial receptor. Moreover, in the same way than chloramphenicol which inhibits the p43 pathway [27], c-Myc overexpression strongly inhibits myogenin expression and the ability of overexpressed myogenic factors to induce terminal differentiation [26].

In addition to this mechanism involving c-Myc, p43 overexpression also increases the expression of the calcium-dependent phosphatase calcineurin via stimulation of mitochondrial activity. Conversely, inhibition of this pathway by chloramphenicol reduces calcineurin expression [164]. As this phosphatase stimulates myogenin expression and myoblast differentiation [180–182], regulation of its expression by the p43 pathway probably represents another mechanism of regulating terminal differentiation. Another consequence for muscle development is related to the ability of calcineurin to favour slow myosin expression and the acquisition of slow contractile features of muscle fibres [180, 183, 184]. In C2C12 myoblasts, upregulation of calcineurin expression by p43 increases the amount of slow relative to fast myosin [164], indicating an involvement in the acquisition of the contractile phenotype of muscle fibres. This is in concordance with the observation that in transgenic mice, p43 overexpression in skeletal muscle changes the more rapid phenotype towards a slower one [165]. Interestingly, preliminary results from the lab suggest that decreasing c-Myc expression induced by p43 also represents an upstream event inducing calcineurin upregulation.

Altogether these data demonstrate that regulation of mitochondrial activity by transcription factors addressed into the organelle plays an important role in controlling the transition between proliferation and differentiation in myoblasts. Moreover, as a decrease in c Myc expression is often considered as an absolute prerequisite for myoblast withdrawal from the cell cycle, downregulation of its expression level likely represents a crucial mechanism in p43 myogenic activity. In line with this possibility, studies have shown that a rise in mitochondrial activity precedes myoblast terminal differentiation, an event that does not occur in differentiation-deficient myoblasts. Interestingly, this event occurs simultaneously to the induction of RXR expression. We have shown that RXRα is partially cleaved by a cytosolic calpain, and that the resulting protein, mt-RXR, is imported into the mitochondrial matrix where it potentiates the transcriptional activity of p43 [163]. Therefore, it can be hypothesized that through this truncated form, RXR expression occurring at the onset of myoblast differentiation could, in this instance, potentiate p43 transcription of the mitochondrial genome and hence its stimulatory influence on mitochondrial activity and terminal differentiation (Fig. 3.7).

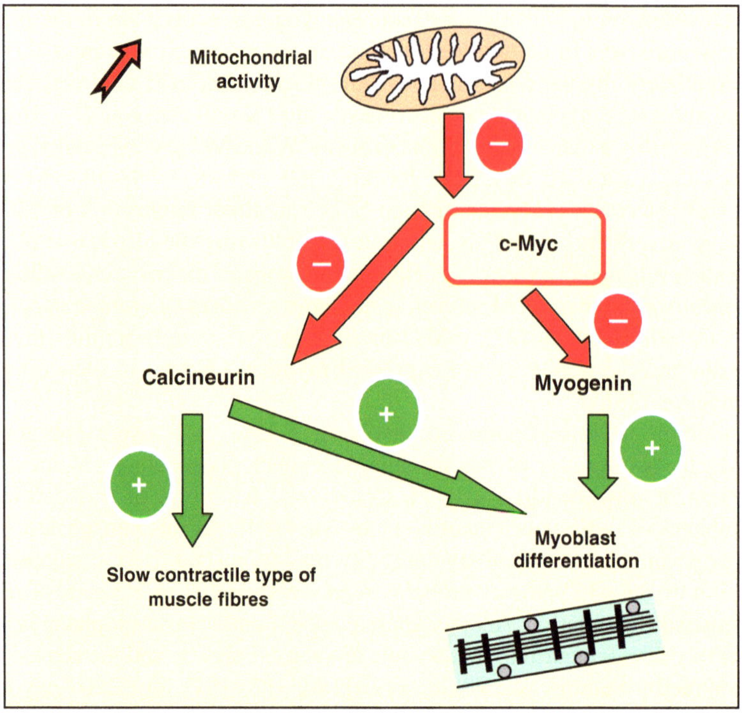

Fig. 3.7 c-Myc is a potent repressor of cell cycle arrest, myogenin expression and myoblast differentiation. Through stimulation of mitochondrial activity, p43 and mt-RXR inhibit c-Myc expression, allowing cell cycle arrest and stimulating myogenin expression and myoblast differentiation. In parallel, this mechanism upregulates calcineurin expression, also increasing myogenin expression and myoblast differentiation. In addition, calcineurin favours the acquisition of a slow oxidative myofibre type

4.5.1 Regulation of Myoblast Differentiation Through the Mitochondrial Nuclear Crosstalk

One intriguing question was to know how a mitochondrial transcription factor could induce changes in nuclear gene expression leading to the onset of the differentiation programme. The involvement of ATP synthesis by mitochondria has been ruled out with the observation that ATP stores are not affected after p43 overexpression [27]. Similarly, inhibition of this pathway by chloramphenicol only transiently decreases ATP stores, which are quickly restored by anaerobic glycolysis [27].

However, mitochondria are considered as important Ca^{2+} accumulation compartments, via captation of this anion through a Ca uniport probably involving uncoupling 2 and 3 proteins (UCP2 and 3) [185]. In parallel, mitochondria independently release Ca^{2+} into the cytosol through a Na^{+}/Ca^{2+} exchanger and the permeability transition pore [186]. These events lead to changes in calcium pulses which govern Ca^{2+} signalling [187]. With respect to the influence of mitochondrial activity on c-Myc expression mentioned above, calcium is known to regulate the expression of this proto-oncogene [188, 189]. In parallel to this, mitochondria produce continuously reactive oxygen species (ROS) through its oxidative activity, considered as signalling molecules, inducing changes in DNA binding and activity of different transcription factors including AP-1, NFκB and p53 [190–192]. In agreement with this organelle signalling, p43 injection in Xenopus eggs stimulates mitochondrial respiration associated with significant changes in calcium pulses [193]. Similarly, in human fibroblasts, p43 expression also stimulates ROS production [179], thereby affecting the mitochondrial nuclear crosstalk. Although the involvement of the ROS pathway in myogenic activity of the organelle remains to be established, recent studies indicate that calcium signalling should probably be considered [194, 195].

This set of data clearly suggests that ligand-dependent mitochondrial transcription factors are important regulators of myoblast differentiation. Overall, these results have led us to propose that mitochondria regulate developmental processes by ensuring coordination between ATP synthesis (needed for the energy-consuming processes of development) and via the induction or downregulation of expression of a set of nuclear genes [196], in particular those involved in muscle development.

5 Conclusions

In this review, we provide evidence that mechanisms allowing myoblast withdrawal from the cell cycle and terminal differentiation involve a functional interaction between several transcription factors. In particular, RXR genes seem to be master genes for the induction of myogenic differentiation. Their expression at a particular step of muscle development reverses the myogenic activity of other transcription factors, initially repressing terminal differentiation. In the presence of RXR, these factors become involved in the induction of myoblast withdrawal from the cell cycle, a key event promoting differentiation.

During myoblast proliferation, in the absence of RXR isoforms, TR and c-Jun inhibit MyoD transcriptional activity via a complex sequestrating the MRF. A major Jun/Fos AP-1 activity then strongly represses myoblast withdrawal from the cell cycle and inhibits MyoD and myogenin expression. It also inhibits the expression of BTG1, a transcriptional coactivator stimulating MRF, TR and RAR activity. High c-Myc levels also contribute to maintaining myoblast proliferation by interfering with proteins involved in cell cycle completion and repressing MyoD and myogenin ability to induce terminal differentiation. RXR expression occurring at the onset of differentiation induces a reversal of these influences. Disruption of the TR/c-Jun/MyoD complex restores full MyoD transcriptional activity. In addition, in the presence of RXR, TR inhibits Jun/Fos activity, an event decreasing c-Jun and Fra2 expression, thus favouring formation of c-Jun/ATF2 complexes which stimulate myogenin expression and myoblast differentiation.

In parallel, the induction of RXR expression is probably associated with the mitochondrial import of mt-RXR, able to stimulate the transcriptional activity of p43 on the mitochondrial genome. This phenomenon could explain the rise in organelle activity reported as preceding the onset of myoblast differentiation [155]. In turn, this leads to a decrease in c-Myc expression allowing myoblast withdrawal from the cell cycle and restoring MRF ability to induce terminal differentiation and myogenin expression (Fig. 3.8).

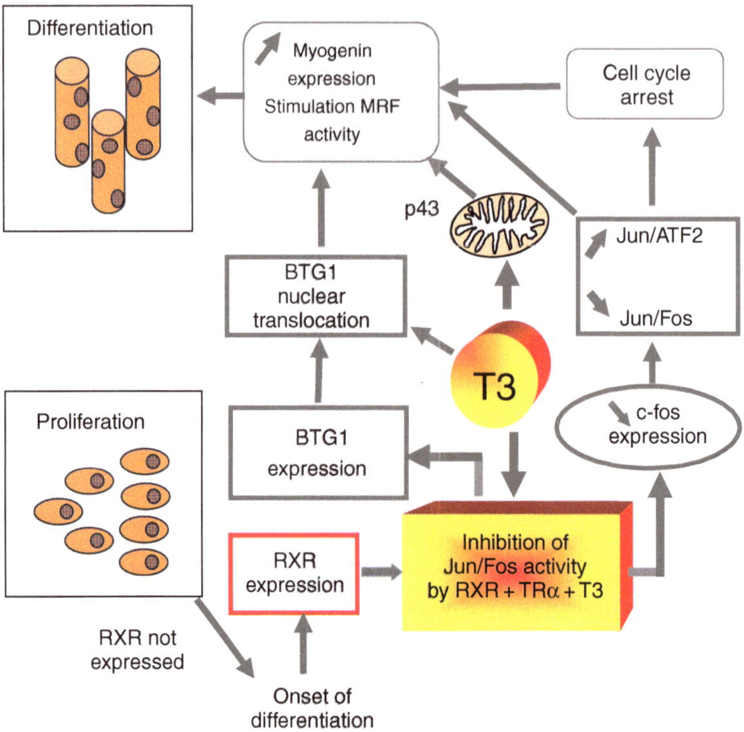

Fig. 3.8 RXR expression is deeply involved in the shift from proliferation to differentiation by reversing TR and c-Jun myogenic influences, and through its truncated mitochondrial form by decreasing c-Myc expression and upregulating calcineurin expression (see comment in the text)

In this respect, induction of RXR expression triggering all these changes appears a major event governing the shift between proliferation and differentiation, with hence considerable involvement in the biological clock of muscle development.

References

1. Buckingham M, Bajard L, Chang T, et al (2003) The formation of skeletal muscle: from somite to limb. J Anat 202(1):59–68

2. Braun T, Buschhausen-Denker G, Bober E, Tannich E, Arnold HH (1989) A novel human muscle factor related to but distinct from MyoD1 induces myogenic conversion in 10T1/2 fibroblasts. Embo J 8(3):701–709
3. Choi J, Costa ML, Mermelstein CS, Chagas C, Holtzer S, Holtzer H (1990) MyoD converts primary dermal fibroblasts, chondroblasts, smooth muscle, and retinal pigmented epithelial cells into striated mononucleated myoblasts and multinucleated myotubes. Proc Natl Acad Sci U S A 87(20):7988–7992
4. Rhodes SJ, Konieczny SF (1989) Identification of MRF4: a new member of the muscle regulatory factor gene family. Genes Dev 3(12B):2050–2061
5. Tapscott SJ, Davis RL, Thayer MJ, Cheng PF, Weintraub H, Lassar AB (1988) MyoD1: a nuclear phosphoprotein requiring a Myc homology region to convert fibroblasts to myoblasts. Science 242(4877):405–411
6. Wright WE, Sassoon DA, Lin VK (1989) Myogenin, a factor regulating myogenesis, has a domain homologous to MyoD. Cell 56(4):607–617
7. Rudnicki MA, Schnegelsberg PN, Stead RH, Braun T, Arnold HH, Jaenisch R (1993) MyoD or Myf-5 is required for the formation of skeletal muscle. Cell 75(7):1351–1359
8. Hasty P, Bradley A, Morris JH, et al (1993) Muscle deficiency and neonatal death in mice with a targeted mutation in the myogenin gene. Nature 364(6437):501–506
9. Nabeshima Y, Hanaoka K, Hayasaka M, et al (1993) Myogenin gene disruption results in perinatal lethality because of severe muscle defect. Nature 364(6437):532–535
10. Bober E, Lyons GE, Braun T, Cossu G, Buckingham M, Arnold HH (1991) The muscle regulatory gene, Myf-6, has a biphasic pattern of expression during early mouse development. J Cell Biol 113(6):1255–1265
11. Kassar-Duchossoy L, Gayraud-Morel B, Gomes D, et al (2004) Mrf4 determines skeletal muscle identity in Myf5:Myod double-mutant mice. Nature 431(7007):466–471
12. Kouzarides T (1995) Transcriptional control by the retinoblastoma protein. Semin Cancer Biol 6(2):91–98
13. Taya Y (1997) RB kinases and RB-binding proteins: new points of view. Trends Biochem Sci 22(1):14–17
14. Mittnacht S (1998) Control of pRB phosphorylation. Curr Opin Genet Dev 8(1):21–27
15. Gu W, Schneider JW, Condorelli G, Kaushal S, Mahdavi V, Nadal-Ginard B (1993) Interaction of myogenic factors and the retinoblastoma protein mediates muscle cell commitment and differentiation. Cell 72(3):309–324
16. Zhang JM, Wei Q, Zhao X, Paterson BM (1999) Coupling of the cell cycle and myogenesis through the cyclin D1-dependent interaction of MyoD with cdk4. Embo J 18(4):926–933
17. Rao SS, Chu C, Kohtz DS (1994) Ectopic expression of cyclin D1 prevents activation of gene transcription by myogenic basic helix-loop-helix regulators. Mol Cell Biol 14(8): 5259–5267
18. Skapek SX, Rhee J, Spicer DB, Lassar AB (1995) Inhibition of myogenic differentiation in proliferating myoblasts by cyclin D1-dependent kinase. Science 267(5200):1022–1024
19. Lassar AB, Davis RL, Wright WE, et al (1991) Functional activity of myogenic HLH proteins requires hetero-oligomerization with E12/E47-like proteins in vivo. Cell 66(2): 305–315
20. Benezra R, Davis RL, Lockshon D, Turner DL, Weintraub H (1990) The protein Id: a negative regulator of helix-loop-helix DNA binding proteins. Cell 61(1):49–59
21. Lemercier C, To RQ, Carrasco RA, Konieczny SF (1998) The basic helix-loop-helix transcription factor Mist1 functions as a transcriptional repressor of myoD. Embo J 17(5):1412–1422
22. Lu J, Webb R, Richardson JA, Olson EN (1999) MyoR: a muscle-restricted basic helix-loop-helix transcription factor that antagonizes the actions of MyoD. Proc Natl Acad Sci U S A 96(2):552–557
23. Hebrok M, Wertz K, Fuchtbauer EM (1994) M-twist is an inhibitor of muscle differentiation. Dev Biol 165(2):537–544

24. Postigo AA, Dean DC (1997) ZEB, a vertebrate homolog of Drosophila Zfh-1, is a negative regulator of muscle differentiation. Embo J 16(13):3935–3943
25. Miner JH, Wold BJ (1991) c-myc inhibition of MyoD and myogenin-initiated myogenic differentiation. Mol Cell Biol 11(5):2842–2851
26. Seyer P, Grandemange S, Busson M, et al (2006) Mitochondrial activity regulates myoblast differentiation by control of c-Myc expression. J Cell Physiol 207(1):75–86
27. Rochard P, Rodier A, Casas F et al (2000) Mitochondrial activity is involved in the regulation of myoblast differentiation through myogenin expression and activity of myogenic factors. J Biol Chem 275(4):2733–2744
28. Sartorelli V, Huang J, Hamamori Y, Kedes L (1997) Molecular mechanisms of myogenic coactivation by p300: direct interaction with the activation domain of MyoD and with the MADS box of MEF2C. Mol Cell Biol 17(2):1010–1026
29. Yuan W, Condorelli G, Caruso M, Felsani A, Giordano A (1996) Human p300 protein is a coactivator for the transcription factor MyoD. J Biol Chem 271(15):9009–9013
30. Puri PL, Sartorelli V, Yang XJ et al (1997) Differential roles of p300 and PCAF acetyltransferases in muscle differentiation. Mol Cell 1(1):35–45
31. Sartorelli V, Puri PL, Hamamori Y et al (1999) Acetylation of MyoD directed by PCAF is necessary for the execution of the muscle program. Mol Cell 4(5):725–734
32. Busson M, Carazo A, Seyer P, et al (2005) Coactivation of nuclear receptors and myogenic factors induces the major BTG1 influence on muscle differentiation. Oncogene 24(10):1698–1710
33. Marchal S, Cassar-Malek I, Magaud JP, Rouault JP, Wrutniak C, Cabello G (1995) Stimulation of avian myoblast differentiation by triiodothyronine: possible involvement of the cAMP pathway. Exp Cell Res 220(1):1–10
34. Rodier A, Marchal-Victorion S, Rochard P et al (1999) BTG1: a triiodothyronine target involved in the myogenic influence of the hormone. Exp Cell Res 249(2):337–348
35. Angel P, Allegretto EA, Okino ST et al (1988) Oncogene jun encodes a sequence-specific trans-activator similar to AP-1. Nature 332(6160):166–171
36. Angel P, Imagawa M, Chiu R, et al (1987) Phorbol ester-inducible genes contain a common cis element recognized by a TPA-modulated trans-acting factor. Cell 49(6):729–739
37. Curran T, Franza BR Jr (1988) Fos and Jun: the AP-1 connection. Cell 55(3):395–397
38. Halazonetis TD, Georgopoulos K, Greenberg ME, Leder P (1988) c-Jun dimerizes with itself and with c-Fos, forming complexes of different DNA binding affinities. Cell 55(5):917–924
39. Hirai S, Yaniv M (1989) Jun DNA-binding is modulated by mutations between the leucines or by direct interaction of fos with the TGACTCA sequence. New Biol 1(2):181–191
40. Sassone-Corsi P, Lamph WW, Kamps M, Verma IM (1988) fos-associated cellular p39 is related to nuclear transcription factor AP-1. Cell 54(4):553–560
41. Hai T, Curran T (1991) Cross-family dimerization of transcription factors Fos/Jun and ATF/CREB alters DNA binding specificity. Proc Natl Acad Sci U S A 88(9):3720–3724
42. Kataoka K, Noda M, Nishizawa M (1994) Maf nuclear oncoprotein recognizes sequences related to an AP-1 site and forms heterodimers with both Fos and Jun. Mol Cell Biol 14(1):700–712
43. Kerppola TK, Curran T (1994) Maf and Nrl can bind to AP-1 sites and form heterodimers with Fos and Jun. Oncogene 9(3):675–684
44. Kouzarides T, Ziff E (1988) The role of the leucine zipper in the fos-jun interaction. Nature 336(6200):646–651
45. Landschulz WH, Johnson PF, McKnight SL (1988) The leucine zipper: a hypothetical structure common to a new class of DNA binding proteins. Science 240(4860):1759–1764
46. Mechta-Grigoriou F, Gerald D, Yaniv M (2001) The mammalian Jun proteins: redundancy and specificity. Oncogene 20(19):2378–2389
47. van Dam H, Duyndam M, Rottier R et al (1993) Heterodimer formation of cJun and ATF-2 is responsible for induction of c-jun by the 243 amino acid adenovirus E1A protein. Embo J 12(2):479–487

48. Hagmeyer BM, Konig H, Herr I et al (1993) Adenovirus E1A negatively and positively modulates transcription of AP-1 dependent genes by dimer-specific regulation of the DNA binding and transactivation activities of Jun. Embo J 12(9):3559–3572
49. Stein B, Angel P, van Dam H, et al (1992) Ultraviolet-radiation induced c-jun gene transcription: two AP-1 like binding sites mediate the response. Photochem Photobiol 55(3):409–415
50. Shin HM, Han TH (1999) CD28-mediated regulation of the c-jun promoter involves the MEF2 transcription factor in Jurkat T cells. Mol Immunol 36(3):197–203
51. Andreucci JJ, Grant D, Cox DM et al (2002) Composition and function of AP-1 transcription complexes during muscle cell differentiation. J Biol Chem 277(19):16426–16432
52. Chen RH, Abate C, Blenis J (1993) Phosphorylation of the c-Fos transrepression domain by mitogen-activated protein kinase and 90-kDa ribosomal S6 kinase. Proc Natl Acad Sci U S A 90(23):10952–10956
53. Deng T, Karin M (1994) c-Fos transcriptional activity stimulated by H-Ras-activated protein kinase distinct from JNK and ERK. Nature 371(6493):171–175
54. Raingeaud J, Gupta S, Rogers JS et al (1995) Pro-inflammatory cytokines and environmental stress cause p38 mitogen-activated protein kinase activation by dual phosphorylation on tyrosine and threonine. J Biol Chem 270(13):7420–7426
55. Papavassiliou AG, Treier M, Bohmann D (1995) Intramolecular signal transduction in c-Jun. Embo J 14(9):2014–2019
56. Arias J, Alberts AS, Brindle P et al (1994) Activation of cAMP and mitogen responsive genes relies on a common nuclear factor. Nature 370(6486):226–229
57. Bannister AJ, Oehler T, Wilhelm D, Angel P, Kouzarides T (1995) Stimulation of c-Jun activity by CBP: c-Jun residues Ser63/73 are required for CBP induced stimulation in vivo and CBP binding in vitro. Oncogene 11(12):2509–2514
58. Musti AM, Treier M, Bohmann D (1997) Reduced ubiquitin-dependent degradation of c-Jun after phosphorylation by MAP kinases. Science 275(5298):400–402
59. Kovary K, Bravo R (1991) The jun and fos protein families are both required for cell cycle progression in fibroblasts. Mol Cell Biol 11(9):4466–4472
60. Riabowol K, Schiff J, Gilman MZ (1992) Transcription factor AP-1 activity is required for initiation of DNA synthesis and is lost during cellular aging. Proc Natl Acad Sci U S A 89(1):157–161
61. Pfarr CM, Mechta F, Spyrou G, Lallemand D, Carillo S, Yaniv M (1994) Mouse JunD negatively regulates fibroblast growth and antagonizes transformation by ras. Cell 76(4): 747–760
62. Bakiri L, Lallemand D, Bossy-Wetzel E, Yaniv M (2000) Cell cycle-dependent variations in c-Jun and JunB phosphorylation: a role in the control of cyclin D1 expression. Embo J 19(9):2056–2068
63. Johnson RS, van Lingen B, Papaioannou VE, Spiegelman BM (1993) A null mutation at the c-jun locus causes embryonic lethality and retarded cell growth in culture. Genes Dev 7(7B):1309–1317
64. Schreiber M, Kolbus A, Piu F et al (1999) Control of cell cycle progression by c-Jun is p53 dependent. Genes Dev 13(5):607–619
65. Lord KA, Abdollahi A, Hoffman-Liebermann B, Liebermann DA (1993) Proto-oncogenes of the fos/jun family of transcription factors are positive regulators of myeloid differentiation. Mol Cell Biol 13(2):841–851
66. Szabo E, Preis LH, Birrer MJ (1994) Constitutive cJun expression induces partial macrophage differentiation in U-937 cells. Cell Growth Differ 5(4):439–446
67. Sark MW, Fischer DF, de Meijer E, van de Putte P, Backendorf C (1998) AP-1 and ets transcription factors regulate the expression of the human SPRR1A keratinocyte terminal differentiation marker. J Biol Chem 273(38):24683–24692
68. Navre M, Ringold GM (1988) A growth factor-repressible gene associated with protein kinase C-mediated inhibition of adipocyte differentiation. J Cell Biol 107(1):279–286

69. Kameda T, Watanabe H, Iba H (1997) C-Jun and JunD suppress maturation of chondrocytes. Cell Growth Differ 8(5):495–503
70. Park K, Chung M, Kim SJ (1992) Inhibition of myogenesis by okadaic acid, an inhibitor of protein phosphatases, 1 and 2A, correlates with the induction of AP1. J Biol Chem 267(15):10810–10815
71. Sulakhe PV, Johnson DD, Phan NT, Wilcox R (1985) Phorbol ester inhibits myoblast fusion and activates beta-adrenergic receptor coupled adenylate cyclase. FEBS Lett 186(2): 281–285
72. Daury L, Busson M, Tourkine N et al (2001) Opposing functions of ATF2 and Fos-like transcription factors in c-Jun-mediated myogenin expression and terminal differentiation of avian myoblasts. Oncogene 20(55):7998–8008
73. Pedraza-Alva G, Zingg JM, Jost JP (1994) AP-1 binds to a putative cAMP response element of the MyoD1 promoter and negatively modulates MyoD1 expression in dividing myoblasts. J Biol Chem 269(9):6978–6985
74. Du R, Chen YF, An XR et al (2005) Cloning and sequence analysis of myostatin promoter in sheep. DNA Seq 16(6):412–417
75. Joulia D, Bernardi H, Garandel V, Rabenoelina F, Vernus B, Cabello G (2003) Mechanisms involved in the inhibition of myoblast proliferation and differentiation by myostatin. Exp Cell Res 286(2):263–275
76. Bengal E, Ransone L, Scharfmann R et al (1992) Functional antagonism between c-Jun and MyoD proteins: a direct physical association. Cell 68(3):507–519
77. Thinakaran G, Bag J (1993) Expression of the protooncogene c-jun is maintained during myogenic differentiation in rat L6 myoblasts. Biochem Cell Biol 71(5–6):260–269
78. Serria MS, Ikeda H, Omoteyama K, Hirokawa J, Nishi S, Sakai M (2003) Regulation and differential expression of the c-maf gene in differentiating cultured cells. Biochem Biophys Res Commun 310(2):318–326
79. Raingeaud J, Whitmarsh AJ, Barrett T, Derijard B, Davis RJ (1996) MKK3- and MKK6-regulated gene expression is mediated by the p38 mitogen-activated protein kinase signal transduction pathway. Mol Cell Biol 16(3):1247–1255
80. van Dam H, Wilhelm D, Herr I, Steffen A, Herrlich P, Angel P (1995) ATF-2 is preferentially activated by stress-activated protein kinases to mediate c-jun induction in response to genotoxic agents. Embo J 14(8):1798–1811
81. Cuenda A, Cohen P (1999) Stress-activated protein kinase-2/p38 and a rapamycin-sensitive pathway are required for C2C12 myogenesis. J Biol Chem 274(7):4341–4346
82. Wu Z, Woodring PJ, Bhakta KS et al (2000) p38 and extracellular signal-regulated kinases regulate the myogenic program at multiple steps. Mol Cell Biol 20(11):3951–3964
83. Zetser A, Gredinger E, Bengal E (1999) p38 mitogen-activated protein kinase pathway promotes skeletal muscle differentiation. Participation of the Mef2c transcription factor. J Biol Chem 274(8):5193–5200
84. Dorman CM, Johnson SE (1999) Activated Raf inhibits avian myogenesis through a MAPK-dependent mechanism. Oncogene 18(37):5167–5176
85. Nesbit CE, Tersak JM, Prochownik EV (1999) MYC oncogenes and human neoplastic disease. Oncogene 18(19):3004–3016
86. Coller HA, Grandori C, Tamayo P et al (2000) Expression analysis with oligonucleotide microarrays reveals that MYC regulates genes involved in growth, cell cycle, signaling, and adhesion. Proc Natl Acad Sci U S A 97(7):3260–3265
87. Grignani F, Lombardi L, Inghirami G, Sternas L, Cechova K, Dalla-Favera R (1990) Negative autoregulation of c-myc gene expression is inactivated in transformed cells. Embo J 9(12):3913–3922
88. Penn LJ, Brooks MW, Laufer EM, Land H (1990) Negative autoregulation of c-myc transcription. Embo J 9(4):1113–1121
89. Xiao Q, Claassen G, Shi J, Adachi S, Sedivy J, Hann SR (1998) Transactivation-defective c-MycS retains the ability to regulate proliferation and apoptosis. Genes Dev 12(24): 3803–3808

90. Claassen GF, Hann SR (1999) Myc-mediated transformation: the repression connection. Oncogene 18(19):2925–2933
91. Amati B, Frank SR, Donjerkovic D, Taubert S (2001) Function of the c-Myc oncoprotein in chromatin remodeling and transcription. Biochim Biophys Acta 1471(3):M135–145
92. Furstoss O, Dorey K, Simon V, Barila D, Superti-Furga G, Roche S (2002) c-Abl is an effector of Src for growth factor-induced c-myc expression and DNA synthesis. Embo J 21(4):514–524
93. Grausz JD, Fradelizi D, Dautry F, Monier R, Lehn P (1986) Modulation of c-fos and c-myc mRNA levels in normal human lymphocytes by calcium ionophore A23187 and phorbol ester. Eur J Immunol 16(10):1217–1221
94. Kim DW, Gazourian L, Quadri SA, Romieu-Mourez R, Sherr DH, Sonenshein GE (2000) The RelA NF-kappaB subunit and the aryl hydrocarbon receptor (AhR) cooperate to transactivate the c-myc promoter in mammary cells. Oncogene 19(48):5498–5506
95. Maki A, Berezesky IK, Fargnoli J, Holbrook NJ, Trump BF (1992) Role of [Ca2+]i in induction of c-fos, c-jun, and c-myc mRNA in rat PTE after oxidative stress. Faseb J 6(3):919–924
96. Reuse S, Maenhaut C, Dumont JE (1990) Regulation of protooncogenes c-fos and c-myc expressions by protein tyrosine kinase, protein kinase C, and cyclic AMP mitogenic pathways in dog primary thyrocytes: a positive and negative control by cyclic AMP on c-myc expression. Exp Cell Res 189(1):33–40
97. Slungaard A, Confer DL, Schubach WH (1987) Rapid transcriptional down-regulation of c-myc expression during cyclic adenosine monophosphate-promoted differentiation of leukemic cells. J Clin Invest 79(5):1542–1547
98. Paulin FE, West MJ, Sullivan NF, Whitney RL, Lyne L, Willis AE (1996) Aberrant translational control of the c-myc gene in multiple myeloma. Oncogene 13(3):505–513
99. West MJ, Sullivan NF, Willis AE (1995) Translational upregulation of the c-myc oncogene in Bloom's syndrome cell lines. Oncogene 11(12):2515–2524
100. Gregory MA, Hann SR (2000) c-Myc proteolysis by the ubiquitin-proteasome pathway: stabilization of c-Myc in Burkitt's lymphoma cells. Mol Cell Biol 20(7):2423–2435
101. Salghetti SE, Kim SY, Tansey WP (1999) Destruction of Myc by ubiquitin-mediated proteolysis: cancer-associated and transforming mutations stabilize Myc. Embo J 18(3):717–726
102. Herbst A, Salghetti SE, Kim SY, Tansey WP (2004) Multiple cell-type-specific elements regulate Myc protein stability. Oncogene 23(21):3863–3871
103. Antonson P, Pray MG, Jacobsson A, Xanthopoulos KG (1995) Myc inhibits CCAAT/enhancer-binding protein alpha-gene expression in HIB-1B hibernoma cells through interactions with the core promoter region. Eur J Biochem 232(2):397–403
104. Li LH, Nerlov C, Prendergast G, MacGregor D, Ziff EB (1994) c-Myc represses transcription in vivo by a novel mechanism dependent on the initiator element and Myc box II. Embo J 13(17):4070–4079
105. Berns K, Martins C, Dannenberg JH, Berns A, te Riele H, Bernards R (2000) p27kip1-independent cell cycle regulation by MYC. Oncogene 19(42):4822–4827
106. Claassen GF, Hann SR (2000) A role for transcriptional repression of p21CIP1 by c-Myc in overcoming transforming growth factor beta -induced cell-cycle arrest. Proc Natl Acad Sci U S A 97(17):9498–9503
107. Wu S, Cetinkaya C, Munoz-Alonso MJ, et al (2003) Myc represses differentiation-induced p21CIP1 expression via Miz-1-dependent interaction with the p21 core promoter. Oncogene 22(3):351–360
108. Yang W, Shen J, Wu M et al (2001) Repression of transcription of the p27(Kip1) cyclin-dependent kinase inhibitor gene by c-Myc. Oncogene 20(14):1688–1702
109. Denis N, Blanc S, Leibovitch MP et al (1987) c-myc oncogene expression inhibits the initiation of myogenic differentiation. Exp Cell Res 172(1):212–217
110. Sejersen T, Sumegi J, Ringertz NR (1985) Density-dependent arrest of DNA replication is accompanied by decreased levels of c-myc mRNA in myogenic but not in differentiation-defective myoblasts. J Cell Physiol 125(3):465–470

111. Yeilding NM, Procopio WN, Rehman MT, Lee WM (1998) c-myc mRNA is down-regulated during myogenic differentiation by accelerated decay that depends on translation of regulatory coding elements. J Biol Chem 273(25):15749–15757
112. Crescenzi M, Crouch DH, Tato F (1994) Transformation by myc prevents fusion but not biochemical differentiation of C2C12 myoblasts: mechanisms of phenotypic correction in mixed culture with normal cells. J Cell Biol 125(5):1137–1145
113. Martinez-Fernandez S, Hernandez-Torres F, Franco D, Lyons GE, Navarro F, Aranega AE (2006) Pitx2c overexpression promotes cell proliferation and arrests differentiation in myoblasts. Dev Dyn 235(11):2930–2939
114. Kataoka Y, Matsumura I, Ezoe S et al (2003) Reciprocal inhibition between MyoD and STAT3 in the regulation of growth and differentiation of myoblasts. J Biol Chem 278(45):44178–44187
115. Zentella A, Massague J (1992) Transforming growth factor beta induces myoblast differentiation in the presence of mitogens. Proc Natl Acad Sci U S A 89(11):5176–5180
116. Datta B, Min W, Burma S, Lengyel P (1998) Increase in p202 expression during skeletal muscle differentiation: inhibition of MyoD protein expression and activity by p202. Mol Cell Biol 18(2):1074–1083
117. Wang H, Liu C, Lu Y et al (2000) The interferon- and differentiation-inducible p202a protein inhibits the transcriptional activity of c-Myc by blocking its association with Max. J Biol Chem 275(35):27377–27285
118. Kronfeld-Kinar Y, Vilchik S, Hyman T, Leibkowicz F, Salzberg S (1999) Involvement of PKR in the regulation of myogenesis. Cell Growth Differ 10(3):201–212
119. Galaktionov K, Chen X, Beach D (1996) Cdc25 cell-cycle phosphatase as a target of c-myc. Nature 382(6591):511–517
120. Prochownik EV, Kukowska J, Rodgers C (1988) c-myc antisense transcripts accelerate differentiation and inhibit G1 progression in murine erythroleukemia cells. Mol Cell Biol 8(9):3683–3695
121. Steiner P, Philipp A, Lukas J et al (1995) Identification of a Myc-dependent step during the formation of active G1 cyclin-cdk complexes. Embo J 14(19):4814–4826
122. Steinman RA, Hoffman B, Iro A, Guillouf C, Liebermann DA, el-Houseini ME (1994) Induction of p21 (WAF-1/CIP1) during differentiation. Oncogene 9(11):3389–3396
123. Kitaura H, Shinshi M, Uchikoshi Y, Ono T, Iguchi-Ariga SM, Ariga H (2000) Reciprocal regulation via protein-protein interaction between c-Myc and p21(cip1/waf1/sdi1) in DNA replication and transcription. J Biol Chem 275(14):10477–10483
124. Lasorella A, Noseda M, Beyna M, Yokota Y, Iavarone A (2000) Id2 is a retinoblastoma protein target and mediates signalling by Myc oncoproteins. Nature 407(6804): 592–598
125. La Rocca SA, Crouch DH, Gillespie DA (1994). c-Myc inhibits myogenic differentiation and myoD expression by a mechanism which can be dissociated from cell transformation. Oncogene 9(12):3499–3508
126. La Rocca SA, Vannucchi S, Pompili M et al (2002). Selective repression of myoD transcription by v-Myc prevents terminal differentiation of quail embryo myoblasts transformed by the MC29 strain of avian myelocytomatosis virus. Oncogene 21(31):4838–4842
127. Beato M, Herrlich P, Schutz G (1995). Steroid hormone receptors: many actors in search of a plot. Cell 83(6):851–857
128. Yen PM (2001). Physiological and molecular basis of thyroid hormone action. Physiol Rev 81(3):1097–1142
129. Chin S, Apriletti J, Gick G (1998) Characterization of a negative thyroid hormone response element in the rat sodium, potassium-adenosine triphosphatase alpha3 gene promoter. Endocrinology 139(8):3423–3431
130. Taylor AH, Wishart P, Lawless DE, Raymond J, Wong NC (1996) Identification of functional positive and negative thyroid hormone-responsive elements in the rat apolipoprotein AI promoter. Biochemistry 35(25):8281–8288

131. Wight PA, Crew MD, Spindler SR (1987) Discrete positive and negative thyroid hormone-responsive transcription regulatory elements of the rat growth hormone gene. J Biol Chem 262(12):5659–5663
132. Drabkin H, Kao FT, Hartz J et al (1988) Localization of human ERBA2 to the 3p22----3p24.1 region of chromosome 3 and variable deletion in small cell lung cancer. Proc Natl Acad Sci U S A 85(23):9258–9262
133. Thompson CC, Evans RM (1989) Trans-activation by thyroid hormone receptors: functional parallels with steroid hormone receptors. Proc Natl Acad Sci U S A 86(10):3494–3498
134. Flamant F, Samarut J (2003) Thyroid hormone receptors: lessons from knockout and knock-in mutant mice. Trends Endocrinol Metab 14(2):85–90
135. Schrader M, Muller KM, Nayeri S, Kahlen JP, Carlberg C (1994) Vitamin D3-thyroid hormone receptor heterodimer polarity directs ligand sensitivity of transactivation. Nature 370(6488):382–386
136. Bogazzi F, Hudson LD, Nikodem VM (1994) A novel heterodimerization partner for thyroid hormone receptor. Peroxisome proliferator-activated receptor. J Biol Chem 269(16): 11683–11686
137. Forman BM, Yang CR, Au M, Casanova J, Ghysdael J, Samuels HH (1989) A domain containing leucine-zipper-like motifs mediate novel in vivo interactions between the thyroid hormone and retinoic acid receptors. Mol Endocrinol 3(10):1610–1626
138. Glass CK, Lipkin SM, Devary OV, Rosenfeld MG (1989) Positive and negative regulation of gene transcription by a retinoic acid-thyroid hormone receptor heterodimer. Cell 59(4): 697–708
139. Sugawara A, Yen PM, Darling DS, Chin WW (1993) Characterization and tissue expression of multiple triiodothyronine receptor-auxiliary proteins and their relationship to the retinoid X-receptors. Endocrinology 133(3):965–971
140. Desbois C, Aubert D, Legrand C, Pain B, Samarut J (1991) A novel mechanism of action for v-ErbA: abrogation of the inactivation of transcription factor AP-1 by retinoic acid and thyroid hormone receptors. Cell 67(4):731–740
141. Zhang XK, Wills KN, Husmann M, Hermann T, Pfahl M (1991) Novel pathway for thyroid hormone receptor action through interaction with jun and fos oncogene activities. Mol Cell Biol 11(12):6016–6025
142. Marchal S, Cassar-Malek I, Pons F, Wrutniak C, Cabello G (1993) Triiodothyronine influences quail myoblast proliferation and differentiation. Biol Cell 78(3):191–197
143. Cassar-Malek I, Marchal S, Rochard P, et al (1996) Induction of c-Erb A-AP-1 interactions and c-Erb A transcriptional activity in myoblasts by RXR. Consequences for muscle differentiation. J Biol Chem 271(19):11392–11399
144. Downes M, Griggs R, Atkins A, Olson EN, Muscat GE (1993) Identification of a thyroid hormone response element in the mouse myogenin gene: characterization of the thyroid hormone and retinoid X receptor heterodimeric binding site. Cell Growth Differ 4(11): 901–909
145. Leblond-Larouche L, Morais R, Zollinger M (1979) Studies of the effect of chloramphenicol, ethidium bromide and camptothecin on the reproduction of Rous sarcoma virus in infected chick embryo cells. J Gen Virol 44(2):323–331
146. Morais R, Gregoire M, Jeannotte L, Gravel D (1980) Chick embryo cells rendered respiration-deficient by chloramphenicol and ethidium bromide are auxotrophic for pyrimidines. Biochem Biophys Res Commun 94(1):71–77
147. van den Bogert C, Spelbrink JN, Dekker HL (1992) Relationship between culture conditions and the dependency on mitochondrial function of mammalian cell proliferation. J Cell Physiol 152(3):632–638
148. Hamai N, Nakamura M, Asano A (1997) Inhibition of mitochondrial protein synthesis impaired C2C12 myoblast differentiation. Cell Struct Funct 22(4):421–431
149. Kaneko T, Watanabe T, Oishi M (1988) Effect of mitochondrial protein synthesis inhibitors on erythroid differentiation of mouse erythroleukemia (Friend) cells. Mol Cell Biol 8(8):3311–3315

150. Korohoda W, Pietrzkowski Z, Reiss K (1993) Chloramphenicol, an inhibitor of mitochondrial protein synthesis, inhibits myoblast fusion and myotube differentiation. Folia Histochem Cytobiol 31(1):9–13
151. Vayssiere JL, Cordeau-Lossouarn L, Larcher JC, Basseville M, Gros F, Croizat B (1992) Participation of the mitochondrial genome in the differentiation of neuroblastoma cells. In Vitro Cell Dev Biol 28A(11–12):763–772
152. Newmeyer DD, Ferguson-Miller S (2003) Mitochondria: releasing power for life and unleashing the machineries of death. Cell 112(4):481–490
153. Larsson NG, Wang J, Wilhelmsson H et al (1998) Mitochondrial transcription factor A is necessary for mtDNA maintenance and embryogenesis in mice. Nat Genet 18(3):231–236
154. Wallace DC (2000) Mitochondrial defects in cardiomyopathy and neuromuscular disease. Am Heart J 139(2 Pt 3):S70–85
155. Rochard P, Cassar-Malek I, Marchal S, Wrutniak C, Cabello G (1996) Changes in mitochondrial activity during avian myoblast differentiation: influence of triiodothyronine or v-erb A expression. J Cell Physiol 168(2):239–247
156. Shadel GS, Clayton DA (1997) Mitochondrial DNA maintenance in vertebrates. Annu Rev Biochem 66:409–435
157. Falkenberg M, Gaspari M, Rantanen A, Trifunovic A, Larsson NG, Gustafsson CM (2002) Mitochondrial transcription factors B1 and B2 activate transcription of human mtDNA. Nat Genet 31(3):289–294
158. McCulloch V, Seidel-Rogol BL, Shadel GS (2002) A human mitochondrial transcription factor is related to RNA adenine methyltransferases and binds S-adenosylmethionine. Mol Cell Biol 22(4):1116–1125
159. Wrutniak C, Cassar-Malek I, Marchal S et al (1995) A 43-kDa protein related to c-Erb A [IMAGE]1 is located in the mitochondrial matrix of rat liver. J Biol Chem 270(27): 16347–16354
160. Casas F, Domenjoud L, Rochard P et al (2000) A 45 kDa protein related to PPARgamma2, induced by peroxisome proliferators, is located in the mitochondrial matrix. FEBS Lett 478(1–2):4–8
161. Wrutniak C, Rochard P, Casas F, Fraysse A, Charrier J, Cabello G (1998) Physiological importance of the T3 mitochondrial pathway. Ann NY Acad Sci 839:93–100
162. Casas F, Rochard P, Rodier A et al (1999) A variant form of the nuclear triiodothyronine receptor c-ErbAalpha1 plays a direct role in regulation of mitochondrial RNA synthesis. Mol Cell Biol 19(12):7913–7924
163. Casas F, Daury L, Grandemange S et al (2003) Endocrine regulation of mitochondrial activity: involvement of truncated RXRalpha and c-Erb Aalpha1 proteins. Faseb J 17(3):426–436
164. Seyer P, Grandemange S, Pessemesse L et al (2006) L'activité mitochondriale est un régulateur majeur de la différenciation des myoblastes et de l'expression des isoformes de myosine. Prod Anim 19:279–286
165. Casas F, Pessemesse L, Grandemange S et al (2008) Overexpression of the T3 mitochondrial receptor in mice induces changes in metabolic and contractile features of muscle fibers. PLoS One 3:e2501
166. Downes M, Mynett-Johnson L, Muscat GE (1994) The retinoic acid and retinoid X receptors are differentially expressed during myoblast differentiation. Endocrinology 134(6): 2658–2661
167. Busson M, Daury L, Seyer P et al (2006) Avian MyoD and c-Jun coordinately induce transcriptional activity of the 3,5,3′-triiodothyronine nuclear receptor c-ErbAalpha1 in proliferating myoblasts. Endocrinology 147(7):3408–3418
168. Baris O, Seyer P, Busson M et al (2007) Identification of mitochondrial activity target genes during myoblast differentiation. FEBS Advanced Lecture Series - Mitochondria in Life, Death and Disease, AUSSOIS (France), 14–16
169. Daury L, Busson M, Casas F, Cassar-Malek I, Wrutniak-Cabello C, Cabello G (2001) The triiodothyronine nuclear receptor c-ErbAalpha1 inhibits avian MyoD transcriptional activity in myoblasts. FEBS Lett 508(2):236–240

170. Casas F, Busson M, Grandemange S et al (2006) Characterization of a novel thyroid hormone receptor alpha variant involved in the regulation of myoblast differentiation. Mol Endocrinol 20(4):749–763
171. Claret FX, Antakly T, Karin M, Saatcioglu F (1996) A shift in the ligand responsiveness of thyroid hormone receptor alpha induced by heterodimerization with retinoid X receptor alpha. Mol Cell Biol 16(1):219–227
172. Flamant F, Samarut J (1998) Involvement of thyroid hormone and its alpha receptor in avian neurulation. Dev Biol 197(1):1–11
173. Li L, Chambard JC, Karin M, Olson EN (1992) Fos and Jun repress transcriptional activation by myogenin and MyoD: the amino terminus of Jun can mediate repression. Genes Dev 6(4):676–689
174. Rimokh R, Rouault JP, Wahbi K et al (1991) A chromosome 12 coding region is juxtaposed to the MYC protooncogene locus in a t(812)(q24q22) translocation in a case of B-cell chronic lymphocytic leukemia. Genes Chromosomes Cancer 3(1):24–36
175. Matsuda S, Rouault J, Magaud J, Berthet C (2001) In search of a function for the TIS21/PC3/BTG1/TOB family. FEBS Lett 497(2–3):67–72
176. Rouault JP, Rimokh R, Tessa C et al (1992) BTG1, a member of a new family of antiproliferative genes. Embo J 11(4):1663–1670
177. Westernacher D, Schaper W (1995) A novel heart derived inhibitor of vascular cell proliferation. Purification and biological activity. J Mol Cell Cardiol 27(8):1535–1543
178. Rodier A, Rochard P, Berthet C et al (2001) Identification of functional domains involved in BTG1 cell localization. Oncogene 20(21):2691–2703
179. Grandemange S, Seyer P, Carazo A et al (2005) Stimulation of mitochondrial activity by p43 overexpression induces human dermal fibroblast transformation. Cancer Res 65(10): 4282–4291
180. Delling U, Tureckova J, Lim HW, De Windt LJ, Rotwein P, Molkentin JD (2000) A calcineurin-NFATc3-dependent pathway regulates skeletal muscle differentiation and slow myosin heavy-chain expression. Mol Cell Biol 20(17):6600–6611
181. Friday BB, Horsley V, Pavlath GK (2000) Calcineurin activity is required for the initiation of skeletal muscle differentiation. J Cell Biol 149(3):657–666
182. Xu Q, Yu L, Liu L et al (2002) p38 Mitogen-activated protein kinase-, calcium-calmodulin-dependent protein kinase-, and calcineurin-mediated signaling pathways transcriptionally regulate myogenin expression. Mol Biol Cell 13(6):1940–1952
183. Chin ER, Olson EN, Richardson JA et al (1998) A calcineurin-dependent transcriptional pathway controls skeletal muscle fiber type. Genes Dev 12(16):2499–2509
184. Meissner JD, Umeda PK, Chang KC, Gros G, Scheibe RJ (2007) Activation of the beta myosin heavy chain promoter by MEF-2D, MyoD, p300, and the calcineurin/NFATc1 pathway. J Cell Physiol 211(1):138–148
185. Trenker M, Malli R, Fertschai I, Levak-Frank S, Graier WF (2007) Uncoupling proteins 2 and 3 are fundamental for mitochondrial Ca2+ uniport. Nat Cell Biol 9(4):445–452
186. Brookes PS, Yoon Y, Robotham JL, Anders MW, Sheu SS (2004) Calcium, ATP, and ROS: a mitochondrial love-hate triangle. Am J Physiol Cell Physiol 287(4):C817–833
187. Rizzuto R, Pinton P, Carrington W et al (1998) Close contacts with the endoplasmic reticulum as determinants of mitochondrial Ca2+ responses. Science 280(5370): 1763–1766
188. Reed JC, Nowell PC, Hoover RG (1985) Regulation of c-myc mRNA levels in normal human lymphocytes by modulators of cell proliferation. Proc Natl Acad Sci U S A 82(12): 4221–4224
189. Tsuda T, Kaibuchi K, West B, Takai Y (1985) Involvement of Ca2+ in platelet-derived growth factor-induced expression of c-myc oncogene in Swiss 3T3 fibroblasts. FEBS Lett 187(1):43–46
190. Bubici C, Papa S, Dean K, Franzoso G (2006) Mutual cross-talk between reactive oxygen species and nuclear factor-kappa B: molecular basis and biological significance. Oncogene 25(51):6731–6748

191. Martindale JL, Holbrook NJ (2002) Cellular response to oxidative stress: signaling for suicide and survival. J Cell Physiol 192(1):1–15
192. Nakano H, Nakajima A, Sakon-Komazawa S, Piao JH, Xue X, Okumura K (2006) Reactive oxygen species mediate crosstalk between NF-kappaB and JNK. Cell Death Differ 13(5):730–737
193. Saelim N, John LM, Wu J et al (2004) Nontranscriptional modulation of intracellular Ca2+ signaling by ligand stimulated thyroid hormone receptor. J Cell Biol 167(5):915–924
194. Biswas G, Adebanjo OA, Freedman BD et al (1999) Retrograde Ca2+ signaling in C2C12 skeletal myocytes in response to mitochondrial genetic and metabolic stress: a novel mode of inter-organelle crosstalk. Embo J 18(3):522–533
195. Luo Y, Bond JD, Ingram VM (1997) Compromised mitochondrial function leads to increased cytosolic calcium and to activation of MAP kinases. Proc Natl Acad Sci U S A 94(18):9705–9710
196. Wrutniak-Cabello C, Casas F, Cabello G (2001) Thyroid hormone action in mitochondria. J Mol Endocrinol 26(1):67–77

Part II
Stem Cells

Chapter 4
The Neural Stem Cells

Stefano Pluchino, Marco Bacigaluppi, Elena Brini, Erica Butti, Chiara Cossetti, Melania Cusimano, Lucia Zanotti, and Gianvito Martino

Abstract Neural stem cells represent a heterogeneous population of mitotically active, self-renewing and multipotent cells of both the developing and the adult central nervous system (CNS) showing complex patterns of gene expression that may vary in both space and time. Endogenous stem cells residing within CNS germinal niches might concur to nervous system repair owing to their ability to drive neurogenesis and gliogenesis during adulthood. Nevertheless, self-renewal, proliferation, migration and differentiation of CNS stem cells may significantly vary upon different types (e.g. acute vs chronic, focal vs multifocal) of CNS injury. In this chapter we address several aspects of neural stem cell pathophysiology.

Keywords Neural progenitor cells · Neurogenesis · Neurons · Glia · Neural stem cell transplantation · Nervous system disorders

Neural stem/precursor cells (NPCs) represent a heterogeneous population of mitotically active, self-renewing and multipotent cells of both the developing and the adult central nervous system (CNS) showing complex patterns of gene expression that may vary in both space and time [1–3]. In the late 1960s, proliferating neural cells – possibly representing newly generated neurons – were identified in the adult rat brain [4, 5]. Since then, NPCs have been isolated virtually from the entire embryonic as well as the adult mammalian CNS. The ganglionic eminence(s), in the mouse embryo, and both the subventricular zone (SVZ) of the lateral ventricles and the sub-granular zone (SGZ) of the hippocampus dentate gyrus (DG), in adult rodents, have been shown to consistently contain stem-like cells capable of driving neurogenesis and gliogenesis [2, 6]. Cells with the structural and molecular characteristics of astrocytes (e.g. expression of glial fibrillary acidic protein [GFAP]) are considered as the bona fide CNS stem cells (or type B cells) in both the SVZ and the

S. Pluchino (✉)
CNS Repair Unit, DIBIT2 and Institute of Experimental Neurology (INSPE), San Raffaele Scientific Institute, Via Olgettina 58, 20123 Milan, Italy; Present address: Cambridge Centre for Brain Repair, Department of Clinical Neurosciences, ED Adrian Building, Forvie Site, Robinson Way, Cambridge, CB2 0PY, UK
e-mail: pluchino.stefano@hsr.it; spp24@cam.ac.uk

A. Giordano, U. Galderisi (eds.), *Cell Cycle Regulation and Differentiation in Cardiovascular and Neural Systems*, DOI 10.1007/978-1-60327-153-0_4,

SGZ [7–9]. In the SVZ, these $GFAP^+$ cells lie in intimate contact with other SVZ cell types, such as the rapidly dividing transit amplifying cells (type C cells) and the lineage-committed (post-mitotic) neuroblasts (type A cells). The cell lineage differentiation pathway in the SVZ goes from type B through type C to type A cells, with the type B cells believed to be the self-renewing primary neural stem cells [7] and the type A cell the neuroblasts migrating into chains to the olfactory bulbs (OB) [10] (Fig. 4.1). Differently from rodents, the adult human brain contains a single ribbon of SVZ astrocytes, which may proliferate in vivo and behave as multipotent progenitor cells in vitro [11, 12]. Nonetheless, conflicting evidence of newly generated neuroblasts from the human SVZ migrating towards the rostral pathway to the OB is provided so far [13, 14].

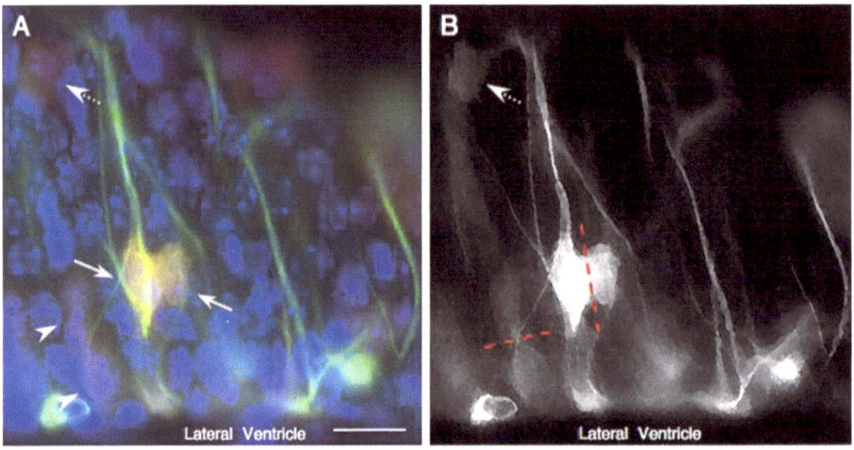

Fig. 4.1 NPCs in the SVZ of the adult mouse brain. (**a**) At 72 h after GFP-retroviral labelling, GFP-expressing SVZ cells (*red*) are positive for the radial glial marker nestin (*green*), visible in the radial processes. (**b**) Symmetry of cell divisions by GFP-expressing cells is defined by the orientation of the cleavage plane with respect to the epithelial surface. Mitotic divisions with the cleavage plane (*red dashed lines*) parallel to the epithelium (*horizontal*) are often asymmetrical (*arrowheads*), while mitotic divisions with a cleavage plane orthogonal to the epithelium (*vertical*) are generally symmetrical (*arrows*). Scale bar: 20 μm

Also in the SGZ, the GFAP-expressing astrocytes function as (type B) CNS stem cells, undergo certain self-renewal and proliferation and turn into (type D) transit amplifying cells, and then differentiate into lineage-committed (type G) migratory granule neurons [15, 16]. Type B cells in the SVZ interdigitate with both the basal lamina and the blood vessels, while in the SGZ bursts of endothelial cell division are spatially and temporally related to clusters of neurogenesis [15]. Indeed, the histopathological analysis of post-mortem brain tissue from patients undergoing bromodeoxyuridine (BrdU) treatment reveals that new neurons – as defined by NeuN, calbindin or neuron-specific enolase (NSE) – are generated from dividing progenitor cells in the DG of the human hippocampus [17], thus suggesting that the human hippocampus retains the ability to generate neurons throughout life.

The maintenance and the differentiation of neural stem cells in CNS germinal niches depend on their physical contact to the basal lamina that acts as a scaffold, sequesters and/or modulates the release of cytokines and growth factors from local cells (e.g. ependymal cells, fibroblasts, macrophages and pericytes) [18]. The rostral migration of at least half of newly generated type A cells also depends on their physical contact with the extracellular matrix and perivascular astrocyte end feet, as they associate with the vasculature in the granule cell layer of the OB and use blood vessels as a scaffold for their migration (vasophilic migration) [19].

1 Endogenous Neural Stem Cells and CNS Diseases

Endogenous stem cells residing within CNS germinal niches might concur to nervous system repair owing to their ability to drive neurogenesis and gliogenesis during adulthood [20]. Nevertheless, self-renewal, proliferation, migration and differentiation of CNS stem cells may significantly vary upon different types (e.g. acute vs chronic, focal vs multifocal) of CNS injury [21–23].

Increased numbers of nestin-expressing proliferating neural progenitor cells as well as of doublecortin-reactive neuronal progenitors are detected at the boundaries of the injury site as early as 1 week after experimental acute focal inflammatory CNS disorders, such as spinal cord injury (SCI) and stroke [23–26]. Experimental acute stroke in rodents triggers neurogenesis and migration of newborn neurons from their sites of origin into ischaemic brain regions [27].

The transient occlusion of the middle cerebral artery in the rat increases the incorporation of BrdU into neural cells in the SGZ of the DG, the effect correlating with activation of the cAMP response element-binding protein (CREB) [28]. Neural cells labelled with BrdU coexpress the immature neuronal markers doublecortin and proliferating cell nuclear antigen (PCNA), while they do not express the more mature cell markers neuronal nuclear (NeuN) and Hu, thus suggesting that they are nascent neurons [29]. The acute stroke is associated to a shortened length of the cell cycle, a decreased G1 phase and an increased cell cycle length of SVZ-resident neural progenitors [30] regulating a transient increase in both (terminal) symmetric cell division and generation of neuronal progenitors migrating through the ischaemic striatum towards the damage, closely associated with blood vessels [25, 31]. Also in patients with stroke, neural cells that express markers associated with newborn neurons are present in the ischaemic penumbra surrounding cerebral cortical infarcts and preferentially localize in the vicinity of blood vessels [32].

On the other hand, following acute SCI in the mouse, neural progenitors in the ependymal zone (EZ) of the central canal mobilize and migrate vigorously towards the direction where the contusion injury is generated – the most favourable migration occurring in the adjacent region close to the epicentre of the lesion – and differentiate optimally into neuronal nuclear (NeuN)-immunoreactive neurons, while not into astrocytes or oligodendrocytes [33]. After cervical SCI in the adult rhesus monkey, BrdU-based analysis of cell proliferation in vivo reveals an increase of $\geq$80-fold in the number of newly divided cells in the spinal cord. By 7 months after injury, 15% of these newly generated neural cells express markers of mature oligodendrocytes while 12% express astrocytic markers. These newly born oligodendrocytes are present in zones of injury-induced demyelination and appear to ensheath or remyelinate host axons [34].

In experimental models of chronic multifocal inflammatory demyelinating disorders such as experimental autoimmune encephalomyelitis (EAE), the animal model of MS, mitotically active neural progenitor cells, which reside either in the SVZ of the brain or in the EZ of central canal of the spinal cord, subvert their physiological destiny – the rostral migration to the OB or the radial migration to the lateral columns of the spinal cord – and migrate into areas of demyelination where they differentiate into glial cells [21, 22].

Though accumulating evidence indicates that endogenous neurogenesis and gliogenesis occur as part of an 'intrinsic' self-repair process during inflammatory CNS disorders, there are no convincing explanations about the overall incapacity of the endogenous CNS stem cell compartment to promote full and long-lasting CNS repair. Recent data suggest that chronic brain inflammation, induced by myelin-specific immune cells, irreversibly alters the proliferative and migratory properties of subventricular zone (SVZ)-resident endogenous NPCs in vivo. This effect is generally sustained by a pro-inflammatory cytokine-dependent inhibition of cell cycle progression leading to significant accumulation of non-migratory neuroblasts within the SVZ germinal niche. In parallel, quantitative reduction of the putative brain stem cells is also observed. Extensive in vitro culturing of neurospheres from mice with chronic brain inflammation completely reverses the impairment, thus suggesting that the hostile chronically inflamed brain microenvironment may sustain a non-cell-autonomous dysfunction of the endogenous NPCs [35].

Furthermore, during sub-acute lipopolysaccharide (LPS)-induced brain inflammation, interleukin (IL)-6 released by microglia significantly impairs neurogenesis in the hippocampus in vivo, the impairment being fully restored when non-steroidal anti-inflammatory drugs (such as indomethacin) are used [36]. In vitro generation of new neurons and oligodendrocyte from NPCs is induced and supported by mouse microglia that have encountered T-cell-associated cytokines (such as interferon-g and IL-4), but blocked by those that have encountered endotoxins (such as LPS) [37]. More recently, hippocampal neurogenesis induced by an enriched environment has been associated with the recruitment of brain-derived neurotrophic factor (BDNF) releasing T cells and the activation of microglia in the DG. When studied in immune-deficient mice, hippocampal neurogenesis has been found markedly impaired and not enhanced by environmental enrichment, while restored and boosted by T cells recognizing a specific CNS antigen, such as myelin basic protein (MBP) [38].

Taken together, these results might suggest that certain common immune-associated mechanism(s) would underlie different aspects of structural plasticity and cell renewal in the adult CNS. Further, we cannot exclude that (at least) in certain chronic CNS inflammatory disorders (such as MS), some regional tropism of blood-borne inflammatory cells for major germinal niches might occur as a consequence of the capacity of the different cell components of the niches to secrete molecules preferentially attracting inflammatory cells. This, in turn, supports the ensuing idea that some CNS diseases might be provocatively viewed as the consequence of a dysfunction of stem cells rather than the upshot of an uncontrolled, and still undiscovered, pathogenic alien(s).

2 Neural Stem Cell Transplantation and CNS Diseases

Soon after the in vivo identification of stem cells from the CNS, different procedures have been developed in order to safely expand and maintain these cells in chemically defined media for years [39]. As a consequence, protocols to obtain in vitro a large number of NPCs have been established, thus supporting the concept that these cells might represent a renewable source of uncommitted ready-to-use cells for transplantation purposes [40]. NPC-based therapies for nervous system disorders – for example, stroke, Parkinson's disease (PD), Huntington's disease (HD), MS, SCI – have been successfully developed. While most of these attempts have succeeded in experimental models, there are still important issues that need to be solved before envisaging any potential human applications of such promising therapies. Not only the ideal cell source for transplantation (e.g., embryonic vs somatic) but also the best route of cell administration (e.g., local vs systemic) should be determined. However, the putative mechanism(s) sustaining both repair capabilities and long-term functional integration of NPCs upon transplantation is also unclear. Although indications that stem cells can reach the target organ and differentiate into the appropriate lineage exist, there is still scarce evidence that transplanted NPCs can reconstruct the three-dimensional brain architecture and give rise to large numbers of properly functioning cells integrating into the brain circuitries.

The route of cell administration represents a major constrain for NPC transplantation and appears to be very much depending on the CNS lesion site(s) (focal vs. multifocal). The anatomo-pathological features of focal CNS disorders, such as PD or acute SCI (also stroke and brain trauma), might suggest that direct local (*intralesional*) cell transplantation might facilitate tissue regeneration, while the multifocality of certain CNS disorders – such as MS and epilepsy – would represent a major limitation for *intralesional* cell transplantation approaches. Following the first observation in experimental brain tumours, in multifocal CNS disorders, systemic (e.g. intravenous and intrathecal) transplantation of NPCs can be therapeutically efficacious owing to the ability of transplanted cells to follow, via the blood stream or cerebrospinal fluid circulation, a gradient of chemoattractants (e.g. pro-inflammatory cytokines and chemokines) occurring at the site of inflammatory lesions [41, 42]. Specific homing of transplanted neural stem cells has been shown, in SCI, epilepsy and stroke. However, the exact molecular mechanism sustaining this phenomenon has been detailed, so far, only in EAE. Tethering, rolling and firm adhesion to inflamed endothelial cells and extravasation into inflamed CNS areas are sequentially mediated by the constitutive expression of functional cell adhesion molecules (CAM) (e.g. CD44), integrins (e.g. α4, β1) and chemokine receptors (e.g. CCR1, CCR2, CCR5, CXCR3, CXCR4) on neural stem cell surface [41, 42]. Irrespective from the characteristics of the experimental disease (e.g. disease course [acute vs. chronic], neuropathological features [focal vs. multifocal] and type of inflammation [primary vs. reactive]), functional recovery obtained by neural stem cell transplantation scarcely correlates with absolute numbers of transplant-derived newly generated terminally differentiated neuronal cells. Transplantation of neural stem cells into rodents with experimental PD or HD very scarcely differentiate into tyrosine hydroxylase (TH)-immunoreactive neurons despite significant behavioural improvement. Similarly, mice with SCI, acute stroke and intracerebral haemorrhage do improve despite pathological evidence of preferential astroglial fate of transplanted NPCs. The large majority of NPCs injected into mice with experimental cerebral haemorrhage or with acute ischaemic stroke express markers of undifferentiation, such as nestin, when surrounding damaged CNS areas. In EAE, very low differentiation of transplanted neural stem cells into myelin-forming oligodendrocytes is accompanied by neurophysiological evidence of axonal protection and remyelination. In the very same context, more than 20% of transplanted cells reaching inflammatory demyelinated areas do not express differentiation markers. This scarce and inappropriate terminal differentiation and the propensity of maintaining an undifferentiated phenotype within the host tissue suggest that transplanted neural stem cells might be therapeutic efficacious via a bystander mechanism(s) alternative to cell replacement. Therefore, transplanted NPCs reduce the scar formation and/or increase survival and function(s) of endogenous glial and neuronal progenitors surviving to the pathological insult. This neuroprotective effect is accompanied by increased in vivo bioavailability of major neurotrophins [e.g. nerve growth factor (NGF), brain-derived growth factor (BDNF), ciliary neurotrophic growth factor (CNTF), glial-derived neurotrophic growth factor (GDNF)]. Moreover, transplanted neural stem cells promote bystander immunomodulation as they release soluble molecules (e.g. cytokines and chemokines), express immune-relevant receptors (e.g. chemokine receptors, CAMs), capable of profoundly altering the inflammatory environment, and up-regulate membrane expression of certain functional death receptor ligands (e.g. FasL, TRAIL, Apo3L) by which they induce programmed cell death (apoptosis) of inflammatory T lymphocytes [41]. Furthermore, transplanted NSCs also significantly and specifically contribute to down-regulate effector functions of inflammatory T cells and macrophages within both the target tissue and draining lymph nodes [43].

All together these results consistently challenge the sole and limited view that stem cells therapeutically work exclusively throughout cell replacement. As a matter of fact, NPC transplantation may also promote CNS repair via intrinsic *neuroprotective* bystander capacities, mainly exerted by undifferentiated stem cells releasing, at the site of tissue damage, a milieu of *neuroprotective* molecules once temporally and spatially orchestrated by environmental needs. The intrinsic nature (*pleiotropism and redundancy*) of these molecules and their '*constitutive*' characteristics represent a *stem cell signature* that also reconcile data showing that other sources of somatic stem cells (e.g. HSCs, MSCs), with very low capabilities of neural (trans) differentiation, may efficiently promote CNS repair. Thus, cell plasticity can also be viewed as the capacity of somatic stem cells to adapt their fate and function(s) to specific environmental needs occurring as a result of different pathological conditions (*therapeutic plasticity*).

The exact knowledge and the potential impact of *non-conventional* stem cell-mediated therapeutic mechanisms might result, in certain circumstances, in more efficacious curative alternatives.

Acknowledgments This work was supported in part by the Italian Multiple Sclerosis Foundation (FISM, 2004/R/15 to S.P. and 2005/R/15 to G.M.), the National Multiple Sclerosis Society (NMSS, RG 3591-A-1 to G.M.), the Italian Ministry of Research and University (MIUR), Italian Ministry of Health, BMW Italy Group and Banca Agricola Popolare di Ragusa (BAPR).

References

1. Temple S (2001) The development of neural stem cells. Nature 414(6859):112–117
2. Gage FH (2000), Mammalian neural stem cells. Science 287(5457):1433–1438
3. Ivanova NB et al (2002) A stem cell molecular signature. Science 298(5593):601–604
4. Altman J, Das GD (1965) Autoradiographic and histological evidence of postnatal hippocampal neurogenesis in rats. J Comp Neurol 124(3):319–335
5. Altman J (1969) Autoradiographic and histological studies of postnatal neurogenesis. 3. Dating the time of production and onset of differentiation of cerebellar microneurons in rats. J Comp Neurol 136(3):269–293
6. Doetsch F (2003) The glial identity of neural stem cells. Nat Neurosci 6(11):1127–1134
7. Doetsch F et al (1999) Subventricular zone astrocytes are neural stem cells in the adult mammalian brain. Cell 97(6):703–716
8. Laywell ED et al (2000) Identification of a multipotent astrocytic stem cell in the immature and adult mouse brain. Proc Natl Acad Sci U S A 97(25):13883–13888
9. Garcia AD et al (2004) GFAP-expressing progenitors are the principal source of constitutive neurogenesis in adult mouse forebrain. Nat Neurosci 7(11):1233–1241
10. Lois C, Garcia-Verdugo JM, Alvarez-Buylla A (1996) Chain migration of neuronal precursors. Science 271(5251):978–981
11. Quinones-Hinojosa A et al (2006) Cellular composition and cytoarchitecture of the adult human subventricular zone: a niche of neural stem cells. J Comp Neurol 494(3):415–434
12. Sanai N et al (2004) Unique astrocyte ribbon in adult human brain contains neural stem cells but lacks chain migration. Nature 427(6976):740–744
13. Curtis MA et al (2007) Human neuroblasts migrate to the olfactory bulb via a lateral ventricular extension. Science 315(5816):1243–1249
14. Sanai N et al (2007) Comment on "Human neuroblasts migrate to the olfactory bulb via a lateral ventricular extension". Science 318(5849):393; author reply 393
15. Palmer TD, Willhoite AR, Gage FH (2000) Vascular niche for adult hippocampal neurogenesis. J Comp Neurol 425(4):479–494
16. Seri B et al (2001) Astrocytes give rise to new neurons in the adult mammalian hippocampus. J Neurosci 21(18):7153–7160
17. Eriksson PS et al (1998) Neurogenesis in the adult human hippocampus. Nat Med 4(11): 1313–1317
18. Mercier F, Kitasako JT, Hatton GI (2002) Anatomy of the brain neurogenic zones revisited: fractones and the fibroblast/macrophage network. J Comp Neurol 451(2):170–188
19. Bovetti S et al (2007) Blood vessels form a scaffold for neuroblast migration in the adult olfactory bulb. J Neurosci 27(22):5976–5980

20. Ming GL, Song H (2005) Adult neurogenesis in the mammalian central nervous system. Annu Rev Neurosci 28:223–250
21. Brundin L et al (2003) Neural stem cells: a potential source for remyelination in neuroinflammatory disease. Brain Pathol 13(3):322–328
22. Picard-Riera N et al (2002) Experimental autoimmune encephalomyelitis mobilizes neural progenitors from the subventricular zone to undergo oligodendrogenesis in adult mice. Proc Natl Acad Sci U S A 99(20):13211–13216
23. Yagita Y et al (2001) Neurogenesis by progenitor cells in the ischemic adult rat hippocampus. Stroke 32(8):1890–1896
24. Tureyen K et al (2004) Transient focal cerebral ischemia-induced neurogenesis in the dentate gyrus of the adult mouse. J Neurosurg 101(5):799–805
25. Zhang R et al (2004) Stroke transiently increases subventricular zone cell division from asymmetric to symmetric and increases neuronal differentiation in the adult rat. J Neurosci 24(25):5810–5815
26. Carmichael ST (2003) Gene expression changes after focal stroke, traumatic brain and spinal cord injuries. Curr Opin Neurol 16(6):699–704
27. Zhang R et al (2004) Activated neural stem cells contribute to stroke-induced neurogenesis and neuroblast migration toward the infarct boundary in adult rats. J Cereb Blood Flow Metab 24(4):441–448
28. Zhu DY et al (2004) Activation of cAMP-response-element-binding protein (CREB) after focal cerebral ischemia stimulates neurogenesis in the adult dentate gyrus. Proc Natl Acad Sci U S A 101(25):9453–9457
29. Jin K et al (2001) Neurogenesis in dentate subgranular zone and rostral subventricular zone after focal cerebral ischemia in the rat. Proc Natl Acad Sci U S A 98(8):4710–4715
30. Zhang RL et al (2006) Reduction of the cell cycle length by decreasing G1 phase and cell cycle reentry expand neuronal progenitor cells in the subventricular zone of adult rat after stroke. J Cereb Blood Flow Metab 26(6):857–863
31. Thored P et al (2007) Long-term neuroblast migration along blood vessels in an area with transient angiogenesis and increased vascularization after stroke. Stroke 38:3032–3039
32. Jin K et al (2006) Evidence for stroke-induced neurogenesis in the human brain. Proc Natl Acad Sci U S A 103(35):13198–13202
33. Ke Y et al (2006) Early response of endogenous adult neural progenitor cells to acute spinal cord injury in mice. Stem Cells 24(4):1011–1019
34. Yang H et al (2006) Endogenous neurogenesis replaces oligodendrocytes and astrocytes after primate spinal cord injury. J Neurosci 26(8):2157–2166
35. Pluchino S et al (2008) Persistent inflammation alters the function of the endogenous brain stem cell compartment. Brain 131:2564–2578
36. Monje ML, Toda H, Palmer TD (2003) Inflammatory blockade restores adult hippocampal neurogenesis. Science 302(5651):1760–1765
37. Butovsky O et al (2006) Microglia activated by IL-4 or IFN-gamma differentially induce neurogenesis and oligodendrogenesis from adult stem/progenitor cells. Mol Cell Neurosci 31(1):149–160
38. Ziv Y et al (2006) Immune cells contribute to the maintenance of neurogenesis and spatial learning abilities in adulthood. Nat Neurosci 9(2):268–275
39. Pluchino S et al (2005) Neural stem cells and their use as therapeutic tool in neurological disorders. Brain Res Brain Res Rev 48(2):211–219
40. Vescovi AL, Snyder EY (1999) Establishment and properties of neural stem cell clones: plasticity in vitro and in vivo. Brain Pathol 9(3):569–598
41. Pluchino S et al (2005) Neurosphere-derived multipotent precursors promote neuroprotection by an immunomodulatory mechanism. Nature 436(7048):266–271
42. Martino G, Pluchino S (2006) The therapeutic potential of neural stem cells. Nat Rev Neurosci 7(5):395–406
43. Einstein O et al (2007) Neural precursors attenuate autoimmune encephalomyelitis by peripheral immunosuppression. Ann Neurol 61(3):209–218

Chapter 5
Cardiac Stem and Progenitor Cells

Ronglih Liao and Regina L. Sohn

Abstract Since the early days of cardiovascular biology, it has been believed that mammalian adult cardiomyocytes exit from the cell cycle soon after birth, with the total number of cardiomyocytes being pre-determined. Recently, the identification of resident cardiac stem/progenitor cells by several independent laboratories has challenged this long-held paradigm and has provoked an exponential increase in the number of investigations. As a consequence, emerging evidence now supports a new theory in which the mammalian heart represents an organ at a dynamic cellular steady rate, with a constant, albeit low, rate of cellular turnover and the intrinsic ability to regenerate lost cells. If this is indeed the case, not only does it re-define myocardial biology, but it also suggests the potential to regenerate lost or diseased myocardium. To date, there is general agreement that adult hearts contain a population(s) of primitive cells that are capable of differentiating into functional myocytes, smooth muscle cells, and endothelial cells; however, what remains to be determined is the number, distribution, and origin of these cells. What also needs to be addressed is the relationship and/or overlap among cardiac stem/progenitor cell populations published thus far. Hopefully, a consensus in these regards will be reached with continued investigation. Note that this is an emerging field of investigation that is evolving at a rapid pace. Herein, we discuss the current views and up-to-date literature describing cardiac stem/progenitor cells with the understanding that this knowledge base will continue to advance and be refined in the days to come.

Keywords Cardiac stem cells · Regeneration · Cardiomyocytes · SP · c-kit · Sca1 · Differentiation · Self-renewal

---

R. Liao (✉)
Cardiac Muscle Research Laboratory, Cardiovascular Division, Department of Medicine, Brigham and Women's Hospital, Harvard Medical School, 77 Avenue Louis Pasteur, New Research Building, NRB 431, Boston, MA 02115, USA
e-mail: rliao@rics.bwh.harvard.edu

A. Giordano, U. Galderisi (eds.), *Cell Cycle Regulation and Differentiation in Cardiovascular and Neural Systems*, DOI 10.1007/978-1-60327-153-0_5,

1 Introduction

Regenerative medicine is an emerging field, involving both basic and clinical investigations, stemming from an increased understanding of stem cell biology and cellular plasticity. Since the first bone marrow transplantation to treat a patient with leukemia by Thomas and colleagues in 1956 [1], countless numbers of patients have benefited from this ground breaking treatment strategy. Indeed, adult stem cell therapy is considered one of the major achievements of modern medicine. Expanding the use of such stem cells for extra-bone marrow regeneration remains one of the next great frontiers of stem cell therapy.

Prior concepts have suggested that unlike the bone marrow, skin, and intestinal epithelium, the heart is a terminally differentiated organ, lacking self-renewal capacity. As a consequence, cardiomyocytes undergo hypertrophy, rather than hyperplasia, in response to hemodynamic stress. This long-held tenant has recently been challenged by the discovery of potential stem/progenitor cells in the heart. The identification of cardiac stem/progenitor cells was first reported in 2002, with the initial observation rapidly confirmed by different identifiers (for details see below), and by many independent laboratories. Such stem/progenitor cells may also explain much earlier observations of mitotic cardiac cells in adult myocardium following stress or injury [2, 3]. Furthermore, the presence of an endogenous pool of stem/progenitor cells has challenged the classical teaching that the mammalian heart contains a pre-determined number of cardiomyocytes after birth devoid of any mechanism to regenerate lost cells. Mounting evidence now suggests that the mammalian heart has the inherent capacity to self-renew and exhibit cellular turnover. Moreover, recent publications by Keating and colleagues even suggest that existing cardiomyocytes may have the potential to divide and proliferate [2–4]. It is, however, important to point out that the renewal capacity of these resident cardiac/progenitor cells is limited to the maintenance of the basal cellular turnover under physiologic conditions. With a larger cardiac insult or injury, such as that occurring during myocardial infarction, the scale of cell death generally exceeds the capacity for endogenous cell regeneration; and thus, the myocardium is left with a region of scar rather than fully regenerated cardiac muscle. Interestingly, such scar formation is also common in organ systems, such as the GI tract, which have greater regenerative capacity following local injury.

While the identification of resident cardiac stem/progenitor cells raises the distinct potential of using these cells for functionally relevant cardiac regeneration, this potential is hindered by our limited understanding of the molecular mechanisms regulating cardiac stem/progenitor cell fate and function. In this chapter, we will begin by reviewing the identification and characterization of currently recognized resident cardiac stem or progenitor cells. We will also review the current knowledge base regarding the origin of these cardiac stem cells and their resident cellular niches in myocardium. Additionally, we will discuss the role of these cardiac stem/progenitor cells in cardiac repair and regeneration as well as the effects of aging on these cardiac stem/progenitor cells. Finally, we will conclude with a brief discussion on current questions and future perspectives.

2 Identification of Cardiac Stem/Progenitor Cells

Emerging data have suggested that the heart does indeed possess endogenous progenitor or stem cells, capable of differentiation into cardiomyocytes. These cardiac stem or resident cells have been identified and characterized by their distinctive surface markers by several laboratories worldwide. Whether these different populations of cardiac resident stem/progenitor cells are truly unique and distinct from each other or represent a spectrum of cells in the developmental pathway has yet to be determined. A hypothesized hierarchy schematic is shown in Fig. 5.1. Nevertheless,

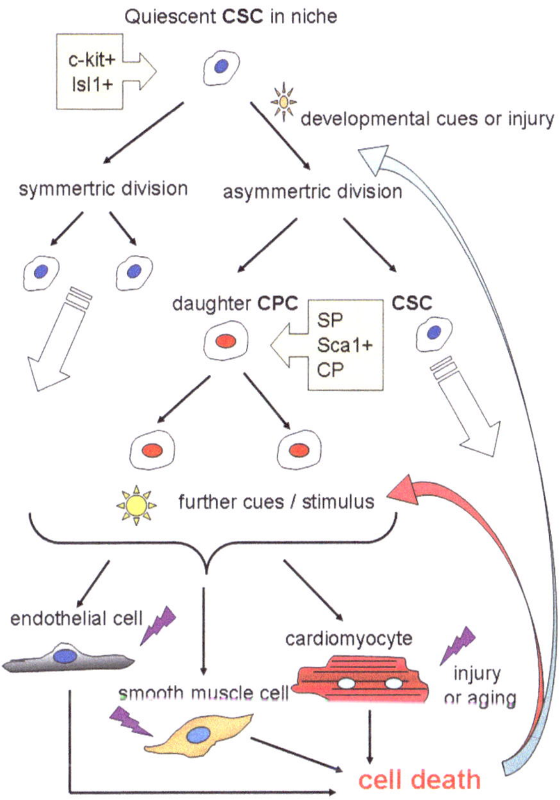

Fig. 5.1 The heart is a dynamic organ capable of cellular turnover. The quiescent cardiac stem cell is capable of a low basal level of cell division to maintain a pool of cardiac stem cells. These cells respond to developmental or injury by undergoing enhanced cell division (both symmetric and asymmetric to allow for renewal) to form daughter cardiac progenitor cells that respond to stimuli to differentiate further into cardiomyocytes, endothelial cells, and smooth muscle cells. The cardiac progenitor cells include cardiac SP, Sca1+, and cardiospheres as described. Whether these progenitor cells are distinct from each other or represent a developmental continuum is unknown. The fully differentiated and mature cardiac cells eventually succumb to aging or injury by undergoing cell death and/or stimulating the existing pools of cardiac stem and progenitor cells

these resident cardiac stem/progenitor cells have generated enormous hope for therapeutic cardiac regeneration. Herein, we discuss and summarize the current reported cardiac stem/progenitor cell populations (Table 5.1).

2.1 Cardiac SP Cells

The report by Hierlihy and colleagues [5] in 2002 was the first to identify the existence of a stem cell pool in post-natal murine hearts based on their specific ability to efflux Hoechst dye. The ability to actively efflux the DNA binding dye, Hoechst 33342, was first introduced to identify highly enriched hematopoietic stem cell populations, termed side population or SP cells, from bone marrow in 1996 by Goodel and colleagues [6]. This Hoechst efflux property can be blocked by verapamil, a reagent known for its ability to inhibit calcium channel activity as well as the multi-drug resistance proteins, Abcg2/Bcrp1 and Abcb1/Mdr1 [7]. Recently, this methodology has been widely utilized to identify tissue-specific stem/progenitor cells in various adult organs, including the pancreas, pituitary, testis, mammary gland, skeletal muscle, liver, lung, and heart [5–15] .

The experiments by Hierlihy et al. [5] stemmed from exciting reports suggesting that bone marrow-derived stem cells undergo cardiomyogenic differentiation when implanted into injured myocardium [16, 17]. In addition, Hierlihy and colleagues hypothesized that almost all organs harbor a resident pool of stem cell-like cells that are capable of maintaining cellular homeostasis during basal states or following injury. Therefore, they sought to identify a similar type of resident progenitor cell in post-natal myocardium. In so doing, they observed that adult hearts may indeed contain a resident pool of stem cell-like cells defined by the side population (SP) phenotype. They found that cardiac SP cells make-up approximately 1% of cardiomyocyte-depleted mono-nuclear cells and this population was low in expression for CD34, c-kit, Sca-1, Flk-2, Thy1.1. Of particular interest is that these cardiac SP cells can form colonies and differentiate into beating cardiomyocytes when cultured with adhered primary neonatal cardiomyocytes. To examine the dynamics of the cardiac SP pool, Hierlihy and colleagues compared the number of cardiac SP cells found in wild-type mice and in mice with cardiac-specific over-expression of a dominant negative form of Mef2c (Mef2cdn). These mice have hypoplastic hearts post-natally due to the small size of individual cardiomyocytes [18]. Compared to wild-type hearts, the percent of cardiac SP cells in hearts from Mef2cdn mice was reproducibly decreased. The Mef2cdn hearts also had a greater total number of cardiomyocytes. No causal relationship was established between the decrease in SP cell number and the increase in cardiomyocyte number in Mef2cdn hearts; nevertheless, it is attractive to speculate that SP cells in Mef2cdn hearts were recruited to undergo cardiomyogenic differentiation and contribute to the increased cardiomyocyte number. The existence of a resident pool of cardiac stem/progenitor cells was later verified by two independent laboratories by their c-kit [19] and Sca-1 [20] expression, as discussed below.

Table 5.1 Comparative data of resident cardiac stem/progenitor cells isolated form different laboratories

	Species	Major surface markers							Proliferative capacity	Cardiac differentiation	Other remarks
		c-kit	Sca1	CD45	CD34	CD31	MDR1	BCRP1			
SP	Mouse [10] (Garry Lab)	–	+	–	–	–	ND	+	Yes	Yes	
	Mouse [11] (Liao Lab)	–	+	–	–	±	+	+	Yes	Yes	CD31- are cardiomyogenic
	Rat [30] (Komuro Lab)	ND	ND	–	–	±	ND	+	Yes	Yes	
c-kit	Mouse, human [19] (Anversa Lab)	+	+	–	–	–	±	ND	Yes	Yes	
Sca1	Mouse [20] (Schneider and Oh Labs)	–	+	–	–	+	ND	+	Yes	Yes	
CP	Human [46] (Marban Lab)	+	±*	–	+	+	–	ND	Yes	Yes	Also CD105+, CD90+, CD133-
Isl1	Mouse [41] (Chien Lab)	–	ND	–	–	–	ND	–*	Yes	Yes	*via dye efflux capacity

SP, side population; CP, cardiosphere; ND, not determined

In 2004, Martin and colleagues examined the expression of Abcg2 (also known as Bcrp1 for breast cancer resistance protein), an ATP binding cassette transporter, in embryonic and adult murine hearts [10]. As in the first report by Hierlihy and colleagues [5], Martin et al. found that Abcg2-expressing SP cells are able to differentiate into α-actinin positive cells. Abcg2 has previously shown to be the molecular determinant for the SP phenotype (Hoechst dye efflux ability) of bone marrow SP cells [21, 22]. Using in situ hybridization and immunohistochemical technologies, the authors elegantly document the expression of Abcg2 during embryonic and cardiac development, with a restricted expression pattern during embryogenesis followed by robust expression in the E8.5 heart and a decrease during midgestation (E11.5 and E13.5). Interestingly, however, while the Abcg2 has been shown to be the molecular determinant for the bone marrow SP phenotype, we have observed that Abcg2 is not required for the SP phenotype in cardiac cells by using an animal model with genetic deletion of Abcg2. Rather, we found that two ABC transporters, Abcg2/Bcrp1 and Abcb1/Mdr1, regulate the dye efflux property in cardiac SP cells in an age-dependent fashion [23]. Moreover, Abcg2 (or ABCG2 in human) has also been found to be expressed in the endothelial cells of veins and capillaries in the heart [24–29]. Taken together, these data have raised caution as to whether Abcg2/Bcrp1 gene or protein expression can be used alone as a molecular marker to identify cardiac SP cells in tissue sections. It is important to highlight that, at the current time, identification of cardiac SP cells is solely based on their ability to efflux Hoechst dye, rather than by a molecular marker. The role of Abcg2 in cardiac SP cells in the regulation of SP phenotype and function remains to be investigated.

Despite the lack of specific molecular/surface markers, Martin et al. also reported a comprehensive transcriptional analysis of cardiac SP cells, even though these cells represent a heterogenous population [10]. These SP cells were freshly isolated and FACS sorted, rather than obtained from a single clone. Gene expression in cardiac SP cells was compared to expression in embryonic stem cells (SM-1, passage 7) and adult cardiomyocytes. This analysis revealed that a distinct molecular program was associated with cardiac SP cells. Most interestingly, several transcription regulators were upregulated in cardiac SP cells, including capsulin, Meox2, Mef2a, and Mef2c, as well as transcripts associated with cell stress, including chaperones, DNA repair, protein degradation, detoxifying enzymes, and oxidative stress genes. Interestingly, Martin and colleagues also observed induction of TGFβ and Notch signaling pathways which may be involved in the regulation of self-renewal and maintenance of cardiac SP cells. The exact role of these genes is unclear at the present time and future study is needed to define their role in regulating the cell fate of cardiac SP cells.

Data from our group have demonstrated that cardiac SP have the capacity for biochemical, and more importantly, functional, differentiation [11]. We found that cardiac SP cells isolated from adult mouse hearts are able to express several cardiac-specific proteins; however, cardiomyogenic differentiation required coupling with primary cardiomyocytes (Fig. 5.2). Using the intracellular calcium indicator, fura-2, we found that cardiac SP cells not only exhibit synchronized contraction and

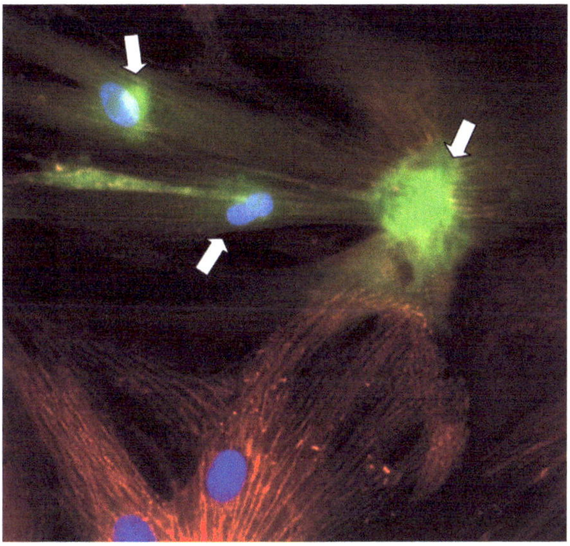

Fig. 5.2 Cardiac SP cells are capable of biochemical and functional differentiation into mature cardiomyocytes. In order for functional differentiation to occur, cardiac SP (GFP, *green*, pointed by *arrows*) couple with primary adult cardiomyocytes (stained for α-actinin, *red*). Mature differentiation is noted by the presence of sarcomeric striations, synchronized contraction and relaxation with adjacent co-cultured adult cardiomyocytes. Nuclei are stained *blue* with DAPI

relaxation with adjacent co-cultured adult cardiomyocytes but also demonstrate intracellular calcium transient indistinguishable from neighboring adult cardiomyocytes. Moreover, we found that the greatest capacity for cardiac differentiation appears to be limited to the Sca1+/CD31– sub-population SP cells. Of interest, cardiac SP cells isolated in Garry's laboratory are primarily negative for CD31 expression, and this discrepancy between groups is mostly likely attributed to the difference in SP isolation procedures. Note that even subtle difference in experimental protocol may yield significant variation. Despite these differences, however, all groups have consistently found that CD31– SP cells retain a strong cardiomyogenic potential.

While the in vitro cardiomyogenic differentiation of cardiac SP cells has been demonstrated by several groups, thus far, less is known about the ability of these cells to undergo cardiomyogenic differentiation in vivo. Recently, Komuro and colleagues studied the homing and differentiation efficiency of intravenously injected cardiac SP cells in a myocardial cryoinjury rat model [30, 31]. Neonatal rat cardiac SP cells were isolated from GFP+ transgenic rats and injected intravenously into normal rats and into rats with cryoinjured hearts. After 4 weeks of transplantation, GFP+ cells (from donor cardiac SP) were found to home to areas of injured myocardium and differentiate into cardiomyocytes, endothelial cells, smooth muscle cells, and fibroblasts.

The dynamic changes of cardiac SP cells during cardiac injury also remain largely unknown. In 2005, our group was the first to show that, upon myocardial injury, the resident cardiac SP population is acutely depleted and then returns to baseline levels 7 days following MI [32]. The reconstitution of cardiac SP cells back to baseline occurred via cell proliferation of endogenous cardiac SP cells, as shown by increased expression of the cell cycle marker Ki67, as well as by homing of bone marrow-derived stem cells, presumably SP cells, to the injured heart [32]. After homing, GFP bone marrow-derived stem cells undergo phenotypic conversion to become cardiac SP cells, as demonstrated by the loss of CD45 expression in the donor-derived GFP cells. Interestingly, the contribution of bone marrow-derived stem cells to the pool of cardiac SP cells was not observed during normal physiologic growth from neonatal to adulthood, as no bone marrow-derived stem cells were detected when GFP-expressing bone marrow was transplanted into lethally irradiated neonatal animals [32]. While we have observed the inability of freshly isolated bone marrow SP cells to undergo cardiomyogenic differentiation, it remains to be determined whether bone marrow-derived stem cells gain a cardiomyogenic capacity after residing in a myocardial niche following ischemic-induced homing. Several important questions still exist, including those regarding, but not limited to, the signals and/or chemoattractants that are essential for the homing of bone marrow-derived stem cells to injured hearts, and whether bone marrow-derived stem cells can attain cardiomyogenic potential in vitro and can be used for cardiac regeneration. All of these questions are the subject of ongoing investigations.

2.2 c-kit Positive Cardiac Cells

In 2003, Beltrami et al. methodically described a population of cardiac stem cells defined by the expression of the stem cell factor, c-kit+, and found them in clusters among cardiomyocytes in adult hearts [19]. This is the most thoroughly characterized cardiac progenitor cell population to date, and the only population, that represent true stem cells, as defined by clonogenicity, self-renewal, and multipotentiality. These c-kit+ cardiac stem cells are negative for blood cell lineages (CD34, CD45, CD20, CD8, and TER-119) and negative for cardiomyocyte, endothelial, and smooth muscle cell markers. However, roughly 10% of c-kit+ cells are positive for early cardiac transcription factors (Nkx2.5, GATA-4, and MEF2) [19]. In vitro, cardiac c-kit+ cells appear to be clonogenic and capable of self-renewal and differentiation into cardiac cell lineages (cardiomyocytes, endothelial, and smooth muscle cell). To determine their ability to regenerate functional myocardium after myocardial infarction, BrdU c-kit+ cells were injected into the borders of infarcted rat hearts 5 h following coronary ligation. Analysis of treated and untreated control animals by immunohistochemistry, echocardiography, hemodynamics, and sarcomeric mechanics revealed that c-kit+ cardiac stem cells were capable of reconstituting lost myocardium with functional cardiomyocytes, arterioles, and capillaries. These in vivo data have significant clinical implications and raise the possibility of regenerating lost myocardium with resident cardiac stem/progenitor cells.

c-kit+ cardiac stem cells have also been detected in hearts explanted from humans with chronic ischemic cardiomyopathy following transplant or lethal MI. This suggests that c-kit+ cardiac stem cells exist in human myocardium. Moreover, these cells may be utilized over time and differentiate into cardiomyocytes, smooth muscle cells, and endothelial cells in vitro [33]. More recently, Anversa and colleagues have demonstrated not only that c-kit+ cardiac stem cells can be isolated and expanded in vitro but also that they can be re-implanted, survive, and integrate in a rodent model of myocardial infraction [34]. Strikingly, these human c-kit positive cardiac stem cells were able to differentiate into cardiomyocytes and vascular cells in injured rodent myocardium, in vivo. These reports further raise the exciting possibility of isolating and expanding cardiac stem cells from myocardial biopsies in patients with chronic ischemic cardiomyopathy for autologous cell transplantation, thus, avoiding the need for immunosuppression.

Houser's laboratory has recently described the isolation and characterization of c-kit+ cardiac stem/progenitor cells from adult feline hearts [35]. The authors documented a detailed study of cardiomyocyte development from purported resident cells and hypertrophy from existing terminally differentiated cardiomyocytes during adolescent growth, a period of enormous cardiac growth. They hypothesized that if indeed cardiac progenitor cells contribute to the homeostasis of cardiomyocytes, the pool of cardiomyocytes should be composed of a combination of newly formed immature cardiomyocytes and terminally differentiated cardiomyocytes. Their extensive analysis included comparing cardiomyocytes isolated from adolescent (11 weeks) and adult (22 weeks) feline hearts by assessing cardiomyocyte volume, numbers, cell cycling activity, telomerase activity, cellular contractility, calcium transients, and electrophysiology. Indeed, Chen et al. found that the heart mass increased 101%, while the volume of individual cardiomyocytes increased only 77% [35]. Therefore, to increase the mass of the adolescent heart, there must also be generation of new cardiomyocytes in addition to hypertrophy. While the authors found that adolescent and adult feline hearts contain a similar percentage of mono-nucleated smaller cardiomyocytes ($\sim$12%) and binucleated larger cardiomyocytes ($\sim$87%), the mono-nucleated cardiomyocytes were functionally immature, whereas the binucleated cardiomyocytes were functionally competent (as assessed by cellular contractility, calcium transients, and electrophysiology). A greater portion of the BrdU positive cells, the majority of Ki67 positive cardiomyocytes, and the majority of high telomerase activity were found in population of smaller, mono-nuclear cardiomyocytes; this suggests again that the growing heart contains cardiomyocytes capable of dividing and forming new cardiomyocytes. The binucleated, larger cardiomyocytes, on the other hand, were terminally differentiated and showed signs of senescence (more $p16^{INK4a}$ positive cells were binucleated). Thus, in order for cardiomyocyte maturation to occur, the newly formed cardiomyocytes must exit from the cell cycle and become terminally differentiated. This study provides data supporting the concept that the heart is not a terminally differentiated organ and that the heart does have the capacity for partial self-renewal.

Another study by Rota et al. confirms the co-existence of a heterogenous population of cardiomyocytes. Here, the authors performed a similarly detailed analysis of murine hearts at 3 months and found cardiomyocytes displaying a spectrum of age and function [36]. Again, this work supports the new tenet that the heart is not a terminally differentiated organ, and it promotes the need for re-evaluation of current hypotheses in myocardial biology. If continued cardiomyocyte regeneration is a normal physiological process during early post-natal development, further understanding of this regulation would have profound contributions to both basic science and translational impact for the treatment of heart disease.

2.3 Sca1 + Cardiac Progenitor Cells

In the same year that c-kit+ cardiac stem cells were identified, Oh et al. employed a different stem cell marker, stem cell antigen (Sca-1), to identify yet another population of resident cardiac progenitor cells in adult hearts [20]. These Sca1+ cells accounted for approximately 14–17% of cardiomyocyte-depleted mono-nuclear cells. As shown in skeletal muscle [8], these Sca1+ cells were found to be small interstitial cells neighboring the basal lamina. They typically coexpress CD31 or its receptor, CD38. Moreover, cardiac Sca1+ cells lack blood lineage markers, including CD8, B220, Gr-1, Mac-1, and TER119, hematopoietic stem cell markers (CD45 and CD 34), c-kit, Flt-1, Flk-1, VE-cadherin, and vWF. Notably, cardiac SP cells are highly enriched for Sca1, and in fact, Sca1+ SP cells contain a higher cardiomyogenic potential than Sca1+ non-SP cells [11].

Cardiac Sca1+ cells express some of cardiac-specific transcription factors, GATA4 and MEF2c but no cardiac structural genes, such as myosin heavy chain or cardiac actin. Like other cardiac progenitor cells, cardiac Sca1+ cells are capable of undergoing cardiomyogenic differentiation, as demonstrated by immunohistochemical staining for sarcomeric α-actin and cardiac troponin I upon exposure to 5-azacytadine in vitro [20]. Oh and colleagues were the first to employ a Cre/Lox donor/reporter system to demonstrate the homing, differentiation, and fusion of cardiac Sca1+ cell in injured hearts. When injected intravenously into mice subjected to ischemic/reperfusion injury, Sca1+ cells were found to home to areas of damaged myocardium and to differentiate into cardiomyocytes, with and without fusion to host cells. Interestingly, engrafted cells were still present in the infarct border zone 2 weeks following implantation and seem to have proliferated as well. Like the extra-cardiac-derived stem/progenitor cells, cardiac Sca1+ cells, though originally isolated form myocardium, were not able to home to uninjured myocardium.

In a recent elegant study designed to identify cardiac stem cells without the use of cell surface markers, Tateishi and colleagues found that Sca1 played a vital role in the cardiac stem cell identity and function [37]. They used an un-biased approach and isolated GFP+ cells from heart homogenate of GFP transgenic mice. The only selection criterion applied in this study was the ability of these isolated cells to expand clonally in serum-free medium. Three colonies were obtained from 9541 single cell cultures. Characterization of these cells revealed that 70% of those clonally expanded cells are Sca1 positive and c-kit negative. In good agreement with most, if not all, reports of cardiac stem/progenitor cells, these Sca1+/c-kit– clonally expanded cells did not express the hematopoietic makers CD45 and CD34 and were negative for the endothelial cell marker, CD31, as well. In addition to expressing Sca1, these cells also express a high degree of CD29, CD105, CD44, and CD106 and, to a much less degree, CD13, CD73, and CD90. In terms of gene expression, these clonally isolated and expanded cells express Bcrp1, Bmi1, TERT (telomerase reverse transcriptase), Nanog and nestin as well as high telomerase activity (all commonly found in various progenitor cells). The authors went on to show that cardiac stem cells isolated form TERT promoter-driven eGFP mice contain Sca1 positive populations. However, these TERT-eGFP expressing cells rarely expressed detectable levels of c-kit, CD45, or CD31. Therefore, they suggested that in adult hearts, Sca1 expression is critically linked to elevated telomerase activity of TERT-expressing cells.

To determine the function of Sca1 in cardiac stem cells, the authors utilized ds-Sca1 RNA driven by RNA polymerase II promoter to generate a mouse model of Sca1 knockdown in which Sca1 proteins expression was reduced in the heart. Their data suggested that deficiency of Sca1 expression significantly affects the proliferation and survival of cardiac stem cells but not their ability to differentiate, in vitro. Intriguingly, the cardiac protective effects of cardiac stem cells were lost when the Sca1-KD cardiac stem cells were used to engraft in post-MI hearts as compared to cells isolated from non-transgenic control. They further suggested that this Sca1-regulated cardiac protection effect may be mediated via PI3K-Akt signaling pathway. While the protein reactive to antibody against Sca1 has been reported both using immunohistochemistry and Western blot in mouse, rat, dog, and human [38], it is unclear if Sca1 is expressed in human cardiac tissue. Nevertheless, given the concern of whether Sca1 is expressed in human hearts, the immediate clinical application of either using Sca1+ cardiac progenitor cells or targeting Sca1+ for optimizing cell-based therapy in human remains to be determined.

2.4 Islet 1 (Isl1) Positive Cardiac Cells

In 2003, Cai and colleagues were the first to report the possibility of using Isl1, a LIM homeodomain transcription factor, as a marker to isolate cardiac progenitor cells. This intriguing study stemmed from their analysis of homozygous Isl1 null mice showing embryonic lethality and died at E10.5 with absence of outflow tract, right ventricle, and much of the atria [39]. These cardiac structures do not arise from the cardiogenic mesoderm derived from the embryo's primitive streak. Instead, Isl1 defines the secondary heart field, providing precursors that proliferate prior to differentiation and migration to the appropriate positions in the heart. This report offered an important perspective regarding the existence of the primary and secondary heart fields and suggests the presence of two sets of cardiogenic progenitor cell populations: one population expressing and requiring Isl1 progenitors to contribute to the outflow tract, right ventricle, atria, some left ventricle; the other population does not express Isl1 to become the progenitors of the majority of the left ventricle.

Further lineage studies using an inducible Isl1-Cre showed that in addition to contributing to the myocardial lineages of the outflow tract, atrial septum, and right ventricle, isl1-derived cells are present in the sinoatrial and atrioventricular nodes, endothelial cells, and vascular smooth muscle (including that of the coronary vessels) [40]. The role of Isl1 in the specification of these distinct lineages remains to be determined.

Subsequently, Laugwitz et al. reported the identification of Isl1+ cardiac progenitor cells, albeit in a very small number, from post-natal rat, mouse, and human myocardium [41]. The localization of Isl1 in cardiac tissue (atrial muscle wall, intra-atrial septum, conus muscle, and right ventricle) was consistent among species, at least in mouse, rat, and human. These Isl1+ progenitor cells were initially identified during early cardiac development, and a small number of Isl1+ cells have been found in the post-natal murine heart (primarily in the outflow tract, atria, and right ventricle). In contrast to the previously discussed cardiac progenitor cells, the Isl1+ progenitor cells do not express Sca1 or c-kit. These Isl1+ progenitors are also capable of differentiating into cardiomyocytes in vitro upon co-culture with neonatal cardiomyocytes and can be expanded in culture on a feeder layer of mesenchymal cells. It is important to point out that Isl1+ cells were found in the animal and human hearts at a very young age. The general lack of Isl1 expression in later stages of life makes it difficult, if not impossible, to isolate these progenitor cells from adult patients for therapeutic regeneration. However, Isl1 may serve as an ideal molecular marker for the identification of cardiac progenitors from ES cells.

Indeed, Moretti et al. reported the isolation of Isl1+ cells from ES cells in 2007 [42]. These cells are called multipotent Isl1+ cardiovascular progenitors (MICP) and have the transcriptional signature of Isl1+/Nkx2.5+/Flk1+. They are clonal, multipotent, and, most importantly, capable of differentiating into striated muscle cells and smooth muscle cells, as well as a subset of non-muscle cells. In their model, Isl1+/Flk1+ cells represent a subset of downstream progenitors, responsible for undergoing endothelial cell differentiation; whereas Isl1+/Nkx2.5+ cells were similar to the Isl1+ progenitors isolated in post-natal hearts and could generate cardiac or smooth muscle cells. Given that the Isl1 progenitors are capable of contributing to multiple cell lineages within the heart, Moretti and colleagues suggest that Isl1+/Nkx2.5/Flk1 cells are part of a hierarchy of cellular programs controlling lineage specification. This is reminiscent of the developmental hierarchy proposed in the bone marrow where a single hematopoietic stem cell can regenerate all of the blood lineages [43]. While this working model establishes an initial framework for the hierarchy of stem/progenitors regulating lineage specification in the second heart field, future studies are necessary to confirm and fine-tune this exciting finding.

2.5 Cardiospheres

Messina and colleagues described yet another interesting method for isolating undifferentiated cardiac progenitor cells from human biopsy specimens. These cells were cultured in vitro, in self-adherent clusters, and are, therefore, termed cardiospheres [44]. These sphere-generating cells can be isolated not only from human atrial or ventricular biopsy but also from embryo, fetal, and post-natal mouse hearts. These cardiospheres generally expressed KDR, CD31, CD34, c-kit, and Sca1 (mouse) and have been shown to be clonogenic, as well as capable of cardiomyogenic differentiation into beating cells when co-cultured. Moreover, cardiospheres isolated and cultured from human hearts were found to regenerate infarcted mouse myocardium, attenuate infarct size, and improve cardiac function as determined by echocardiography. Similarly, Tomita and colleagues found the generation of neurosphere-like clusters, also referred to as "cardiospheres," from neonatal cardiac SP cells [45]. The cardiospheres derived from cardiac SP cells have been shown to be clonogenic and possess multi-lineage differentiation potential to expressing cardiac, smooth muscle, and neuronal genes and proteins [45].

The identification of cardiospheres from very small human biopsy specimens is significant for it allows for the growth of cells in numbers that would permit human cell therapy. It would also allow for autologous cell transplantation, obviating the need for immunosuppression. Indeed, a more recent study by Marban's laboratory demonstrated that cardiosphere cells can be expanded from percutaneous endomyocardial biopsy specimens and when implanted into mouse model of myocardial infarction, these cardiospheres were able to promote cardiac regeneration and improve cardiac function [46]. This study provided proof of concept for the use of cardiac stem cells isolated from human myocardium biopsy specimens for cell therapy. While it is encouraging and provides motivation for future therapeutic application in patients, several open questions remain. Of most importance, it is not clear whether the quality and quantity of cardiac stem/progenitor cells would be affected by the stage of the heart disease or age of the patient requiring cell therapy.

3 Origin of Cardiac Stem Cells

In theory, the stem/progenitor cells isolated from myocardium could either originate endogenously during development or home from exogenous bone marrow or other tissues. Data from several laboratories suggest that the cardiac stem/progenitor cells are distinct from bone marrow-derived stem cells, while many others demonstrate these cardiac stem/progenitor cells may be of extra cardiac origin, such as bone marrow. The bone marrow is known to contain the largest reservoir of stem/progenitor cells; and, through circulation and homing mechanisms, the bone marrow replenishes the stem/progenitors pool of various organs [47–50]. The homing of bone marrow-derived stem cells to other organs following injury has been unambiguously documented; however, the migration of bone marrow-derived stem cell to non-injured organs remains controversial. Using an animal model of parabiosis, circulating hematopoietic stem cells do not home to the heart in the absence of cardiac injury [51–53]. Using a cardiac α-actin promoter-driven EGFP, the cardiac histogenesis patterns have been documented and suggest that fetal and post-natal cardiomyocytes are derived from a common progenitor cell [54].

Tomita et al. used double transgenic mice encoding protein 0 (Schwann cell myelin marker)-cre/Floxed-EGFP and revealed that cardiac neural crest-derived stem cells migrate into the heart during development and remain in the myocardium, presumably in certain niches as dormant cardiac stem cells. Upon stimulation, these cells can then differentiate into cardiomyocytes as well as neurons, glia, and smooth muscle cells [45]. While our laboratory is unable to track the origin of cardiac SP cells at the current time, we concur with the notion that cardiac stem/progenitor cells are derived during development and reside in the myocardium to contribute to the physiologic cellular homeostasis of cardiac cells with limited involvement of extra-cardiac-derived stem cells. In contrast, bone marrow-derived stem cells contribute significantly to the maintenance of the pool of cardiac SP cells only following cardiac injury.

Some researchers, however, have suggested that cardiac stem/progenitor cells are derived from bone marrow. Using a model of GFP-labeled bone marrow transplant, Li and colleagues [55] found that, under their experimental conditions, the majority of the c-kit+ cells in the heart were also GFP positive. While these experiments were not designed to lineage track the origin of the c-kit+ cardiac stem cells, the data suggest that the c-kit+ cardiac stem cells may be of bone marrow origin. In summary, the origin of these stem and progenitor cells isolated from adult myocardium remains highly debatable and largely unanswered due to the lack of specific molecular markers to definitively identify the cardiac stem/progenitor cells. Equally important to the origin of cardiac stem/progenitor cell, it remains unknown whether the various cardiac stem/progenitor cells isolated using different surface markers or phenotype are in fact from the same progeny and reflect stem/progenitor cells at different stages in their development. Further investigation is necessary to ascertain the answers to these questions.

4 Cardiac Stem Cell Niche

To date, the most studied stem cell microenvironments, or niches, are those of the bone marrow, skin, and hair follicle [56]. Little is known, however, regarding the cardiac stem cell niche. Stem cell niches would harbor microenvironments in which stem/progenitor cells can exist in a quiescent state, and then, when activated, differentiate into mature cells. This would require a tightly regulated supporting network. However, since the fundamental properties of various organs differ substantially, the microenvironment where stem/progenitors reside, renew, and differentiate may be different. In principle, the niches or microenvironments are the discrete environments in the interstitium where primitive cells, including stem cells and early linage committed progenitor cells, are stored, renewed, and differentiated. To investigate the microenvironments or niches where stem cells reside, a molecular marker of stem cells and exquisite immunohistochemistry techniques are necessary. Leri and colleagues were the first to demonstrate the existence of cardiac stem cell niches in the rodent myocardium [57] (see Fig. 5.3).

Using a combination of immunohistochemistry and long- and short-term BrdU pulse/chase assay, Urbanek et al, demonstrated the existence of cardiac niches

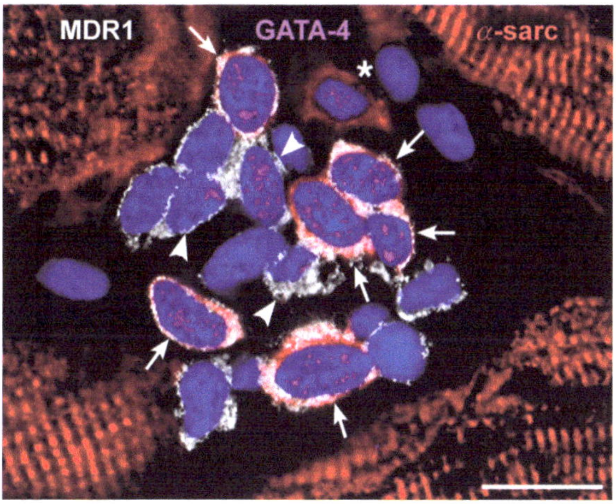

Fig. 5.3 The putative cardiac niche and its surrounding network: In this immunohistochemistry analysis of the apical myocardium, a cluster of progenitor cells is observed. Shown are 14 Abcb2/Mdr1+ cells (*white*). Of these 14, nine express GATA-4 (*magenta dots*) in the nuclei (*stained blue* with propidium iodide), and six of these cells express α-sarcomeric actin in the cytoplasm (*red; arrows*; myocyte precursors). Cardiac progenitors are the three cells positive for GATA-4 only (*arrow heads*). Five cardiac stem cells are identified by the lone expression of Abcb2/Mdr1; they are also lineage negative. There is also one small early developing myocyte which no longer has Abcb2/Mdr1 expression (*asterix*). Bar = 10 μm. (Photo courtesy of Drs P. Anversa, A. Leri, and J. Kajstura)

where clusters of cardiac stem cells or early lineage-committed cells resided [57]. By the authors' definition, cardiac stem cells are primitive cells expressing c-kit, Abcb1/Mdr1, and Sca1 without tissue-specific transcription factor and/or proteins of cardiac cells. Similarly, the early lineage-committed progenitor cells are defined by the coexpression of stem cells markers (i.e., c-kit, Abcb1/Mdr1, or Sca1) and cardiomyocyte, endothelial cells, or smooth muscle cell-specific transcription factor(s). While cardiac niches can be identified in the atria, ventricular base-midregion, and apex, the number of cardiac stem cells and/or progenitor cells is much higher within the atria relative to other regions. These cardiac stem/progenitor cells are interwoven with neighboring cardiomyocytes and fibroblasts via gap junction proteins, including connexins 43 and 45 and N- and E-cadherin. Interestingly, by using a calcium translocation assay, it was found that cardiac stem/progenitor cells are connected and supported by surrounding cardiomyocytes and fibroblasts but not endothelial cells. Moreover, integrin receptors, α4-integrin together with fibronectin and the α2 chain of laminin, seem to play a critical role in maintaining the homeostasis of cardiac stem/progenitor cells. β1-integrin is usually found on many primitive and committed progenitor cells, while α4-integrin, which is linked to the renewal of hematopoietic stem cells [58, 59], is exclusively expressed in the not-yet-committed stem cells. This suggests that α4-integrin may be critical for maintenance of cardiac stem cells at their undifferentiated and most "stem" stage. Upon activation, the α2 chain of laminin and fibronectin (both are ligands for α4β1-integrin receptor) is thought to serve to transducer signal from the extracellular compartment to the α4β1-integrin receptor and initiate the downstream signaling cascade to direct stem cells to undergo lineage-committed differentiation. Numb and α-adaptin have been shown to interact to permit asymmetric cell division, providing for different daughter cells (stem cell and committed cell) [60–63]. Based on the expression pattern of Numb and α-adaptin, Urbanek et al. note that the cardiac stem cells undergo twofold greater asymmetric division vs. symmetric division and the ratio between the events of symmetric and asymmetric division is independent of the location of cardiac niches (atrial vs. apical). Though the proposed thesis is attractive, future experiments from this and other groups are certainly needed to confirm the existence of cardiac niches and to address their regulation and function.

In steps toward this end, Komuro and colleagues also performed immunohistochemistry to identify the location and distribution of cardiac SP cells [31]. In this study, they stained neonatal rat hearts with Abcg2/Bcrp1 and CD31 antibodies and found several Abcg2/Bcrp1 positive and CD31 negative cells (potentially representing the SP populations in the heart) existed in both the perivascular and interstitial spaces among mature cardiomyocytes [31]. While they did not find any significant difference in cellular distribution of these stem/progenitor niches among apex, mid, and base of left ventricle, as well as no difference between left and right ventricle, they discovered that these ABCG2/Bcrp1 positive and CD31 negative cells communicate with neighboring cardiomyocytes via CD29 and N-cadherin. This finding is similar to those described by Leri and colleagues [57], suggesting that, indeed, cardiac stem/progenitor cells are localized in discrete areas in the myocardium and supported by surrounding cardiomyocytes.

While the adult myocardium is no longer felt to be a terminally differentiated organ, the degree of cardiomyocyte turnover, either from existing cardiomyocyte cycling or differentiation from resident stem/progenitor cells, remains largely unclear. Utilizing BrdU long-term label-retaining assay (6 days labeling and 10 weeks chasing), Urbanek et al. showed that cardiomyocytes are continually turning over and the majority of the cycling cardiomyocytes are differentiated from the pool of stem/progenitor cells. Intriguingly, the half-life of a cardiomyocyte depends on the particular location: cardiomyocytes located in the base-mid section exhibit double the half-life than those in the atrial and apical regions [57].

More recently, Lee and colleagues used a genetic "pulse-chase" strategy with an inducible Cre-loxP system to track the percent regenerated cardiomyocytes from stem/progenitor cells [64]. In this system, mice with the cardiomyocyte-specific α-myosin heavy chain promoter driving the expression of a tamoxifen-inducible Cre-recombinase are mated with mice containing a reporter such that β-galactosidase expression is replaced by GFP expression after excision of a stop codon flanked by loxP sites. Treatment with tamoxifen ("pulse") labels cardiomyocytes with GFP. The percent of GFP positive cardiomyocytes are assessed during the course of normal aging over 1 year and cardiac injury ("chase"). They found that resident stem/progenitor cells indeed contribute to the myocyte regeneration in the setting of post-injury, including myocardial infarction and pressure overload, but not in the non-injured basal condition during aging [64]. These results differ from those found by Urbanek et al; however, the discrepancy in basal cardiomyocyte turn over could be due, in part, to the different methodologies used. Further research will be necessary to resolve the controversy of the degree of cellular turnover at the basal level. In addition, whether or not the same degree of cellular turnover occurs in humans also needs to be addressed.

Though dividing cardiomyocytes and resident cardiac stem cells appear to be present, their ability to regenerate myocardium after injury is insufficient. Thus, to repair myocardial damage, one could determine the factors that increase the population of dividing cardiomyocytes. Work by Kuhn et al. addresses this possibility by using recombinant periostin [4]. Periostin is part of the extracellular matrix [65]; associated with the epithelial–mesenchymal transition during cardiac development [66, 67]; and, it is re-expressed in adult tissues after cardiac, skeletal muscle, vascular, and bone injury [68–72]. Kuhn et al. found that recombinant periostin induced increased cell division in neonatal and mature cardiomyocytes to undergo mitosis in vitro, to similar degrees as treatment with FGF [73]. Such induction required α_v, β_1, β_3, and β_5 integrins and an intact phosphotidylinositol-3-OH (PI3) kinase pathway. To test the effects of periostin in vivo after myocardial infarction, recombinant periostin was delivered via Gelfoam epicardial patches at the time of infarction. The authors found that 12 weeks after injury, periostin improved cardiac function by echo and assessment of pressure–volume loops; reduced fibrosis; and increased angiogenesis. They also found increased BrdU incorporation by cardiomyocytes in the border zone. While this work is promising, it is in contrast to work by Oka et al. showing that periostin is involved in the hypertrophic remodeling response [74, 75]. A periostin knockout mouse was more likely to have ventricular rupture after MI; while the surviving mice had less fibrosis and better left ventricular function. Mice overexpressing periostin, on the other hand, exhibited hypertrophy with normal aging and appeared protected against ventricular rupture after MI. Oka et al. found that the function of fibroblasts is altered with periostin expression. Other groups using adenoviral and liposomal delivery of periostin to cardiomyocytes found hypertrophy or dilation, respectively, of the hearts [76, 77]. Clearly further research is necessary to clarify the role of periostin in the heart. Nevertheless, the research above shows that modulation of the cardiac extracellular environment may provide the stimulus needed for cardiac repair. More research is necessary, however, before it can be added to the treatment armamentarium for cardiac damage.

5 Aging and Cardiac Progenitor Cells

With the average lifespan and the aging population continuously increasing in the Western world, so is the average age of patient populations with cardiovascular disease. It is commonly acknowledged that the regenerative potential of any organ declines with age, and this impairment could be associated with the senescence of the stem/progenitor cells in the given organ. The heart is no exception, and it is particularly vulnerable to the aging process. A new paradigm has been evolving and demonstrates that the heart contains populations of stem/progenitor cells which may contribute to the renewal of cardiac cells. As we have reviewed above, this new hypothesis suggests that the mammalian heart consists of non-dividing and dividing cardiomyocytes. Over the individual's life span, the death of senescent cardiomyocytes is balanced by the proliferation of existing cardiomyocytes as well as the repopulation of cardiomyocytes, at least in the early post-natal life, by resident stem/progenitor cells (review see [78]). Unfortunately, this self-renewal capacity is a slow process and it is not sufficient to counteract the loss of cardiomyocytes over one's life span or following cardiac injury. As a consequence, the heart is left with either poorly functioning senescent cardiomyocytes or fast growing fibrosis to replace lost cardiomyocytes.

As the adult human heart ages, the number of cardiomyocytes decreases in males more than in females: in male hearts from 17 to 89 years of age, there is 45 million myocytes fewer per year in the left ventricle and 19 million fewer in the right ventricle [79–81]; the loss of myocytes is compensated by myocyte hypertrophy. In females, on the other hand, the numbers of myocytes in the left and right ventricles is essentially unchanged in 20–95-year-old hearts. Though the number of myocytes is unchanged, the female hearts do have cardiomyocyte death [82, 83]; therefore, to maintain the number of cells, there must be proliferation of cardiomyocytes. Thus, this supports the noted new paradigm that the heart is, indeed, a dynamic organ, capable of some degree of renewal.

Telomerase activity is necessary to maintain telomere lengths in the dividing cells; as the cells become senescent, the telomeres shorten and the cell may go through apoptosis [84]. Analysis reveals that telomerase activity is present in cardiomyocytes of both young and old hearts [85], though it is much less in the older hearts of men. Again, the adult heart is not a static organ.

However, despite the documented presence of proliferating cardiomyocytes and the resident cardiac stem/progenitor cells, it is not enough to offset the age-associated loss of cardiomyocytes in men, nor is it enough to repair damaged hearts. We have presented above potential avenues to improve the regenerative potential of the heart. However, whether the cardiac progenitors of older hearts are still able to "perform" as well as their counterparts in the young hearts remain to be determined. Subtle changes in the milieu or stem cell niche may affect the ability of the cardiomyocyte cell or precursor to respond. For example, in skeletal muscle, satellite cells are primarily responsible for regeneration after injury; this regenerative capacity is reduced during aging and there is increased myocyte death and fibrosis [86]. Though aged satellite cells are less functional, they do retain the potential for repair: when exposed to a young environment in a parabiosis model, aged skeletal muscle exhibits increased regeneration and less fibrosis after cryoinjury; these processes are mediated, in part, by restoring signaling of the Notch pathway and by suppressing Wnt activity[86–88]. A similar phenomenon may be present in the aging heart and should be considered as a potential obstacle in future therapies.

6 The Role of Resident Cardiac Stem Cell in Repair and Regeneration

Since the first cell-based therapy in human using autologous skeletal myoblast took place in 2000 [89], a rather significant number of clinical trials, from small open-labeled safety trails to a relatively larger double-blinded, placebo-controlled efficacy trails have been conducted worldwide [90–104]. Currently, there are more than 10 clinical trials in the United States either ongoing or about to start to determine the potential use of stem cell therapy in a host of different cardiac diseases [105]. This rather rapid translation from bench to bed side was fueled by the observation that some degree of beneficial effects in animal models always occurred regardless of the cell type tested, which ranged from somatic to embryonic-derived cells, as well as the initial encouraging clinical results. It is intuitively appealing to use resident cardiac stem/progenitor cells for cell-based myocardial repair and regeneration. However, the biggest hurdle to overcome is the difficulty in the acquisition of cardiac biopsies and isolation and expansion of these resident stem/progenitor cells. Significant efforts have been put forth in this regard, and recent publications have detailed such isolation and expansion of cardiac stem/progenitor cells from human hearts or biopsies [34, 46]. Alternatively, these resident cardiac stem/progenitor cells can be activated locally in situ by growth factors and cytokines to promote their proliferation and differentiation [4]. In any case, further understanding of the biology and the regulation of these resident cardiac stem/progenitor cells is a prerequisite for the successful translation from animal models to human. We are heading into an exciting era of regenerative medicine in both basic science research and clinical application. With the continued advances in stem cell biology, molecular biology, and tissue bioengineering, we may soon to be able to regenerate damaged hearts with normal and functional cardiac tissue with an integrated cardiomyocyte, vasculature, and supporting matrix network.

7 Conclusion and Future Prospective

In summary, the recognition of the existence of endogenous cardiac stem/progenitor cells in the adult mammalian heart has raised the distinct possibility for true cardiac regeneration. No longer can the heart be considered a post-mitotic organ, as the heart does indeed have the capacity and potential for repair/regeneration. As currently stands, however, these resident cardiac stem cells are insufficient to overcome the tissue loss due to aging or severe injury such as myocardial infarction. The difficulty and complications involved in regenerating normal and functional cardiac tissue in a sick heart are daunting; however, with continued progress in understanding stem cell biology, its enormous promise as therapy is becoming more and more a reality. It is still unclear as to the what ideal cell type is for such therapies and the ideal patient population to be targeted. Further intensive basic science investigations, in conjunction with carefully designed randomized, double-blinded, placebo-controlled clinical trials, is obligatory to advance the practicality of cardiovascular regenerative medicine into routine clinical practice.

Acknowledgments This work was supported by research grants funding from the National Institutes of Health (RL).

References

1. Thomas ED, Lochte HL Jr, Lu WC, Ferrebee JW (1957) Intravenous infusion of bone marrow in patients receiving radiation and chemotherapy. N Engl J Med 257(11):491–496
2. Beltrami AP, Urbanek K, Kajstura J et al (2001) Evidence that human cardiac myocytes divide after myocardial infarction. N Engl J Med 344(23):1750–1757
3. Kajstura J, Leri A, Finato N, Di Loreto C, Beltrami CA, Anversa P (1998) Myocyte proliferation in end-stage cardiac failure in humans. Proc Natl Acad Sci U S A 95(15):8801–8805
4. Kuhn B, del Monte F, Hajjar RJ et al (2007) Periostin induces proliferation of differentiated cardiomyocytes and promotes cardiac repair. Nat Med 13(8):962–969
5. Hierlihy AM, Seale P, Lobe CG, Rudnicki MA, Megeney LA (2002) The post-natal heart contains a myocardial stem cell population. FEBS Lett 530(1–3):239–243
6. Goodell MA, Brose K, Paradis G, Conner AS, Mulligan RC (1996) Isolation and functional properties of murine hematopoietic stem cells that are replicating in vivo. J Exp Med 183(4):1797–1806
7. Asakura A, Rudnicki MA (2002) Side population cells from diverse adult tissues are capable of in vitro hematopoietic differentiation. Exp Hematol 30(11):1339–1345
8. Asakura A, Seale P, Girgis-Gabardo A, Rudnicki MA (2002) Myogenic specification of side population cells in skeletal muscle. J Cell Biol 159(1):123–134
9. Jackson KA, Majka SM, Wang H et al (2001) Regeneration of ischemic cardiac muscle and vascular endothelium by adult stem cells. J Clin Invest 107(11):1395–1402
10. Martin CM, Meeson AP, Robertson SM et al (2004) Persistent expression of the ATP-binding cassette transporter, Abcg2, identifies cardiac SP cells in the developing and adult heart. Dev Biol 265(1):262–275
11. Pfister O, Mouquet F, Jain M et al (2005) CD31- but Not CD31+ cardiac side population cells exhibit functional cardiomyogenic differentiation. Circ Res 97(1):52–61
12. Summer R, Kotton DN, Sun X, Ma B, Fitzsimmons K, Fine A (2003) Side population cells and Bcrp1 expression in lung. Am J Physiol Lung Cell Mol Physiol 285(1):L97–104
13. Gussoni E, Soneoka Y, Strickland CD et al (1999) Dystrophin expression in the mdx mouse restored by stem cell transplantation. Nature 401(6751):390–394
14. Torrente Y, Tremblay JP, Pisati F et al (2001) Intraarterial injection of muscle-derived CD34(+)Sca-1(+) stem cells restores dystrophin in mdx mice. J Cell Biol 152(2):335–348
15. Sampaolesi M, Torrente Y, Innocenzi A et al (2003) Cell therapy of alpha-sarcoglycan null dystrophic mice through intra-arterial delivery of mesoangioblasts. Science 301(5632): 487–492
16. Orlic D, Kajstura J, Chimenti S et al (2001) Bone marrow cells regenerate infarcted myocardium. Nature 410(6829):701–705
17. Kocher AA, Schuster MD, Szabolcs MJ et al (2001) Neovascularization of ischemic myocardium by human bone-marrow-derived angioblasts prevents cardiomyocyte apoptosis, reduces remodeling and improves cardiac function. Nat Med 7(4):430–436
18. Kolodziejczyk SM, Wang L, Balazsi K, DeRepentigny Y, Kothary R, Megeney LA (1999) MEF2 is upregulated during cardiac hypertrophy and is required for normal post-natal growth of the myocardium. Curr Biol 9(20):1203–1206
19. Beltrami AP, Barlucchi L, Torella D et al (2003) Adult cardiac stem cells are multipotent and support myocardial regeneration. Cell 114(6):763–776

20. Oh H, Bradfute SB, Gallardo TD et al (2003) Cardiac progenitor cells from adult myocardium: homing, differentiation, and fusion after infarction. Proc Natl Acad Sci U S A 100(21):12313–12318
21. Zhou S, Schuetz JD, Bunting KD et al (2001) The ABC transporter Bcrp1/ABCG2 is expressed in a wide variety of stem cells and is a molecular determinant of the side-population phenotype. Nat Med 7(9):1028–1034
22. Bunting KD (2002) ABC transporters as phenotypic markers and functional regulators of stem cells. Stem Cells 20(1):11–20
23. Pfister O, Oikonomopoulos A, Sereti KI, et al. (2008) Role of the ATP-binding cassette transporter Abcg2 in the phenotype and function of cardiac side population cells. Circ Res. 103(8):825–835
24. Maliepaard M, Scheffer GL, Faneyte IF et al (2001) Subcellular localization and distribution of the breast cancer resistance protein transporter in normal human tissues. Cancer Res 61(8):3458–3464
25. Aronica E, Gorter JA, Redeker S et al (2005) Localization of breast cancer resistance protein (BCRP) in microvessel endothelium of human control and epileptic brain. Epilepsia 46(6):849–857
26. Asashima T, Hori S, Ohtsuki S et al (2006) ATP-binding cassette transporter G2 mediates the efflux of phototoxins on the luminal membrane of retinal capillary endothelial cells. Pharm Res 23(6):1235–1242
27. Sisodiya SM, Martinian L, Scheffer GL et al (2006) Vascular colocalization of P-glycoprotein, multidrug-resistance associated protein 1, breast cancer resistance protein and major vault protein in human epileptogenic pathologies. Neuropathol Appl Neurobiol 32(1):51–63
28. Zhang W, Mojsilovic-Petrovic J, Andrade MF, Zhang H, Ball M, Stanimirovic DB (2003) The expression and functional characterization of ABCG2 in brain endothelial cells and vessels. Faseb J 17(14):2085–2087
29. Meissner K, Heydrich B, Jedlitschky G et al (2006) The ATP-binding cassette transporter ABCG2 (BCRP), a marker for side population stem cells, is expressed in human heart. J Histochem Cytochem 54(2):215–221
30. Wada H, Matsuura K, Oyama T, Iwanaga K, Takahashi T, Goto T, Nagai T, Komuro I (2005) Homing and differentiation of intravenously transplanted cardiac side population cells in injured myocardium. Circulation 111:801
31. Oyama T, Nagai T, Wada H et al (2007) Cardiac side population cells have a potential to migrate and differentiate into cardiomyocytes in vitro and in vivo. J Cell Biol 176(3): 329–341
32. Mouquet F, Pfister O, Jain M et al (2005) Restoration of cardiac progenitor cells after myocardial infarction by self-proliferation and selective homing of bone marrow-derived stem cells. Circ Res 97(11):1090–1092
33. Torella D, Ellison GM, Mendez-Ferrer S, Ibanez B, Nadal-Ginard B (2006) Resident human cardiac stem cells: role in cardiac cellular homeostasis and potential for myocardial regeneration. Nat Clin Pract Cardiovasc Med 3 Suppl 1:S8–13
34. Bearzi C, Rota M, Hosoda T et al (2007) Human cardiac stem cells. Proc Natl Acad Sci U S A 104(35):14068–14073
35. Chen X, Wilson RM, Kubo H et al (2007) Adolescent feline heart contains a population of small, proliferative ventricular myocytes with immature physiological properties. Circ Res 100(4):536–544
36. Rota M, Hosoda T, De Angelis A et al (2007) The young mouse heart is composed of myocytes heterogeneous in age and function. Circ Res 101(4): 387–399
37. Tateishi K, Ashihara E, Takehara N et al (2007) Clonally amplified cardiac stem cells are regulated by Sca-1 signaling for efficient cardiovascular regeneration. J Cell Sci 120 (Pt 10):1791–1800

38. Zhang XB, Schwartz JL, Humphries RK, Kiem HP (2007) Effects of HOXB4 overexpression on ex vivo expansion and immortalization of hematopoietic cells from different species. Stem Cells 25(8):2074–2081
39. Cai CL, Liang X, Shi Y et al (2003) Isl1 identifies a cardiac progenitor population that proliferates prior to differentiation and contributes a majority of cells to the heart. Dev Cell 5(6):877–889
40. Sun Y, Liang X, Najafi N et al (2007) Islet 1 is expressed in distinct cardiovascular lineages, including pacemaker and coronary vascular cells. Dev Biol 304(1):286–296
41. Laugwitz KL, Moretti A, Lam J et al (2005) Postnatal isl1+ cardioblasts enter fully differentiated cardiomyocyte lineages. Nature 433(7026):647–653
42. Moretti A, Caron L, Nakano A et al (2006) Multipotent embryonic Isl1(+) progenitor cells lead to cardiac, smooth muscle, and endothelial cell diversification. Cell 127(6):1151–1165
43. Kiel MJ, Yilmaz OH, Iwashita T, Yilmaz OH, Terhorst C, Morrison SJ (2005) SLAM family receptors distinguish hematopoietic stem and progenitor cells and reveal endothelial niches for stem cells. Cell 121(7):1109–1121
44. Messina E, De Angelis L, Frati G et al (2004) Isolation and expansion of adult cardiac stem cells from human and murine heart. Circ Res 95(9):911–921
45. Tomita Y, Matsumura K, Wakamatsu Y et al (2005) Cardiac neural crest cells contribute to the dormant multipotent stem cell in the mammalian heart. J Cell Biol 170(7):1135–1146
46. Smith RR, Barile L, Cho HC et al (2007) Regenerative potential of cardiosphere-derived cells expanded from percutaneous endomyocardial biopsy specimens. Circulation 115(7):896–908
47. Theise ND, d'Inverno M (2004) Understanding cell lineages as complex adaptive systems. Blood Cells Mol Dis 32(1):17–20
48. Grove JE, Bruscia E, Krause DS (2004) Plasticity of bone marrow-derived stem cells. Stem Cells 22(4):487–500
49. Lakshmipathy U, Verfaillie C (2005) Stem cell plasticity. Blood Rev 19(1):29–38
50. Quesenberry PJ, Abedi M, Aliotta J et al (2004) Stem cell plasticity: an overview. Blood Cells Mol Dis 32(1):1–4
51. Wagers AJ, Sherwood RI, Christensen JL, Weissman IL (2002) Little evidence for developmental plasticity of adult hematopoietic stem cells. Science 297(5590):2256–2259
52. Calvi LM, Adams GB, Weibrecht KW et al (2003) Osteoblastic cells regulate the haematopoietic stem cell niche. Nature 425(6960):841–846
53. Wagers AJ, Weissman IL (2004) Plasticity of adult stem cells. Cell 116(5):639–648
54. Eberhard D, Jockusch H (2005) Patterns of myocardial histogenesis as revealed by mouse chimeras. Dev Biol 278(2):336–346
55. Fazel S, Cimini M, Chen L et al (2006) Cardioprotective c-kit+ cells are from the bone marrow and regulate the myocardial balance of angiogenic cytokines. J Clin Invest 116(7):1865–1877
56. Fuchs E, Tumbar T, Guasch G (2004) Socializing with the neighbors: stem cells and their niche. Cell 116(6):769–778
57. Urbanek K, Cesselli D, Rota M et al (2006) Stem cell niches in the adult mouse heart. Proc Natl Acad Sci U S A 103(24):9226–9231
58. Priestley GV, Scott LM, Ulyanova T, Papayannopoulou T (2006) Lack of alpha4 integrin expression in stem cells restricts competitive function and self-renewal activity. Blood 107(7):2959–2967
59. Shinoda G, Umeda K, Heike T et al (2007) alpha4-Integrin(+) endothelium derived from primate embryonic stem cells generates primitive and definitive hematopoietic cells. Blood 109(6):2406–2415
60. Rhyu MS, Jan LY, Jan YN (1994) Asymmetric distribution of numb protein during division of the sensory organ precursor cell confers distinct fates to daughter cells. Cell 76(3):477–491
61. Jan YN, Jan LY (1998) Asymmetric cell division. Nature 392(6678):775–778

62. Berdnik D, Torok T, Gonzalez-Gaitan M, Knoblich JA (2002) The endocytic protein alpha-adaptin is required for numb-mediated asymmetric cell division in Drosophila. Dev Cell 3(2):221–231
63. Santolini E, Puri C, Salcini AE et al (2000) Numb is an endocytic protein. J Cell Biol 151(6):1345–1352
64. Hsieh PC, Segers VF, Davis ME et al (2007) Evidence from a genetic fate-mapping study that stem cells refresh adult mammalian cardiomyocytes after injury. Nat Med 13(8): 970–974
65. Takeshita S, Kikuno R, Tezuka K, Amann E (1993) Osteoblast-specific factor 2: cloning of a putative bone adhesion protein with homology with the insect protein fasciclin I. Biochem J 294 (Pt 1):271–278
66. Litvin J, Zhu S, Norris R, Markwald R (2005) Periostin family of proteins: therapeutic targets for heart disease. Anat Rec A Discov Mol Cell Evol Biol 287(2):1205–1212
67. Butcher JT, Norris RA, Hoffman S, Mjaatvedt CH, Markwald RR (2007) Periostin promotes atrioventricular mesenchyme matrix invasion and remodeling mediated by integrin signaling through Rho/PI 3-kinase. Dev Biol 302(1):256–266
68. Wang D, Oparil S, Feng JA et al (2003) Effects of pressure overload on extracellular matrix expression in the heart of the atrial natriuretic peptide-null mouse. Hypertension 42(1):88–95
69. Lindner V, Wang Q, Conley BA, Friesel RE, Vary CP (2005) Vascular injury induces expression of periostin: implications for vascular cell differentiation and migration. Arterioscler Thromb Vasc Biol 25(1):77–83
70. Stanton LW, Garrard LJ, Damm D et al (2000) Altered patterns of gene expression in response to myocardial infarction. Circ Res 86(9):939–945
71. Goetsch SC, Hawke TJ, Gallardo TD, Richardson JA, Garry DJ (2003) Transcriptional profiling and regulation of the extracellular matrix during muscle regeneration. Physiol Genomics 14(3):261–271
72. Nakazawa T, Nakajima A, Seki N et al (2004) Gene expression of periostin in the early stage of fracture healing detected by cDNA microarray analysis. J Orthop Res 22(3):520–525
73. Engel FB, Schebesta M, Duong MT et al (2005) p38 MAP kinase inhibition enables proliferation of adult mammalian cardiomyocytes. Genes Dev 19(10):1175–1187
74. Norris RA, Damon B, Mironov V et al (2007) Periostin regulates collagen fibrillogenesis and the biomechanical properties of connective tissues. J Cell Biochem 101(3):695–711
75. Oka T, Xu J, Kaiser RA et al (2007) Genetic manipulation of periostin expression reveals a role in cardiac hypertrophy and ventricular remodeling. Circ Res 101(3):313–321
76. Litvin J, Blagg A, Mu A et al (2006) Periostin and periostin-like factor in the human heart: possible therapeutic targets. Cardiovasc Pathol 15(1):24–32
77. Katsuragi N, Morishita R, Nakamura N et al (2004) Periostin as a novel factor responsible for ventricular dilation. Circulation 110(13):1806–1813
78. Anversa P, Rota M, Urbanek K et al (2005) Myocardial aging--a stem cell problem. Basic Res Cardiol 100(6):482–493
79. Olivetti G, Cigola E, Maestri R et al (1996) Aging, cardiac hypertrophy and ischemic cardiomyopathy do not affect the proportion of mononucleated and multinucleated myocytes in the human heart. J Mol Cell Cardiol 28(7):1463–1477
80. Olivetti G, Giordano G, Corradi D et al (1995) Gender differences and aging: effects on the human heart. J Am Coll Cardiol 26(4):1068–1079
81. Olivetti G, Melissari M, Capasso JM, Anversa P (1991) Cardiomyopathy of the aging human heart. Myocyte loss and reactive cellular hypertrophy. Circ Res 68(6):1560–1568
82. Olivetti G, Abbi R, Quaini F et al (1997) Apoptosis in the failing human heart. N Engl J Med 336(16):1131–1141
83. Guerra S, Leri A, Wang X et al (1999) Myocyte death in the failing human heart is gender dependent. Circ Res 85(9):856–866
84. Blasco MA, Hahn WC (2003) Evolving views of telomerase and cancer. Trends Cell Biol 13(6):289–294

85. Leri A, Malhotra A, Liew CC, Kajstura J, Anversa P (2000) Telomerase activity in rat cardiac myocytes is age and gender dependent. J Mol Cell Cardiol 32(3):385–390
86. Conboy IM, Conboy MJ, Smythe GM, Rando TA (2003) Notch-mediated restoration of regenerative potential to aged muscle. Science 302(5650):1575–1577
87. Conboy IM, Conboy MJ, Wagers AJ, Girma ER, Weissman IL, Rando TA (2005) Rejuvenation of aged progenitor cells by exposure to a young systemic environment. Nature 433(7027):760–764
88. Brack AS, Conboy MJ, Roy S et al (2007) Increased Wnt signaling during aging alters muscle stem cell fate and increases fibrosis. Science 317(5839):807–810
89. Menasche P, Hagege AA, Scorsin M et al (2001) Myoblast transplantation for heart failure. Lancet 357(9252):279–280
90. Assmus B, Honold J, Schachinger V et al (2006) Transcoronary transplantation of progenitor cells after myocardial infarction. N Engl J Med 355(12):1222–1232
91. Assmus B, Schachinger V, Teupe C et al (2002) Transplantation of progenitor cells and regeneration enhancement in acute myocardial infarction (TOPCARE-AMI). Circulation 106(24):3009–3017
92. Chen SL, Fang WW, Qian J et al (2004) Improvement of cardiac function after transplantation of autologous bone marrow mesenchymal stem cells in patients with acute myocardial infarction. Chin Med J (Engl) 117(10):1443–1448
93. Fuchs S, Baffour R, Zhou YF et al (2001) Transendocardial delivery of autologous bone marrow enhances collateral perfusion and regional function in pigs with chronic experimental myocardial ischemia. J Am Coll Cardiol 37(6):1726–1732
94. Janssens S, Dubois C, Bogaert J et al (2006) Autologous bone marrow-derived stem-cell transfer in patients with ST-segment elevation myocardial infarction: double-blind, randomised controlled trial. Lancet 367(9505):113–121
95. Lunde K, Solheim S, Aakhus S et al (2006) Intracoronary injection of mononuclear bone marrow cells in acute myocardial infarction. N Engl J Med 355(12):1199–1209
96. Lunde K, Solheim S, Aakhus S, Arnesen H, Abdelnoor M, Forfang K (2005) Autologous stem cell transplantation in acute myocardial infarction: The ASTAMI randomized controlled trial. Intracoronary transplantation of autologous mononuclear bone marrow cells, study design and safety aspects. Scand Cardiovasc J 39(3):150–158
97. Meyer GP, Wollert KC, Lotz J et al (2006) Intracoronary bone marrow cell transfer after myocardial infarction: eighteen months' follow-up data from the randomized, controlled BOOST (BOne marrOw transfer to enhance ST-elevation infarct regeneration) trial. Circulation 113(10):1287–1294
98. Schachinger V, Erbs S, Elsasser A, Haberbosch W, Holschermann H, Yu J, Corti R, Mathey D, Susellbeck T, Assmus B,Tonn T, Dimmeler S, Zeiher AM (2006) Intracoronary infusion of bone-marrow-derived progenitor cells is associated with improved clinical outcome in patients with acute myocardial infarction: 12 months follow up of the REPAR-AMI trial. Circulation 114:II-787
99. Schachinger V, Assmus B, Britten MB et al (2004) Transplantation of progenitor cells and regeneration enhancement in acute myocardial infarction: final one-year results of the TOPCARE-AMI Trial. J Am Coll Cardiol 44(8):1690–1699
100. Schachinger V, Erbs S, Elsasser A et al (2006) Intracoronary bone marrow-derived progenitor cells in acute myocardial infarction. N Engl J Med 355(12):1210–1221
101. Schachinger V, Tonn T, Dimmeler S, Zeiher AM (2006) Bone-marrow-derived progenitor cell therapy in need of proof of concept: design of the REPAIR-AMI trial. Nat Clin Pract Cardiovasc Med 3 Suppl 1:S23–28
102. Stamm C, Kleine HD, Westphal B et al (2004) CABG and bone marrow stem cell transplantation after myocardial infarction. Thorac Cardiovasc Surg 52(3):152–158
103. Stamm C, Westphal B, Kleine HD et al (2003) Autologous bone-marrow stem-cell transplantation for myocardial regeneration. Lancet 361(9351):45–46

104. Strauer BE, Brehm M, Zeus T et al (2002) Repair of infarcted myocardium by autologous intracoronary mononuclear bone marrow cell transplantation in humans. Circulation 106(15):1913–1918
105. Jolicoeur EM, Granger CB, Fakunding JL et al (2007) Bringing cardiovascular cell-based therapy to clinical application: perspectives based on a National Heart, Lung, and Blood Institute Cell Therapy Working Group meeting. Am Heart J 153(5):732–742

Chapter 6
Muscle Stem Cells

Shihuan Kuang and Michael A. Rudnicki

Abstract Accumulating evidence suggests that there are two categories of stem cells in skeletal muscle: (1) satellite cells, the default muscle stem cells that are responsible for muscle growth and regeneration under physiological conditions and (2) other multipotent stem cells that are capable of myogenic differentiation during muscle regeneration induced by injury or diseases. The latter category includes different cell populations isolated by various researchers using several techniques. When used in tissue engineering applications these stem cells have been demonstrated to possess promising potential for the regeneration and repair of muscle. Here, we review the origin, localization, isolation, and myogenic functions of muscle stem cells with particular interests in their practical and potential implications, as well as challenges, in cell-based therapies for muscle diseases.

Keywords CD34 · Mesoangioblast · Muscular dystrophy · Myogenesis · Pax7 · Pericyte · regeneration · Satellite cell · Sca-1 · Side population · Stem cell therapy

1 Introduction

1.1 Skeletal Muscle Regeneration as a Model to Study Stem Cell Function

Skeletal muscles are the most abundant tissue in human body and are responsible for all body movements. Mature muscle cells, commonly known as muscle fibers or myofibers, are long cylindrical shaped multinuclear cells. These multinuclear

M.A. Rudnicki (✉)
The Sprott Center for Stem Cell Research, Ottawa Health Research Institute, Regenerative Medicine Program, 501 Smyth Road, Ottawa, ON, Canada, K1H 8L6
e-mail: mrudnicki@ohri.ca

A. Giordano, U. Galderisi (eds.), *Cell Cycle Regulation and Differentiation in Cardiovascular and Neural Systems*, DOI 10.1007/978-1-60327-153-0_6,

muscle fibers are formed through fusion of hundreds of mononuclear muscle cells called myocytes. The capability of myocytes to fuse with each other or with existing muscle fibers provides an effective mechanism for muscle growth and repair during development and regeneration [1]. Conversely, poor muscle regenerations under certain pathological conditions, such as various forms of muscular dystrophy, often reflect an inadequate supply of myocytes and compromised muscle stem cell function.

1.2 Transcriptional Control of Muscle Development and Regeneration

During development, most skeletal muscles are derived from embryonic somites, paired segmental structures sitting bilaterally along the neural tube. The somites are originally cleaved off the presomitic mesoderm in an anterior to posterior fashion with the segmentation of body plan. Primitive somites are ball-like structures containing a central sclerotome that eventually gives rise to the skeleton and a peripheral portion that undergoes mesenchyme-to-epithelium transition to transform into a dorsal–lateral sheet called dermomyotome [2]. Within the dermomyotome, a population of progenitors, marked by their expression of the paired domain transcription factors $Pax3^+$ and $Pax7^+$, eventually give rise to most skeletal muscles as well as adult satellite cells [3], the primary stem cells that are responsible for muscle growth and repair.

During postnatal muscle growth or regeneration, quiescent satellite cells and other muscle resident stem cells are activated and proliferate to give rise to myoblasts, myogenic precursor cells that express the basic helix-loop-helix (bHLH) family transcription factors Myf5 and MyoD. Some myoblasts subsequently withdraw from cell cycle, up-regulate the expression of two other bHLH myogenic regulatory factors myogenin and Mrf4, then differentiate into myocytes. Myocytes are terminally differentiated myoblasts that express myosin heavy chain (MHC) and muscle creatine kinase (MCK) [1].

The molecular regulation of quiescence, activation, proliferation, and differentiation of embryonic and adult muscle stem cells is strikingly similar [2]. First, myogenic specification of multipotent stem cells is regulated by Pax7 and Pax3. The $Pax3^+/Pax7^+$ progenitor cells in the embryonic dermomyotome are shown to give rise to dermal, adipogenic, and myogenic lineages, probably mediated by asymmetric cell divisions. Wnt and Shh signalings are positive regulators of myogenic differentiation, whereas BMP signaling is shown to inhibit myogenic differentiation and promote alternative lineage differentiation [2]. In adult, Pax7 is the master regulator of muscle lineage specification and stem cell function, whereas Pax3 seems to be insufficient, if not dispensable, for effective muscle regeneration [4–6]. Second, myogenic lineage commitment of muscle stem cells is regulated by Myf5 and MyoD. Quiescent satellite cells heterogeneously express Myf5, suggesting that they are composed of both myogenic committed and non-committed cells [7]. MyoD is not expressed by quiescent satellite cells, but is rapidly up-regulated in the activated satellite cells and proliferating myoblasts. Third, differentiation of myoblasts into fusion competent myocytes is regulated by the myogenin and MRF4. Together, the "Pax3/Pax7 → Myf5/MyoD → Myogenin/MRF4" axis defines the transcription framework regulating embryonic muscle development and adult muscle regeneration. Emerging knowledge supports the notion that essentially all extracellular molecules and intracellular signaling pathways act upon this transcriptional network to regulate muscle differentiation.

1.3 Various Types of Stem Cells Contributing to Muscle Regeneration

In adult, skeletal muscles have remarkable capability for remodeling and regeneration upon muscle injuries, due to myogenic differentiation of satellite cells and other types of muscle-specific stem cells. Muscle stem cells are a broad term referring to various types of myogenic cells that are isolated from skeletal muscles and are capable of long-term proliferation and myogenic differentiation (Table 6.1). They can be roughly classified into four categories: (1) satellite cells; (2) muscle side populations; (3) vessel-associated myogenic cells; and (4) other muscle resident stem cells mostly found in the interstitial connective tissues. Whether all these categories of cells are true adult stem cells is still a debated topic. The commonly accepted defining features of adult stem cells are self-renewal and differentiation, often mediated by asymmetric cell divisions. Except for satellite cells, there is still a lack of solid evidence to demonstrate that a single cell, from category 2 or 3 or 4, is capable of both self-renewal and myogenic differentiation in vivo.

These different types of muscle-derived stem cells have been isolated by various ways. For example, satellite cells are identified by their sublaminar localization and expression of specific markers. Some of the markers, such as CD34, α7 integrin, and Vcam1, have been successfully used to prospectively isolate satellite cells by FACS [7–9]. Several other populations of stem cells are also isolated by FACS based on specific marker expression, including pericytes by ALP expression, side population cells by Hoechst33442 dye exclusion, circulating AC133$^+$ cells, and myoendothelial cells [10–12]. In addition, stem cells have also been isolated by culture techniques. Bone marrow-derived stem cells, mesoangioblasts, and muscle-derived stem cells (MDSC) are examples of culture-derived stem cells.

2 Muscle Satellite Cells

2.1 Identification and Molecular Markers

Satellite cells were first observed in 1961 by Mauro with electron microscope and so named because of their intimate association with mature muscle fibers [13]. He

Table 6.1 Myogenic stem cells

Cell type	Origin	Localization	Molecular markers	Myogenic potential
Satellite cells	Somite	Sublaminar, adjacent to myofibers	Pax7, CD34, Myf5, M-Cad, CD56	Spontaneous High efficiency [9, 27]
SP cells	Somite and unknown	Interstitium? derived by FACS	Sca1	Inducible by co-culture [47, 53]
Pericytes	Mesenchymal	Associated with microvasculatures	ALP, PDGFRβ	Spontaneous, efficient [47, 53]
Mesoangioblasts	Mesenchymal	Dorsal aorta, blood vessel in muscle	CD34, Flk1, VE-Cad	Efficient [54, 55]
Myoendothelial cells	?	Interstitium	CD34	Spontaneous, efficient [57, 58]
AC133+ cells	? Bone marrow	Circulation?	AC133	Inducible by Wnt or co-culture with myoblast [12]
CD45Sca1	?, derived by FACS	?, muscle resident	CD45, Sca1	Inducible by Wnt [59]
BM-derived cells	Mesoderm	Bone marrow	Hematopoietic stem cell markers	Inducible [60–64]
MDSC	?, culture-derived	?, muscle resident	CD34, Sca1	Spontaneous [39]
ES cell derivatives	ES cells	Culture-derived	CD73, CD56	Spontaneous [65]

?, indicates unknown

noticed the satellite cells as being wedged between two membranes: the plasma membrane of muscle fiber and the basal membrane that wraps up the muscle fiber along its whole length. If the identification of satellite cells is a natural by-product of the application of electron microscope in biology, then his prediction of the origin and biological function of satellite cell is far more outreaching. Prior to his work, it was known that skeletal muscles have remarkable regenerative capacity and that some non-fused "free" cells are probably responsible for the regeneration. However, the source of these "free" cells was unclear or thought to be derived from the surviving nuclei of the degenerated muscle cell. In his seminal work, Mauro predicted that satellite cells are more likely to be remnants of embryonic myoblasts that have remained dormant and are "ready to recapitulate embryonic development of skeletal muscle fiber" upon damage [13]. He therefore provided a mechanism that was "pertinent to the vexing problem of skeletal muscle regeneration" and speculated that the lack of regenerative capacity in cardiac muscle was due to the lack of satellite cells [13]. His interpretation still holds true today: It has been generally accepted that satellite cells are indeed the main source of progenitors for adult muscle repair and regeneration.

Satellite cells are known to express various transcription factors and cell surface markers. Pax7 is not only important for the formation and function of satellite cells (elaborated below), it is also a specific and reliable marker for all satellite cells. Other markers, including CD34, Myf5, M-Cad, NCAM, VCAM-1, Syndecan3, and Syndecan4, are also expressed by satellite cells with certain heterogeneity [1]. In the past few years, several novel markers have been identified to be expressed by satellite cells. These include α7-integrin, β1-integrin, CXCR4, caveolin-1, Sox8, Sox15, sphingomyelin, VAMP-2, calcitonin receptor (CTR), MEGF10, VE-Cad, ICAM1, claudin 5, ESAM, and PCDHB9 [14].

2.2 *Role of Pax7 in Satellite Cell Function*

The paired box homeodomain transcription factor Pax7 plays multiple roles in satellite cell generation, maintenance, and function. The perinatal ablation and further age-dependent decline of satellite cells in Pax7 mutants demonstrate an essential role of Pax7 in the specification and subsequent self-renewal or survival of satellite cells [4–6, 15]. Surviving Pax7 mutant satellite cells are apoptotic in vivo and do not proliferate due to cell cycle arrest upon activation [4, 5]. In contrast, in vitro gain-of-function studies confirmed that Pax7 promotes proliferation and self-renewal but inhibits differentiation of satellite cells [16]. Furthermore, Pax7 mutant muscles fail to regenerate following injury, consistent with the observation that Pax7 is required for the myogenic progression of not only satellite cells but also other adult stem cells [4, 15]. Indeed, Pax7 is necessary for the myogenic conversion of $Sca1^+CD45^+$ muscle resident stem cells and $Pax3^+$ interstitial progenitor cells [4, 17]. Interestingly, the function of Pax7 cannot be compensated by the closely related Pax3 protein [5], suggesting a unique requirement of Pax7 in the regenerative production of myogenic progenitors.

2.3 *Regulation of Activation, Proliferation, Self-Renewal, and Differentiation*

Quiescent satellite cells are activated to enter cell cycle in response to growth signals or muscle injuries. Presently it is not clear how satellite cell quiescence is maintained, but a number of molecular cues are known to induce satellite cell activation. Of these, the hepatocyte growth factor (HGF) and nitric oxide (NO) signaling plays a critical role in the initial activation phase [18, 19]. NO has been shown to regulate the release of HGF, which subsequently acts through c-Met receptor to activate ERK and p38 family of mitogen-activated protein kinases (MAPKs) signaling pathways [20]. Inhibition of p38α/β promotes cell cycle withdrawal and prevents differentiation of proliferating myoblasts, suggesting that p38α/β reversibly regulates the quiescent state of the skeletal muscle satellite cell. Activation of satellite cells and p38α/β MAPKs occurs concomitantly, providing further support that these MAPKs function as a molecular switch for satellite cell activation [20].

Recent studies also indicate that sphingosine-1-phosphate (S1P) signaling is involved in the activation of satellite cells and proliferation of mesoangioblasts [21, 22]. S1P is expressed in the membrane invaginations of quiescent satellite cells and released by stress to trigger the sphingomyelin signaling. In addition, NF-κB signaling is also involved in activation of satellite cells. Depletion of IKK2, an inhibitor of NF-κB kinases, facilitates skeletal muscle regeneration through enhanced satellite cell activation [23].

Proliferation of satellite cells is regulated by various signaling pathway. Of prime importance is the bFGF signaling [1]. Interestingly, p38 MAPKs, whose role in promoting activation is discussed above, also functions downstream of bFGF to stimulate activated satellite cell proliferation [20, 24]. This finding suggests that p38 regulates distinct targets depending on its temporal kinetics, resulting in separate functions during myogenesis. Other growth factors and cytokines, including IGF, EGF, BDNF VEGF, PDGF, TWEAK, IL-6, and LIF, have also been documented to play roles in regulating satellite cell proliferation and differentiation [1, 25, 26].

Recent studies strongly suggest that some satellite cells are capable of self-renewal to maintain the homeostasis of the stem/progenitor cells in vivo [7, 27]. Several lines of evidence suggest that Notch signaling balances the cell fate choice between self-renewal and differentiation. Perturbation of Notch signaling through genetic mutation or pharmacological treatment results in depletion of muscle progenitor pool, indicative of compromised self-renewal [7, 28, 29]. Conversely, enhancement of Notch signaling restores the regenerative capacity of aged muscle [30]. Together, these results indicate that Notch signaling positively regulate satellite cell self-renewal.

2.4 Multipotency of Satellite Cells

In addition to their self-renewal capacity, satellite cells have also been shown to possess potentials for multi-lineage differentiation. First, myogenic differentiation of satellite cells has been well documented. In vivo tracing of radio isotope labeled satellite cells indicate that satellite cells can differentiate into myonuclei during muscle growth and regeneration [31–34]. In vitro cultivation of physically and/or enzymatically isolated single myofibers containing only satellite cells and differentiated myonuclei suggests that satellite cells can spontaneous differentiate to form multinuclear myotubes [35]. In addition, cultured satellite cells can also undergo adipogenic, osteogenic, fibroblastic, and even neural and endothelial differentiation in vitro and in vivo [36–44]. However, it is unclear how efficient satellite cells undergo alternative osteogenic and adipogenic differentiation in vivo. It is also unclear whether multi-lineage differentiation capacity is a property of all satellite cells or a sub-population of primitive satellite cells.

2.5 Satellite Cells as a Candidate for Stem Cell-Based Therapies

Due to their extraordinary efficiency in spontaneous muscle differentiation, satellite cell-derived myoblasts have been used in numerous clinical trials for muscular dystrophies [45]. However, several hurdles greatly slowed down efforts to harness the great potential of the satellite cell-based therapy. The main problems include host immune rejection, the poor survival, self-renewal, and migration of donor cells after local intramuscular injection, and their incompatibility with systemic delivery through circulation.

Recent studies indicate that freshly isolated satellite cells, compared to cultured myoblasts, are far more effective in restoration of dystrophin expression in mdx mice and give rise to functional satellite cells in host muscle [9, 27]. Remarkably, the number of cells required is several magnitudes lower as compared to conventional myoblast transfer. For example, a single wild-type myofiber carrying only seven to eight satellite cells can robustly contribute to the restoration of dystrophin in hundreds of mdx myofibers and give rise to self-renewed satellite cells at the same time [27]. Interestingly, MyoD mutant myoblasts engraft host muscle more efficiently due to their improved survival upon grafting, suggesting that inhibiting MyoD activity represent a strategy to improve efficiency of myoblast transplantation [46]. These results not only provide a promising direction for future development of stem cell-based therapy in muscular dystrophies but also suggest that the current conditions for cultivation of satellite cells result in alterations in the functionality satellite cells. Since in vitro culture is necessary for genetic manipulation of satellite cells to be used for autologous transplantation, future research should focus on culture conditions that maintain the high myogenicity of satellite cells. A thorough understanding of the molecular control of satellite cell activation, proliferation, and self-renewal would lead to such a solution.

Poor migration of focally injected myoblasts is another factor that hinders the application of satellite cell-based therapy for muscular dystrophies. It would be interesting to explore the feasibility of systemic delivery of fresh or cultured satellite cells. Several other cell types, including SP cells, pericytes, and mesoangioblasts, are capable of migration into muscles through the circulation system (see below). A challenging task would be to genetically engineer satellite cells so that they can also be used for systemic delivery. Finally, to ensure the long-term contribution of transplanted satellite cells to host muscle regeneration, a functional pool of satellite cells must be present or restored after transplantation. We have recently shown that sub-population of $Myf5^-$ satellite cells are particularly efficiently in the self-renewal of satellite cell compartment upon injection into the host muscle in mice [7]. Future studies should focus on developing strategies to purify equivalent self-renewing satellite cells from humans and on the understanding of the signaling mechanisms controlling the self-renewal process.

3 Other Myogenic Stem Cells Resident to the Muscle

3.1 Side Population Cells

Side population (SP) cells are originally isolated from bone marrow-derived hematopoietic stem cells based on Hoechst 33342 dye exclusion. They represent ~1% of all cells, as opposed to the rest of the main population (MP) cells [11]. SP cells are subsequently isolated from muscle and other tissues [47–49]. Interestingly, significant heterogeneity exists among SP cells from different tissues and even among SP cells from the same tissue [49]. SP cells from bone marrow are all $CD45^+$ but vary in Sca1 expression (roughly half positive and half negative). In contrast, SP cells derived from muscles are predominantly $Sca1^+CD45^-$ with a minor fraction being $Sca1^-CD45^-$. Although a rare population (<1%) of muscle SP cells are $Cd45^+$ and capable of hematopoietic differentiation, it is unknown whether these are contaminating SP cells from bone marrow through peripheral blood.

The origin, localization, and normal function of SP cells remain unclear. Recent genetic lineage analysis indicates that roughly 50% of muscle SP cells are derived from Pax3 expressing cells in embryonic dermomyotome [50]. Muscle SP cells are thought to be localized in the interstitium. Intriguingly, in human fetal muscles, SP cells express BMP4, probably functioning to induce proliferation and myogenic differentiation of neighboring MP cells that express BMP receptor 1a [51].

Both bone marrow- and muscle-derived SP cells have dual hematopoietic and myogenic potentials. They have been shown to reconstitute the hematopoietic system, give rise to dystrophin expressing muscle fibers and satellite cells after transplantation into mdx mice [47, 48, 52]. However, the contribution of transplanted SP cells to muscle regeneration is extremely low. One advantage of using SP cells is that they can be systemically delivered via intra-arterial or intra-venous injection. To this end, muscle SP cells that have been cultured and transduced with lentivirus expressing minidystrophin were intra-arterial injected into mdx mice. Overall, these donor SP cells are capable of myogenic differentiation, as indicated by Pax7 and desmin expression, resulting in 5–8% dystrophin-positive muscle fibers in the host [53]. These results represent a step toward the improvement of cell-based therapies for DMD and other myogenic disorders. Since SP cells do not express myogenic markers and are capable of myogenic differentiation only when co-cultured with myogenic cells, an intriguing possibility is that co-injecting SP cells with satellite cells may further improve their efficiency in muscle engraftment.

3.2 Vessel-Associated Myogenic Progenitors

Pericytes and mesoangioblasts are derived from mesenchymal lineage and found to be associated with blood vessels. Pericytes are mainly found at the surface of microvessels where their main function is to regulate blood flow through interaction with vascular smooth muscle cells. Pericytes are also involved in the maintenance of capillary structure and remodeling of microvasculatures. Surprisingly, a recent study reveals a role of pericytes in skeletal muscle regeneration [10].

Pericytes isolated from human skeletal muscle are capable of spontaneous differentiation into muscle. Prior to muscle differentiation, however, pericytes do not express classical myogenic markers such as Pax7, Myf5, or MyoD, suggesting that they represent a distinct population from satellite cells. Instead, pericytes express alkaline phosphatase (ALP) and PDGFRβ and can be prospectively isolated based on ALP expression [10]. Importantly, pericytes express β2 and α4 integrins that enable them to adhere to and across the endothelium (leukocyte also express these integrins), a feature that makes them an ideal candidate for systemic delivery [10]. Indeed, intra-arterial injections of wild type or genetically modified dystrophin-mutant pericytes into *mdx/SCID* mice resulted in restoration of dystrophin expression, with much higher efficiency than satellite cells. The injected pericytes also gave rise to satellite cells in the host muscle [10]. These results open a new avenue to exploring the feasibility of pericytes as a promising candidate for cell-based therapy to enhance muscle regeneration in human.

Mesoangioblasts are associated with embryonic and fetal aorta, and with adult blood vessels within the muscle [54, 55]. They are culture-derived cells meaning that their exact origin and anatomical localization are unclear. Phenotypic analysis indicates that cultured mesangioblasts are positive for CD34, Flk1, and VE-cadherin, markers that are not expressed by pericytes [10, 56]. This suggests that mesoangioblasts and pericytes are phenotypically distinct populations while sharing certain common behavioral and functional characteristics. Like pericytes, mesoangioblasts are capable of penetrating the endothelium of blood vessels and are thus amenable to systemic delivery through intra-arterial injection. Mesoangioblasts are also capable of efficient myogenic differentiation in vitro and in vivo, though their normal physiological function is unknown. Strikingly, intra-arterial delivery of wild-type or genetically corrected mutant mesoangioblasts results in an extensive recovery of dystrophin expression and muscle function in dystrophic mice and dogs [54, 55].

Pericytes and mesoangioblasts represent two exciting candidates for cell-based therapies, each having their own advantages and disadvantages. One disadvantage of the mesoangioblasts is that there is a lack of method to purify them directly using FACS. Currently they are derived through a prolonged culture procedure, which may render potential problems when they are considered for clinical usage. On the other hand, they have a superior proliferation capacity, making it possible to genetically modify autologous patient cells in vitro and amplify sufficient number of cells for subsequent treatment. In contrast, pericytes can be directly isolated with FACS but are less potential in vitro proliferation. Future clinical trials in human will prove the usage of these endothelial-associated myogenic progenitors in the treatment of muscular dystrophies.

3.3 Interstitial Myogenic Progenitors

The interstitial myogenic progenitors include several types of cells resident to the muscle. First, myogenic endothelial (myoendothelial) cells that are closely associated with muscle fibers but localized outside the myofiber basal lamina represent a class of interstitial stem cells [57, 58]. These cells have been identified in mice and human and express both endothelial and myogenic markers. Human myoendothelial stem cells express CD56 (NCAM), CD34, and CD144, whereas those of mice are $CD34^+$ and $Sca1^+$, but $CD144^-$, $CD45^-$, $CD31^-$, c-kit^-, and $FLK1^-$ [57, 58]. This marker expression profile suggests that myoendothelial cells are distinct from vessel-associated myogenic progenitors and satellite cells. The myoendothelial cells can form colony under clonal culture conditions and possess multi-lineage differentiation potential, including endothelial and myogenic differentiation. When myoendothelial cells were injected into injured muscles in SCID mice, they efficiently differentiated into myofibers that expressed donor-specific markers [57, 58]. However, it is unknown if they can also give rise to satellite cells. In addition, several issues remain to be resolved. It is yet unknown whether myoendothelial cells can be delivered through intra-arterial injection. If so, it will be a huge advantage for these cells to be considered as candidates for clinical applications.

Next, $Pax3^+$ interstitial cells have been shown to be capable of myogenic differentiation, with a low efficiency, in Pax7 mutant muscles [4]. It is likely that these $Pax3^+$ interstitial cells are closely related to, or derived from, the pericytes, since both cells express Pax3 [4, 10]. As mentioned above, pericytes are capable of spontaneous myogenic differentiation with high efficiency. The low myogenic potential of the $Pax3^+$ interstitial cells in the absence of Pax7 suggests that Pax7 plays an important role in their myogenic differentiation.

Third, $CD45^+Scal^+$ muscle resident stem cells have been shown to differentiate into myogenic lineage in response to Wnt signaling during muscle regeneration

[59]. Intriguingly, $CD45^{+}Sca1^{+}$ cells from non-regenerating muscles fail to undergo myogenic differentiation even in the presence of inductive signals. Importantly, the myogenic specification of $CD45^{+}Sca1^{+}$ cells also requires Pax7, suggesting a common role of Pax7 in myogenic lineage progression of all adult stem cells [17]. It is unknown where these cells are located within the muscle.

Finally, a population of culture-derived stem cells (MDSC) based on adherence to collagen coating have been shown to be an excellent candidate for cell-based restoration of dystrophin expression in dystrophic muscles [39]. MDSC express MyoD, Desmin, CD34, and Sca1, but do not express M-cad, therefore they are proposed to represent an interstitial cell population distinct from satellite cells. It is also possible that they are derived from a rare fraction of sublaminar satellite cells or blood vessel-associated cells.

3.4 Myogenic Progenitors Non-resident to Muscle

It is worth mentioning that several other types of non-muscle resident stem cells have also been proven to be capable of myogenic differentiation under certain circumstances. These include circulating human $AC133^{+}$ cells [12], bone marrow-derived hematopoietic stem cells, and stromal stem cells [60–64], as well as ES cell-derived myogenic cells [65]. Transplantation studies indicate that these cells also differentiate into muscle in vivo and some adopt satellite cell positions. These stem cells from different resources are important supplement to the muscle resident stem cells as candidates for cell-based therapies to treatment muscle diseases.

3.5 A Possible Developmental Link Among Various Myogenic Stem Cells

Given the heterogeneity of these various myogenic stem cells (Table 6.1), including the side population and vessel-associated cells, it is imperative to examine the relationship among them. It would be important to address whether this heterogeneity reflect a hierarchical composition of cells derived from a common origin (as illustrated in Fig. 6.1) or it rather suggests different developmental origins of these various types of cells. For example, it has been suggested that mesoangioblasts share a common somitic origin with skeletal muscle precursors and give rise to adult pericytes [66, 67]. This hypothesis therefore provides a lineage connection among satellite cells, mesoangioblasts, and pericytes.

Another question is whether these various stem cells are involved in the maintenance and growth of skeletal muscles under normal physiological conditions, or they just act as an emergent reserve for transdifferentiating into a variety of cell types in response to injuries. The latter possibility is supported by recent evidence that progenitors of endothelial origin only undergo myogenic conversion (or transdifferentiation) upon spontaneous fuse with myotubes [68]. In addition, mesoangioblasts

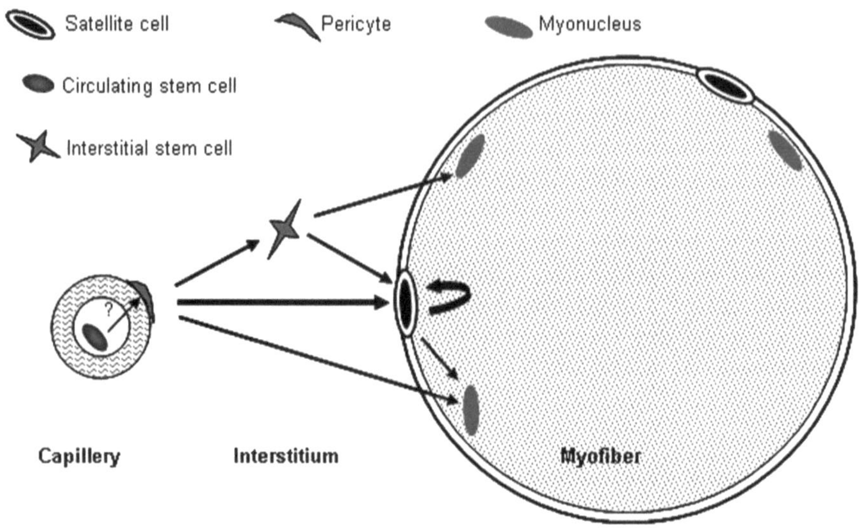

Fig. 6.1 A possible developmental link among various populations of myogenic stem cells. Satellite cells reside beside multinuclear myofibers underneath the basal lamina are the primary stem cells mediating postnatal muscle growth and repair. Developmentally, satellite cells are derived from progenitors located in the embryonic dermomyotome. During muscle regeneration, however, various adult stem cells, including those localized in the interstitium, those associated with blood vessels, and those from circulation may also give rise to satellite cells. These various types of stem cells can also give rise to differentiated myonuclei

migrate toward soluble factors including HMGB1 and SDF1, released from cells undergoing necrosis due to injuries, supporting their role as a reserved population [69].

4 Conclusions and Perspectives

Skeletal muscles are remarkably efficient in regeneration due to myogenic progenitor and stem cells resident to the muscle. These include several different types of cells based on localization, function, and origin. Satellite cells localized adjacent to muscle fibers underneath the basal lamina have been shown to be necessary and sufficient for muscle growth and repair. In addition, various other types of stem cells are also capable of myogenic differentiation upon injuries. Importantly, some cells in this later category, including side population cells, pericytes, and myoendothelial cells, are highly myogenic and amenable to systemic delivery, therefore possessing great potential for applications to the regeneration and repair of diseased muscle. Investigations into the mechanisms regulating the quiescence, activation, proliferation, and differentiation of these various types of myogenic stem cells will lead the way to successful cell-based therapy for muscle diseases.

Given the various types of myogenic stem cells (Table 6.1), one challenge in future development of stem cell-based therapies is to understand the characteristics of these different cells. Obviously, choosing the right cell to use at the right time is particularly important for specific purposes. Another challenge is to delineate the lineage relationships among these various types of stem cells. Are they derived from different lineages and represent cells programmed for specific fate (with certain plasticity)? Do they represent hierarchical progenies of the same lineage at different developmental stages? Since skeletal muscles are composed of muscle cells, neurons and vascular systems as a functional entity, understanding these questions is crucial for choosing the right cells to be used for transplantation. If the various types of muscle stem cells are programmed for specific cell fate, then transplantation of a combinatorial of stem cells would achieve the maximal effect on the synergistic regeneration of muscle, nerve, and vasculatures. In contrast, if these stem cells represent hierarchical descendents of the same lineage, then the most primitive stem cell population should be used for therapies. Finally, it is crucial to understand the interactions among these various stem cells and their interaction with the niche [26]. Active interactions between satellite cells and other muscle resident or circulating cells can positively regulate stem cell function and muscle regeneration.

References

1. Charge SB, Rudnicki MA (2004). Cellular and molecular regulation of muscle regeneration. Physiol Rev 84:209–238
2. Parker MH, Seale P, Rudnicki MA (2003) Looking back to the embryo: defining transcriptional networks in adult myogenesis. Nat Rev Genet 4:497–507
3. Buckingham M, Relaix F (2007) The role of pax genes in the development of tissues and organs: pax3 and pax7 regulate muscle progenitor cell functions. Annu Rev Cell Dev Biol 23:645–673
4. Kuang S, Charge SB, Seale P, Huh M, Rudnicki MA (2006) Distinct roles for Pax7 and Pax3 in adult regenerative myogenesis. J Cell Biol 172:103–113
5. Relaix F, Montarras D, Zaffran S et al (2006) Pax3 and Pax7 have distinct and overlapping functions in adult muscle progenitor cells. J Cell Biol 172:91–102
6. Seale P, Sabourin LA, Girgis-Gabardo A, Mansouri A, Gruss P, Rudnicki MA (2000) Pax7 is required for the specification of myogenic satellite cells. Cell 102:777–786
7. Kuang S, Kuroda K, Le Grand F, Rudnicki MA (2007) Asymmetric self-renewal and commitment of satellite stem cells in muscle. Cell 129:999–1010
8. Fukada S, Uezumi A, Ikemoto M et al (2007) Molecular signature of quiescent satellite cells in adult skeletal muscle. Stem Cells 25:2448–2459
9. Montarras D, Morgan J, Collins C et al (2005) Direct isolation of satellite cells for skeletal muscle regeneration. Science 309:2064–2067
10. Dellavalle A, Sampaolesi M, Tonlorenzi R et al (2007) Pericytes of human skeletal muscle are myogenic precursors distinct from satellite cells. Nat Cell Biol 9:255–267
11. Goodell MA, Brose K, Paradis G, Conner AS, Mulligan RC (1996) Isolation and functional properties of murine hematopoietic stem cells that are replicating in vivo. J Exp Med 183:1797–1806
12. Torrente Y, Belicchi M, Sampaolesi M et al (2004) Human circulating AC133(+) stem cells restore dystrophin expression and ameliorate function in dystrophic skeletal muscle. J Clin Invest 114:182–195

13. Mauro A (1961) Satellite cell of skeletal muscle fibers. J Biophys Biochem Cytol 9:493–495
14. Kuang S, Rudnicki MA (2008) The emerging biology of satellite cells and their therapeutic potentials. Trends Mol Med 14:82–91
15. Oustanina S, Hause G, Braun T (2004) Pax7 directs postnatal renewal and propagation of myogenic satellite cells but not their specification. Embo J 23:3430–3439
16. Olguin HC, Olwin BB (2004) Pax-7 up-regulation inhibits myogenesis and cell cycle progression in satellite cells: a potential mechanism for self-renewal. Dev Biol 275:375–388
17. Seale P, Ishibashi J, Scime A, Rudnicki MA (2004) Pax7 is necessary and sufficient for the myogenic specification of CD45+:Sca1+ stem cells from injured muscle. PLoS Biol 2:E130
18. Tatsumi R, Liu X, Pulido A et al (2006) Satellite cell activation in stretched skeletal muscle and the role of nitric oxide and hepatocyte growth factor. Am J Physiol Cell Physiol 290:C1487–1494
19. Wozniak AC, Anderson JE (2007) Nitric oxide-dependence of satellite stem cell activation and quiescence on normal skeletal muscle fibers. Dev Dyn 236:240–250
20. Jones NC, Tyner KJ, Nibarger L et al (2005) The p38alpha/beta MAPK functions as a molecular switch to activate the quiescent satellite cell. J Cell Biol 169:105–116
21. Donati C, Cencetti F, Nincheri P et al (2007) Sphingosine 1-Phosphate Mediates Proliferation and Survival of Mesoangioblasts. Stem Cells 25:1713–1749
22. Nagata Y, Partridge TA, Matsuda R, Zammit PS (2006) Entry of muscle satellite cells into the cell cycle requires sphingolipid signaling. J Cell Biol 174:245–253
23. Mourkioti F, Kratsios P, Luedde T et al (2006) Targeted ablation of IKK2 improves skeletal muscle strength, maintains mass, and promotes regeneration. J Clin Invest 116:2945–2954
24. Perdiguero E, Ruiz-Bonilla V, Serrano AL, Munoz-Canoves P (2007) Genetic deficiency of p38alpha reveals its critical role in myoblast cell cycle exit: the p38alpha-JNK connection. Cell Cycle 6:1298–1303
25. Girgenrath M, Weng S, Kostek CA et al (2006) TWEAK, via its receptor Fn14, is a novel regulator of mesenchymal progenitor cells and skeletal muscle regeneration. EMBO J 25:5826–5839
26. Christov C, Chretien F, Abou-Khalil R et al (2007) Muscle satellite cells and endothelial cells: close neighbors and privileged partners. Mol Biol Cell 18:1397–1409
27. Collins CA, Olsen I, Zammit PS et al (2005) Stem cell function, self-renewal, and behavioral heterogeneity of cells from the adult muscle satellite cell niche. Cell 122:289–301
28. Schuster-Gossler K, Cordes R, Gossler A (2007) Premature myogenic differentiation and depletion of progenitor cells cause severe muscle hypotrophy in Delta1 mutants. Proc Natl Acad Sci U S A 104:537–542
29. Vasyutina E, Lenhard DC, Wende H, Erdmann B, Epstein JA, Birchmeier C (2007) RBP-J (Rbpsuh) is essential to maintain muscle progenitor cells and to generate satellite cells. Proc Natl Acad Sci USA 104:4443–4448
30. Conboy IM, Conboy MJ, Smythe GM, Rando TA (2003) Notch-mediated restoration of regenerative potential to aged muscle. Science 302:1575–1577
31. Lipton BH, Schultz E (1979) Developmental fate of skeletal muscle satellite cells. Science 205:1292–1294
32. Moss FP, Leblond CP (1971) Satellite cells as the source of nuclei in muscles of growing rats. Anat Rec 170:421–435
33. Snow MH (1978) An autoradiographic study of satellite cell differentiation into regenerating myotubes following transplantation of muscles in young rats. Cell Tissue Res 186:535–540
34. Watt DJ, Lambert K, Morgan JE, Partridge TA, Sloper JC (1982) Incorporation of donor muscle precursor cells into an area of muscle regeneration in the host mouse. J Neurol Sci 57:319–331
35. Rosenblatt JD, Lunt AI, Parry DJ, Partridge TA (1995) Culturing satellite cells from living single muscle fiber explants. In Vitro Cell Dev Biol 31:773–779
36. Asakura A, Komaki M, Rudnicki M (2001) Muscle satellite cells are multipotential stem cells that exhibit myogenic, osteogenic, and adipogenic differentiation. Differentiation 68:245–253

37. Csete M, Walikonis J, Slawny N et al (2001) Oxygen-mediated regulation of skeletal muscle satellite cell proliferation and adipogenesis in culture. J Cell Physiol 189:189–196
38. Ozeki N, Jethanandani P, Nakamura H, Ziober BL, Kramer RH (2007) Modulation of satellite cell adhesion and motility following BMP2-induced differentiation to osteoblast lineage. Biochem Biophys Res Commun 353:54–59
39. Qu-Petersen Z, Deasy B, Jankowski R et al (2002) Identification of a novel population of muscle stem cells in mice: potential for muscle regeneration. J Cell Biol 157:851–864
40. Sarig R, Baruchi Z, Fuchs O, Nudel U, Yaffe D (2006) Regeneration and transdifferentiation potential of muscle-derived stem cells propagated as myospheres. Stem Cells 24:1769–1778
41. Shefer G, Wleklinski-Lee M, Yablonka-Reuveni Z (2004) Skeletal muscle satellite cells can spontaneously enter an alternative mesenchymal pathway. J Cell Sci 117:5393–5404
42. Shefer G, Yablonka-Reuveni Z (2007) Reflections on lineage potential of skeletal muscle satellite cells: do they sometimes go MAD? Crit Rev Eukaryot Gene Expr 17:13–29
43. Sherwood RI, Christensen JL, Conboy IM et al (2004) Isolation of adult mouse myogenic progenitors: functional heterogeneity of cells within and engrafting skeletal muscle. Cell 119:543–554
44. Yada E, Yamanouchi K, Nishihara M (2006) Adipogenic potential of satellite cells from distinct skeletal muscle origins in the rat. J Vet Med Sci 68:479–486
45. Mouly V, Aamiri A, Perie S et al (2005) Myoblast transfer therapy: is there any light at the end of the tunnel? Acta Myol 24:128–133
46. Asakura A, Hirai H, Kablar B et al (2007) Increased survival of muscle stem cells lacking the MyoD gene after transplantation into regenerating skeletal muscle. Proc Natl Acad Sci USA 104:16552–16557
47. Asakura A, Seale P, Girgis-Gabardo A, Rudnicki MA (2002) Myogenic specification of side population cells in skeletal muscle. J Cell Biol 159:123–134
48. Gussoni E, Soneoka Y, Strickland CD et al (1999) Dystrophin expression in the mdx mouse restored by stem cell transplantation. Nature 401:390–394
49. Montanaro F, Liadaki K, Schienda J, Flint A, Gussoni E, Kunkel LM (2004) Demystifying SP cell purification: viability, yield, and phenotype are defined by isolation parameters. Exp Cell Res 298:144–154
50. Schienda J, Engleka KA, Jun S et al (2006) Somitic origin of limb muscle satellite and side population cells. Proc Natl Acad Sci U S A 103:945–950
51. Frank NY, Kho AT, Schatton T et al (2006) Regulation of myogenic progenitor proliferation in human fetal skeletal muscle by BMP4 and its antagonist Gremlin. J Cell Biol 175: 99–110
52. Jackson KA, Mi T, Goodell MA (1999) Hematopoietic potential of stem cells isolated from murine skeletal muscle. Proc Natl Acad Sci U S A 96:14482–14486
53. Bachrach E, Perez AL, Choi YH et al (2006) Muscle engraftment of myogenic progenitor cells following intraarterial transplantation. Muscle Nerve 34:44–52
54. Sampaolesi M, Blot S, D'Antona G et al (2006) Mesoangioblast stem cells ameliorate muscle function in dystrophic dogs. Nature 444:574–579
55. Sampaolesi M, Torrente Y, Innocenzi A et al (2003) Cell therapy of alpha-sarcoglycan null dystrophic mice through intra-arterial delivery of mesoangioblasts. Science 301:487–492
56. Cossu G, Bianco P (2003) Mesoangioblasts--vascular progenitors for extravascular mesodermal tissues. Curr Opin Genet Dev 13:537–542
57. Tamaki T, Akatsuka A, Ando K et al (2002) Identification of myogenic-endothelial progenitor cells in the interstitial spaces of skeletal muscle. J Cell Biol 157:571–577
58. Zheng B, Cao B, Crisan M et al (2007) Prospective identification of myogenic endothelial cells in human skeletal muscle. Nat Biotechnol 25:1025–1034
59. Polesskaya A, Seale P, Rudnicki MA (2003) Wnt signaling induces the myogenic specification of resident CD45+ adult stem cells during muscle regeneration. Cell 113:841–852
60. Camargo FD, Green R, Capetanaki Y, Jackson KA, Goodell MA (2003) Single hematopoietic stem cells generate skeletal muscle through myeloid intermediates. Nat Med 9:1520–1527

61. Corbel SY, Lee A, Yi L et al (2003) Contribution of hematopoietic stem cells to skeletal muscle. Nat Med 9:1528–1532
62. Dezawa M, Ishikawa H, Itokazu Y et al (2005) Bone marrow stromal cells generate muscle cells and repair muscle degeneration. Science 309:314–317
63. Ferrari G, Cusella-De Angelis G, Coletta M et al (1998) Muscle regeneration by bone marrow-derived myogenic progenitors. Science 279:1528–1530
64. LaBarge MA, Blau HM (2002) Biological progression from adult bone marrow to mononucleate muscle stem cell to multinucleate muscle fiber in response to injury. Cell 111:589–601
65. Barberi T, Bradbury M, Dincer Z, Panagiotakos G, Socci ND, Studer L (2007) Derivation of engraftable skeletal myoblasts from human embryonic stem cells. Nat Med 13:642–648
66. De Angelis L, Berghella L, Coletta M et al (1999) Skeletal myogenic progenitors originating from embryonic dorsal aorta coexpress endothelial and myogenic markers and contribute to postnatal muscle growth and regeneration. J Cell Biol 147:869–878
67. Esner M, Meilhac SM, Relaix F, Nicolas JF, Cossu G, Buckingham ME (2006) Smooth muscle of the dorsal aorta shares a common clonal origin with skeletal muscle of the myotome. Development 133:737–749
68. Kirillova I, Gussoni E, Goldhamer DJ, Yablonka-Reuveni Z (2007) Myogenic reprogramming of retina-derived cells following their spontaneous fusion with myotubes. Dev Biol 311:449–463
69. Palumbo R, Galvez BG, Pusterla T et al (2007) Cells migrating to sites of tissue damage in response to the danger signal HMGB1 require NF-{kappa}B activation. J Cell Biol 179:33–40

Chapter 7
Marrow Stromal Mesenchymal Stem Cells

Cynthia B. Ripoll and Bruce A. Bunnell

Abstract The broad definition of a stem cell is a population of cells that has the ability to self-renew and to differentiate into one or more types of specialized terminally differentiated cells. It has become evident that stem cells persist in and can be isolated from many organs postnatally. Stem cells isolated from various sources have been demonstrated to vary in their differentiation capacity or pluripotentiality. Differentiation causes stem cells to adopt the phenotypic, biochemical, and functional properties of more terminally differentiated cells. As such, the newly differentiated cells may replace and/or support cells damaged by disease. Investigators have begun to examine postnatal sources of pluripotent stem cells, such as mesenchymal stem cells from bone marrow and adipose tissue for therapeutic applications. The anti-inflammatory, immunosuppressive, and neurotrophic factors secreted by mesenchymal stem cells have increased interest in developing stem cell-based therapies for many devastating conditions such as myocardial infarction, stroke, Parkinson's disease, and amyotrophic lateral sclerosis. This chapter summarizes the origins, isolation, characterization, cell cycle properties, and differentiation potential of bone marrow stromal mesenchymal stem cells (MSCs).

Keywords Mesenchymal stem cells · Cell cycle · Differentiation · Expansion · Immunophenotype · Regenerative medicine

1 Introduction

The broad definition of a stem cell is a population of cells that has the ability to self-renew and to differentiate into one or more specialized mature cell types [1–4]. The classification of stem cells is dependent on their species and tissue of origin

B.A. Bunnell (✉)
Department of Pharmacology, Division of Gene Therapy, Center for Gene Therapy, Tulane National Primate Research Center, Tulane University Health Sciences Center, 18703 Three Rivers Road, Covington, LA 70433, USA
e-mail: bbunnell@tulane.edu

A. Giordano, U. Galderisi (eds.), *Cell Cycle Regulation and Differentiation in Cardiovascular and Neural Systems*, DOI 10.1007/978-1-60327-153-0_7,

and their differentiation potential into mature cell type(s). Stem cells isolated from various sources have been demonstrated to vary in their differentiation capacity or pluripotentiality. This differentiation capacity, or plasticity, is thought to be an indication of cellular differences in the organisms, the ontogenic stage of the stem cell, or the environmental conditions to which the stem cell is exposed [5]. Two primary groups of stem cells have been defined, to date. The first group, embryonic stem cells (ESCs), is believed to have an indefinite replicative capacity and the ability to generate all somatic cells of the body [6]. The second population of stem cells is denoted as adult stem cells, which have been characterized from adherent and nonadherent fractions of bone marrow, adipose tissue, as well as specific organs [7–9].

1.1 Adult Stem Cells

Tissue-specific stem cells comprise the second group and are derived from specific organs, such as brain, gut, lung, liver, and bone marrow [10–12]. It has become evident that these stem cells persist in adult tissues, although they represent a rare population localized in small niches [13]. Postnatal (adult) stem cells are not totipotent; however, they are pluripotent, and as such they retain a broad differentiation potential but their developmental potential is more restricted than embryonic cells and can be organ specific. Adult stem cells were initially thought to have the differentiation capacity limited to their tissue of origin; however, recent studies have demonstrated that stem cells have the capacity to differentiate into cells of mesodermal, endodermal, and ectodermal origins [8, 14–19]. The plasticity of stem cells most often refers to the inherent ability retained within stem cells to cross lineage barriers and to adopt the phenotypic, biochemical, and functional properties of cells unique to other tissues. For example, studies have demonstrated that adult stem cells can differentiate into cells types of many different organ systems including cardiomyocytes, liver cells, lung cells, myocytes, and neural tissue [8, 14–18].

As a result of this plasticity, stem cells may be useful to clinicians interested in regenerative medicine to effectively repair damaged or failing organs. Ideally, stem cells for regenerative medicinal applications should meet the following criteria: (i) they are found in abundant quantities (millions to billions of cells); (ii) they can be harvested by a minimally invasive procedure; (iii) they can undergo differentiation along multiple cell lineage pathways in a controlled, reproducible manner; and (iv) they can be safely and effectively transplanted to either an autologous or allogeneic host.

1.2 Bone Marrow-Derived MSCs

Mesenchymal stem cells, also known as marrow stromal cells or MSCs, represent a small, nonhematopoietic subpopulation of cells that reside in the bone marrow, that were initially described by Friedenstein and colleagues in the 1970s [20]. MSCs have attracted increasing attention for their potential use in cell and gene therapy because they have several appealing features [8, 18]. It has subsequently been determined that MSCs have the capacity to differentiate into cell types along multiple lineages. Extensive research has focused on the biological characterization of MSCs obtained from several distinct species including rodents, canines, felines, nonhuman primates (baboons), and humans [8, 21–23].

MSCs are readily isolated from a patient by simple bone marrow aspiration under local anesthesia. They can be readily expanded in culture up to a billion fold in 8 weeks [24–27]. Although MSCs can be expanded rapidly, they are not immortal. Therefore, they do not pose a danger of producing tumors as seen with embryonic stem cells and with most immortal cell lines [28]. MSCs can differentiate ex vivo and in vivo into multiple cell lineages. Moreover, they can be readily transduced with genes with the use of viral and nonviral vectors [29–31]. Additionally, MSCs have the remarkable property that they home to sites of tissue injury and repair the tissue either by differentiating into tissue-specific cell phenotypes [32–37] or by creating a milieu that increases the capacity of the endogenous cells to repair the tissue [38, 39].

2 MSC Immunophenotype

Multiple, independent groups have examined the surface immunophenotype of MSCs isolated from human and other species [9, 40–43]. Hematopoietic stem cell markers including CD3, CD4, CD8, CD11b, CD13, and CD45 were found to be negative in all MSCs at all passages. MSCs have been reported to be strongly positive for CD59, a SCA-1 homolog, CD90 (Thy-1), and HLA-1, while they were negative for CD164. The expression profile changes as a function of time in passage and plastic adherence [40, 44]. After two or more successive passages in culture, the MSCs express characteristic adhesion (CD44, CD73, CD90, CD105, CD166) molecules, surface enzymes, extracellular matrix and cytoskeletal proteins, and proteins associated with the stromal cell phenotype. MSCs are positive for MSC-specific surface markers such as CD105, CD106, CD146, and CD161. Despite any differences in the isolation and culture procedures, the immunophenotype is relatively consistent between laboratories.

3 Cell Cycle and Proliferation of MSCs

The cell cycle is a highly ordered process that results in the faithful duplication and transmission of genetic information from one cell generation to the next [45]. Primary mammalian somatic cells can replicate in vitro an estimated 50 cumulative population doublings, after which the cultures stop dividing [46]. This phenomenon is termed Hayflick's limit and is more readily known as replicative senescence.

While it has been demonstrated that MSCs and other stem cell populations continuously grow in vitro for 10–20 passages, it appears that MSCs, similar to all other primary cells, are subject to the Hayflick limit [47, 48]. In vitro, early passages MSCs display a cell doubling time of 24–48 h, depending on the culture medium and passage number [44, 49]. Our group has determined that the duration of the cell cycle increases markedly, to as long as 150 h in aged cultures of human and nonhuman primate MSCs. It is evident that human MSCs derived from the marrow become senescent during protracted culture, as indicated by their decreased differentiation potential, shortening of the mean telomere length, and morphological alterations [50]. We have observed concomitant increases in the levels of the expression of senescence-associated β-galactosidase in all MSC populations as they age; however, our group and others have not observed marked shortening of the telomeres in human MSCs (Izadpanah and Bunnell, unpublished observations).

An important factor involved in cell senescence is the maintenance of mean telomere length as a result of decreased telomerase activity. It has previously been reported by our group that cultures of MSCs derived from the bone marrow and adipose tissue underwent morphological alteration, a decline in multilineage differentiation potential, and a marked decrease in telomerase activity in progressively increasing passages of MSCs [47].

It is unknown whether the MSCs or a subpopulation of MSCs are able to escape cellular senescence in a manner similar to immortalized or transformed cells. There is an increasing body of evidence that MSCs, and other stem cells, can undergo spontaneous transformation to malignant cells [51]. The transformation of MSCs appears to be the direct result of spontaneous genetic alterations that accumulate during extended culture. Murine MSCs have been demonstrated to undergo malignant transformation upon extended culture and form sarcomas upon in vivo transplantation [52, 53]. The mechanisms of transformation observed in one of the studies with murine MSCs were associated with chromosomal abnormalities, increased telomerase activity, and elevated c-myc expression levels. With prolonged passage for >4 months, MSCs from human bone marrow have not been reported to undergo malignant transformation; however, this has been reported for MSCs from human adipose tissue. In at least one laboratory, serially passaged adipose-derived MSCs displayed karyotypic abnormalities at a frequency of >30% and, when implanted into immunodeficient mice, formed tumors at a frequency of 50% [51]. However, extended culture of human bone marrow MSCs failed to reveal any chromosomal alterations. Genome-wide transcriptome comparison of MSCs at early and late passages indicates that the expression of genes involved in cell cycle, protein ubiquitination, and apoptosis was altered (Izadpanah and Bunnell, unpublished observations).

The data presented here demonstrate that MSCs continuously cultured for protracted periods have altered cell cycle progression and chromosomal alterations that result in both cellular senescence and crisis. These findings indicate that caution should be exercised in the manipulation and culture of MSCs.

4 Mechanisms for Therapeutic Utility

Investigators have postulated a number of non-exclusive mechanisms through which MSCs can be used to repair and regenerate tissues. Initially, it was believed that the differentiation of MSCs into terminally differentiated cell phenotypes was the exclusive mechanism of tissue repair mediated by the MSCs. However, more recent data present a paradox in that tissue repair is often observed without significant evidence of either engraftment or differentiation of the MSCs. These data suggest that MSCs may mediate therapeutic efficacy via several disparate mechanisms. Presently, there are at least four mechanisms of action by which MSCs can provide repair, including differentiation, paracrine effects, mediation of immune reactions, and anti-inflammatory effects.

4.1 Differentiation/Fusion

MSCs are able to differentiate into bone, cartilage, fat, and myoblasts in response to certain growth factors and chemical inducers supplied in the culture medium in vitro or bioactive factors found in the local microenvironment in vivo. Adult stem cells were initially thought to have the differentiation capacity limited to their tissue of origin; however, recent studies have demonstrated that stem cells have the capacity to differentiate into cells of mesodermal, endodermal, and ectodermal origins [14–19]. The plasticity of MSCs most often refers to the inherent ability retained within stem cells to cross lineage barriers and to adopt the phenotypic, biochemical, and functional properties of cells unique to other tissues. For example, studies have demonstrated that adult stem cells can differentiate into cell types of many different organ systems including cardiomyocytes, liver cells, lung cells, myocytes, and neural tissue [14–19].

A few years ago, the results of some studies suggested that cell–cell fusion between MSCs and host tissue cells, not transdifferentiation, may be responsible for the co-localization of cell specific markers with transplanted MSCs [54–56]. However, significantly more evidence suggests that cell fusion is an extremely rare event [57– 62]. In reality, the phenomenon of cell fusion both in vitro and in vivo appears to be rare (<1/100,000 cells) and seems to occur most readily in cells such as hepatocytes, skeletal muscle, cardiac muscle, and Purkinje cells where polyploidy is commonly seen [62].

4.2 Paracrine-Mediated Effects

While the inherent differentiation potential of MSCs has been demonstrated in several organ systems, in more recent studies investigators have reported functional improvements in diseased organs with little to no evidence of either long-term engraftment or differentiation. It is becoming evident that MSCs delivered into an injured or diseased tissue may provide a therapeutic benefit through the localized secretion of cytokines and growth factors that stimulate recovery in a paracrine manner. The MSCs modulate the host's "stem cell niche" by stimulating the recruitment of endogenous stem cells to the site and promoting their differentiation along the required lineage pathway. The influence that MSCs exert over neighboring host cells can be referred to as trophic since the MSCs themselves may not differentiate but rather regulate the regeneration, turnover, and differentiation of endogenous stem cells.

In a related manner, MSCs might provide antioxidants chemicals, free radical scavengers, and chaperone/heat shock proteins at an ischemic site. As a result, toxic substances released into the local environment would be removed, thereby promoting recovery of the surviving cells.

MSCs have been shown to secrete a large number of cytokines and chemokines in vitro. In addition, the pattern of cytokines secreted by MSCs changes as the cells engraft into new microenvironments, such as the brain. The paracrine-mediated effects of MSCs are also illustrated in multiple studies involving the application of MSCs as a therapeutic tool in animal models for diseases including spinal cord injury, diseases of the central nervous system (parkinsonism and stroke), and for the repair of cardiac tissue following myocardial infarction. The application of MSCs for spinal cord injury has demonstrated moderate to significant improvements in gait in animals rendered paraplegic. Several groups have shown that MSCs promote de novo neurogenesis, angiogenesis, and functional recovery in rats, rabbits, and nonhuman primates [63–65]. In paralyzed rats, the MSCs were tightly associated with immature astrocytes and formed bundles of cells that bridged the injury [63].

Most of the studies for MSCs and myocardial infarction have demonstrated improvements in left ventricular ejection fraction and decreased ventricular remodeling when MSCs were injected into the coronary artery following myocardial infarction. Several groups have shown co-localization of cardiac markers with transplanted MSCs in cardiac repair models [66–70], and that administered MSCs display upregulation of genes for vascular endothelial growth factor, fibroblast growth factor-2, insulin-like growth factor, thymosin beta 4, various interleukins, macrophage colony-stimulating factor, leukemia inhibitor factor, placental growth factor, and monocyte chemoattractant protein-1 [71]. In the brain, the transplantation of MSCs into the dentate gyrus of the hippocampus of mice markedly enhanced the proliferation migration and differentiation of endogenous neural stem cells [72]. The data published from a clinical trial for *osteogeneis imperfecta* performed by Horwitz and colleagues seem to support paracrine-mediated repair mechanisms. In this study, five children between the age of 4 and 6 years received MSCs from a donor that had previously been used for a bone marrow transplant in these children. The data indicate that, while the children demonstrated significant clinical improvements in their disease, the levels of donor MSCs detected in the bone skin and other tissues less than 1%.

Moreover, the localized production of bioactive molecules may also inhibit scar formation (fibrosis) and apoptosis, stimulate angiogenesis, and promote mitosis and differentiation of the host stem cells if transplanted into a diseased recipient [73]. Taken together, the data from these studies provide a paradigm to explain the significant levels of therapeutic efficacy provided by MSCs in disease models though a limited number of cells engrafted.

4.3 Modulation of Immune Responses

MSCs retain the ability to suppress immune reactions and inhibit T-cell alloreactivity [74–78]. The suppressive activity of MSCs is effective against T cells stimulated with mitogens or alloantigens and is independent of HLA matching [74–78]. The suppression appears to be effective on both naïve and memory T cells and the suppressed T cells do not undergo apoptosis or anergy [74, 75, 77]. While the mechanism of suppression is unknown, it is more than likely a soluble factor(s) because suppression can occur in studies using transwell culture plates (semipermeable membrane culture systems) to separate the cells [79, 80]. Importantly, these data indicate that the MSCs may not elicit a cytotoxic T-cell response in vivo. The immunosuppressive properties of MSCs have prompted the proposal of clinical trials using MSCs to suppress or inhibit graft-versus-host disease (GVHD) in animal models and in transplant patients [78, 81, 82]. The effectiveness of MSCs for their immunomodulation or prevention of GVHD is presently being investigated in numerous human clinical trials.

Moreover, several recent publications support further evaluation of allogeneic MSC transplantation. Independent studies from several laboratories have determined that passaged human MSCs, as opposed to freshly isolated cells, reduce their expression of surface histocompatibility antigens and no longer stimulate a mixed lymphocyte reaction when co-cultured with allogeneic peripheral blood monocytes [40, 51, 83]. The ability to transplant allogeneic MSCs will have a significant impact on their clinical utility by reducing their cost, improving their manufacture, and allowing their application at the point of care.

4.4 Anti-inflammatory Effects

The suppression the inflammatory response by MSCs is a recently described property that has not yet been thoroughly investigated. In a recent study by Ortiz et al., MSCs blocked both the inflammation and fibrosis in a murine model pulmonary fibrosis induced by administration of bleomycin [84]. The mechanism for the inhibition appears to be mediated the MSC-mediated production of interleukin-1 receptor antagonist, which inhibited proliferation of a T-cell line dependent on interleukin-1α and the production of tumor necrosis factor-α (TNF-α) in vitro. A second demonstration of the anti-inflammatory property of MSCs was observed in models of repair for spinal cord injury in rats. Multiple laboratories observed that the injection of MSCs into the lesion site resulted in improved motor function, but only very limited engraftment or differentiation was observed [85]. Recently, Prockop and colleagues observed a marked decrease in the frequency of astrocytes and macrophages at the lesion site 5 weeks after MSC infusion, with only a few MSCs being detected (D. Prockop, personal communication). The MSCs apparently inhibited the chronic inflammatory response observed in response to the injury. It is evident that detailed investigations into the mechanisms associated with the anti-inflammatory responses are required to more fully describe this mode of action.

5 Therapeutic Efficacy in the Heart and CNS

5.1 Central Nervous System

MSCs have also demonstrated the capacity to differentiate neural and glial cells in vitro and in vivo [18, 86–91]. The MSCs express the neuronal-associated markers nestin, NeuN, and intermediate filament. They also express the oligodendrocyte marker, glial fibrillary acidic protein (GFAP). Further studies with murine MSCs have detected the neuronal-associated glutamate receptor subunits NR1 and NR2, MAP2, S-100, and β-III tubulin [27, 92–97]. To date, no studies have clearly demonstrated that MSCs exhibit the electrophysiological profile of mature neuronal cell.

In vivo, MSCs display a beneficial effect in multiple central nervous system injury models. The therapeutic efficacy of MSCs has been assessed in many different models of CNS injury and disease, including cerebral ischemia and traumatic injury of the brain or spinal cord. Hermann et al. showed that human MSCs effectively differentiated into a clonogenic neural cells that grew in clumps of cells resembling neurospheres [86]. The resulting cells were demonstrated to produce dopamine that was released in potassium-dependent manner. The delivery of MSCs by intraventricular, intrastriatal, and intravenous routes of delivery resulted in targeting of the cells to the injured tissue and positive effects on the functional recovery in mice and rats in models of traumatic brain injury [98, 99, 2006; 100, 2006; 101, 2006]. When human MSCs were administered intravenously in rats with brain injury, the cells migrated to the site of injury with a few cells expressing neuronal and astrocytic markers [102–104]. Intraventricular, intrastriatal, and intravenous delivery of MSCs exerted a similar positive effect on the recovery of mice and rats in models of middle cerebral artery occlusion (stroke) [105–112]. Multiple tracking techniques have demonstrated that the MSCs migrate to the ischemic injury site within the brain [112, 113].

A large number of studies investigating the therapeutic efficacy provided by MSCs have been performed in rat and mouse models of traumatic. In another study, it was determined that MSCs migrated to the injury site and displayed biochemical markers of both neuronal and oligodendroglial cells; however, it was suggested that the MSCs provided architecture for axon guidance that may mediate repair

[63, 114]. Determination of the phenotype of differentiated cells is contingent upon morphological, immunophenotypic, and functional criteria. MSCs differentiated into neural or glial cells can be identified by the induction of expression lineage-specific markers, such as NeuN or glial fibrillary acid protein (GFAP), respectively. However, it is important to note that even though various staining protocols or immunodetection procedures help to identify differentiated cells, they do not necessarily provide definitive evidence of their functionality.

With the release of neurotrophic factors, MSCs may also mediate the therapeutic efficacy observed in CNS disease models. The neurotrophic factors secreted by MSCs can mediate neuronal cell survival, induce proliferation of endogenous cells, and promote the regeneration of nerve fibers [72, 101, 115–117]. Human MSCs increased production of BDNF, NGF, VEGF, and HGF when cultured in the presence of supernatant from ischemic brain extracts [102]. The production of the neurotrophic factors by MSCs, including BDNF, NGF, and GDNF, was demonstrated by Arnhold and colleagues [118]. Moreover, Crigler and colleagues demonstrated that MSCs encode several neuro-regulatory factors, as well as brain-derived neurotrophic factor BDNF and β-nerve growth factor (β-NGF). Interrogation of the human MSC transcriptome identified expressed mRNAs encoding various neurite-inducing factors, axon guidance, and neural cell adhesion molecules. Collectively, these studies reveal the existence of MSC subpopulations that co-express neurotrophins and other potent neuro-regulatory molecules. Thus, paracrine factors released by MSCs or stimulated by their presence may account for the beneficial effects of MSC transplants on central nervous system injury models.

The production of neurotrophic factors and neuro-regulatory molecules may directly contribute to MSC-induced effects on elevated levels of neurogenesis, neuronal cell survival, and nerve regeneration. The transplantation of MSCs into a rat model of traumatic brain injury was associated with increased levels of NGF and BDNF [119]. Increased levels of BDNF and NGF were also observed following MSC administration in another traumatic brain injury model study [120]. The injection of MSCs resulted in the induction and migration of new neural cells from the ventricular zone and choroid plexus into the surrounding injured brain tissue [98]. The implantation of human MSCs into the dentate gyrus of the hippocampus of immunodeficient mice resulted in markedly increased proliferation of endogenous neural stem cells [72]. Labeling of the mice at 7 days post-transplantation with BrdUrd demonstrated that endogenous cells migrated throughout the dorsal hippocampus and expressed markers for astrocytes and for neural or oligodendrocyte progenitors. At 30 days after implantation, the newly generated cells expressed markers for more mature neurons and astrocytes. Also, subpopulations of BrdUrd-labeled cells exhibited elaborate processes immunoreactive for ciliary neurotrophic factor, neurotrophin-4/5, nerve growth factor, or VEGF.

Taken together these data raise the possibility of therapeutic applications for neural and glial tissue repair using MSCs, even though they are derived from mesenchymal germ layer instead of the ectodermal germ layer.

5.2 Cardiac Injury and Disease

Upon injury, native cardiomyocytes only have a limited potential to regenerate mature functional heart tissue [121]. Important experimental findings in recent years suggest considerable therapeutic potential for cellular replacement in the context of acute myocardial injury (e.g., infarction) and chronic, progressive cardiac disease (e.g., left ventricular remodeling and heart failure). Orlic and colleagues were among the first to show that immunoselected subpopulations of bone marrow hematopoietic cells (Lin^- $c\text{-}kit^+$) effectively differentiate into cardiomyocytes, as well as endothelial and smooth muscle cells, using a mouse model of myocardial infarction [122–124]. Since that time, may different types of cells have been investigated for cardiac repair, including hematopoietic stem cells, peripheral blood stem cells, cardiac stem cells, endothelial progenitor cells, skeletal myoblasts, and MSCs [125].

Although limited in volume, the existing literature does suggest that MSCs are able to engraft and survive within an infarcted myocardial milieu, acquire phenotypic markers consistent with cardiomyocyte and vascular-related lineages, and positively impact structural and functional endpoints. The differentiation of MSCs into cardiomyocytes has been demonstrated by several groups [57, 71, 126–130]. The strategies for inducing cardiomyocyte differentiation are diverse and involve culture of MSCs in the presence of 5-azacytidine and/or a cocktail of growth factors, or in co-culture with primary cardiomyocytes [127, 131–134]. The use of these protocols efficiently induces the expression of cardiomyocyte-specific markers or even spontaneous contractions. In studies using animal models of myocardial infarction (MI), MSCs have been reported to provide a therapeutic benefit. Kawada and colleagues transplanted $EGFP^+$-MSCs into the bone marrow of lethally irradiated mice, induced the MI and treated animals with granulocyte colony-stimulating factor (G-CSF), and monitored for the presence of EGFP cells in the heart. $EGFP^+/actinin^+$ cells were detected in the heart, indicating cells mobilized and possibly differentiated into cardiomyocytes [69]. Iso and colleagues intravenously administered human MSCs into immune-deficient mice with acute MI. While both fibrosis and cardiac function were markedly improved in these animals, there was no engraftment of MSCs detected 3 weeks after infusion [135]. It is hypothesized the MSCs provided a benefit through paracrine effects. In another study, Tang and colleagues demonstrated that transplanted MSCs could up-regulate neovascularization and several growth factors in the ischemic heart, such as vascular endothelial growth factor (VEGF) [136].

Recently, a few reports presenting the outcomes of human clinical trials using bone marrow-derived mesenchymal stem cells for heart disease have been published [137–140]. In the first study by Chen and colleagues, intracoronary infusion of autologous MSCs in patients within 12 h of suffering a myocardial infarction (MI) resulted in a significant reduction in infarct size, increased left ventricular ejection fraction (LVEF) [137]. In the clinical trial performed by Katritsis and colleagues, patients that suffered an MI underwent transcoronary infusion of a combination of MSCs and endothelial progenitor cells (EPCs) into the infarcted tissue. The results from this trial indicate improvements in the wall movement index, and echocardiography indicated improved myocardial contractility in nonviable tissue [138]. Patients who suffered an MI underwent intracoronary infusion of MSCs combined with EPCs. The results of this clinical trial indicate that the patients receiving the stem cells demonstrated myocardial repair, as indicated by left ventricular wall motion improvements and ablation of tachycardia [139]. Although limited in number, the results of human clinical trials using MSCs for the treatment of cardiac disease are encouraging. Despite these positive outcomes, however, a great deal remains to be learned about cell-based therapies for myocardial damage. In addition to these studies, numerous clinical trials with other stem cells are ongoing [125].

Acknowledgments Financial Support. The research was supported by the National Center for Research Resources, National Institutes of Health, grant number RR00164, and a grant from the State of Louisiana Millennium Health Excellence Fund and the Louisiana Gene Therapy Research Consortium.

References

1. Jackson KA, Majka SM, Wulf GG, Goodell MA (2002) Stem cells: a minireview. J Cell Biochem Suppl 38:1–6
2. Kondo M, Wagers AJ, Manz MG, Prohaska SS, Scherer DC, Beilhack GF, Shizuru JA, Weissman IL (2003) Biology of hematopoietic stem cells and progenitors: implications for clinical application. Annu Rev Immunol 21:759–806
3. Krause DS (2002) Plasticity of marrow-derived stem cells. Gene Ther 9, 754–758
4. Verfaillie CM (2002) Adult stem cells: assessing the case for pluripotency. Trends Cell Biol 12:502–508
5. Wells WA (2002) Is transdifferentiation in trouble? J Cell Biol 157:15–18
6. Jones JM, Thomson JA (2000) Human embryonic stem cell technology. Semin Reprod Med 18:219–223
7. Herzog EL, Chai L, Krause DS (2003) Plasticity of marrow-derived stem cells. Blood 102:3483–3493
8. Prockop DJ (1997) Marrow stromal cells as stem cells for nonhematopoietic tissues. Science 276:71–74
9. Zuk PA, Zhu M, Ashjian P, De Ugarte DA, Huang JI, Mizuno H, Alfonso ZC, Fraser JK, Benhaim P, Hedrick MH (2002) Human adipose tissue is a source of multipotent stem cells. Mol Biol Cell 13:4279–4295
10. Gimble JM, Katz AJ, Bunnell BA (2007) Adipose-derived stem cells for regenerative medicine. Circ Res 100:1249–1260
11. Vieyra DS, Jackson KA, Goodell MA (2005) Plasticity and tissue regenerative potential of bone marrow-derived cells. Stem Cell Rev 1:65–69
12. Wei G, Schubiger G, Harder F, Muller AM (2000) Stem cell plasticity in mammals and transdetermination in Drosophila: common themes? Stem Cells 18:409–414
13. Woodbury D, Reynolds K, Black IB (2002) Adult bone marrow stromal stem cells express germline, ectodermal, endodermal, and mesodermal genes prior to neurogenesis. J Neurosci Res 69:908–917
14. Beltrami AP, Urbanek K, Kajstura J, Yan SM, Finato N, Bussani R, Nadal-Ginard B, Silvestri F, Leri A, Beltrami CA, Anversa P (2001) Evidence that human cardiac myocytes divide after myocardial infarction. N Engl J Med 344:1750–1757
15. Gussoni E, Soneoka Y, Strickland CD, Buzney EA, Khan MK, Flint AF, Kunkel LM, Mulligan RC (1999) Dystrophin expression in the mdx mouse restored by stem cell transplantation. Nature 401:390–394

16. Kotton DN, Fine A (2003) Derivation of lung epithelium from bone marrow cells. Cytotherapy 5:169–173
17. Petersen BE, Bowen WC, Patrene KD, Mars WM, Sullivan AK, Murase N, Boggs SS, Greenberger JS, Goff JP (1999) Bone marrow as a potential source of hepatic oval cells. Science 284:1168–1170
18. Pittenger MF, Mackay AM, Beck SC, Jaiswal RK, Douglas R, Mosca JD, Moorman MA, Simonetti DW, Craig S, Marshak DR (1999) Multilineage potential of adult human mesenchymal stem cells. Science 284:143–147
19. Terskikh AV, Easterday MC, Li L, Hood L, Kornblum HI, Geschwind DH, Weissman IL (2001) From hematopoiesis to neuropoiesis: evidence of overlapping genetic programs. Proc Natl Acad Sci U S A 98:7934–7939
20. Friedenstein AJ, Deriglasova UF, Kulagina NN, Panasuk AF, Rudakowa SF, Luria EA, Ruadkow IA (1974) Precursors for fibroblasts in different populations of hematopoietic cells as detected by the in vitro colony assay method. Exp Hematol 2:83–92
21. Devine SM, Cobbs C, Jennings M, Bartholomew A, Hoffman R (2003) Mesenchymal stem cells distribute to a wide range of tissues following systemic infusion into nonhuman primates. Blood 101:2999–3001
22. Kadiyala S, Young RG, Thiede MA, Bruder SP (1997) Culture expanded canine mesenchymal stem cells possess osteochondrogenic potential in vivo and in vitro. Cell Transplant 6:125–134
23. Perez-Losada J, Sanchez-Martin M, Rodriguez-Garcia A, Sanchez M L, Orfao A, Flores T, Sanchez-Garcia I (2002) Zinc-finger transcription factor Slug contributes to the function of the stem cell factor c-kit signaling pathway. Blood 100:1274–1286
24. Colter DC, Class R, DiGirolamo CM, Prockop DJ (2000) Rapid expansion of recycling stem cells in cultures of plastic-adherent cells from human bone marrow. Proc Natl Acad Sci U S A 97:3213–3218
25. Bantel-Schaal U, Hub B, Kartenbeck J (2002) Endocytosis of adeno-associated virus type 5 leads to accumulation of virus particles in the Golgi compartment. J Virol 76:2340–2349
26. Hung SC, Chen NJ, Hsieh SL, Li H, Ma HL, LoW H (2002) Isolation and characterization of size-sieved stem cells from human bone marrow. Stem Cells 20:249–258
27. Hung SC, Cheng H, Pan CY, Tsai MJ, Kao LS, Ma HL (2002) In vitro differentiation of size-sieved stem cells into electrically active neural cells. Stem Cells 20:522–529
28. Bjorklund A, Kirik D, Rosenblad C, Georgievska B, Lundberg C, Mandel RJ (2000) Towards a neuroprotective gene therapy for Parkinson's disease: use of adenovirus, AAV and lentivirus vectors for gene transfer of GDNF to the nigrostriatal system in the rat Parkinson model. Brain Res 886:82–98
29. Zhang XY, La Russa VF, Bao L, Kolls J, Schwarzenberger P, Reiser J (2002) Lentiviral vectors for sustained transgene expression in human bone marrow-derived stromal cells. Mol Ther 5:555–565
30. Zhang XY, La Russa VF, Reiser J (2004) Transduction of bone-marrow-derived mesenchymal stem cells by using lentivirus vectors pseudotyped with modified RD114 envelope glycoproteins. J Virol 78:1219–1229
31. Peister A, Mellad JA, Wang M, Tucker HA, Prockop DJ (2004) Stable transfection of MSCs by electroporation. Gene Ther 11:224–228
32. Pereira RF, O'Hara MD, Laptev AV, Halford KW, Pollard MD, Class R, Simon D, Livezey K, Prockop DJ (1998) Marrow stromal cells as a source of progenitor cells for nonhematopoietic tissues in transgenic mice with a phenotype of osteogenesis imperfecta. Proc Natl Acad Sci USA 95:1142–1147
33. Ferrari G, Cusella-De Angelis G, Coletta M, Paolucci E, Stornaiuolo A, Cossu G, Mavilio F (1998) Muscle regeneration by bone marrow-derived myogenic progenitors. Science 279:1528–1530
34. Liechty KW, MacKenzie TC, Shaaban AF, Radu A, Moseley AM, Deans R, Marshak DR, Flake AW (2000) Human mesenchymal stem cells engraft and demonstrate site-specific differentiation after in utero transplantation in sheep. Nat Med 6:1282–1286

35. Kotton DN, Ma BY, Cardoso WV, Sanderson EA, Summer RS, Williams MC, Fine A (2001) Bone marrow-derived cells as progenitors of lung alveolar epithelium. Development 128:5181–5188
36. Mackenzie TC, Flake AW (2001) Multilineage differentiation of human MSC after in utero transplantation. Cytotherapy 3:403–405
37. Dezawa M, Takahashi I, Esaki M, Takano M, Sawada H (2001) Sciatic nerve regeneration in rats induced by transplantation of in vitro differentiated bone-marrow stromal cells. Eur J Neurosci 14:1771–1776
38. Chen J, Sanberg PR, Li Y, Wang L, Lu M, Willing AE, Sanchez-Ramos J, Chopp M (2001) Intravenous administration of human umbilical cord blood reduces behavioral deficits after stroke in rats. Stroke 32:2682–2688
39. Ankeny DP, McTigue DM, Jakeman LB (2004) Bone marrow transplants provide tissue protection and directional guidance for axons after contusive spinal cord injury in rats. Exp Neurol 190:17–31
40. Case J, Horvath TL, Howell JC, Yoder MC, March KL, Srour EF (2005) Clonal multilineage differentiation of murine common pluripotent stem cells isolated from skeletal muscle and adipose stromal cells. Ann NY Acad Sci 1044:183–200
41. Gronthos S, Franklin DM, Leddy HA, Robey PG, Storms RW, Gimble JM (2001) Surface protein characterization of human adipose tissue-derived stromal cells. J Cell Physiol 189:54–63
42. Katz AJ, Tholpady A, Tholpady SS, Shang H, Ogle RC (2005) Cell surface and transcriptional characterization of human adipose-derived adherent stromal (hADAS) cells. Stem Cells 23:412–423
43. Yoshimura H, Muneta T, Nimura A, Yokoyama A, Koga H, Sekiya I (2007) Comparison of rat mesenchymal stem cells derived from bone marrow, synovium, periosteum, adipose tissue, and muscle. Cell Tissue Res 327:449–462
44. Safford KM, Rice HE (2005) Stem cell therapy for neurologic disorders: therapeutic potential of adipose-derived stem cells. Curr Drug Targets 6:57–62
45. Israels ED, Israels LG (2001) The cell cycle. Stem Cells 19:88–91
46. Hayflick L, Moorhead PS (1961) The serial cultivation of human diploid cell strains. Exp Cell Res 25:585–621
47. Izadpanah R, Joswig T, Tsien F, Dufour J, Kirijan JC, Bunnell BA (2005) Characterization of multipotent mesenchymal stem cells from the bone marrow of rhesus macaques. Stem Cells Dev 14:440–451
48. Jiang Y, Jahagirdar BN, Reinhardt RL, Schwartz RE, Keene CD, Ortiz-Gonzalez XR, Reyes M, Lenvik T, Lund T, Blackstad M, Du J, Aldrich S, Lisberg A, Low WC, Largaespada DA, Verfaillie CM (2002) Pluripotency of mesenchymal stem cells derived from adult marrow. Nature 418:41–49
49. Izadpanah R, Trygg C, Patel B, Kriedt C, Dufour J, Gimble JM, Bunnell BA (2006) Biologic properties of mesenchymal stem cells derived from bone marrow and adipose tissue. J Cell Biochem 99:1285–1297
50. Bonab MM, Alimoghaddam K, Talebian F, Ghaffari SH, Ghavamzadeh A, Nikbin B (2006) Aging of mesenchymal stem cell in vitro. BMC Cell Biol 7:14
51. Rubio D, Garcia-Castro J, Martin MC, de la Fuente R, Cigudosa JC, Lloyd AC, Bernad A (2005) Spontaneous human adult stem cell transformation. Cancer Res 65:3035–3039
52. Miura M, Miura Y, Padilla-Nash HM, Molinolo AA, Fu B, Patel V, Seo BM, Sonoyama W, Zheng JJ, Baker CC, Chen W, Ried T, Shi S (2006) Accumulated chromosomal instability in murine bone marrow mesenchymal stem cells leads to malignant transformation. Stem Cells 24:1095–1103
53. Tolar J, Nauta AJ, Osborn MJ, Panoskaltsis Mortari A, McElmurry RT, Bell S, Xia L, Zhou N, Riddle M, Schroeder TM, Westendorf JJ, McIvor RS, Hogendoorn PC, Szuhai K, Oseth L, Hirsch B, Yant SR, Kay MA, Peister A, Prockop DJ, Fibbe WE, Blazar BR (2007) Sarcoma derived from cultured mesenchymal stem cells. Stem Cells 25:371–379

54. Alvarez-Dolado M, Pardal R, Garcia-Verdugo JM, Fike JR, Lee HO, Pfeffer K, Lois C, Morrison SJ, Alvarez-Buylla A (2003) Fusion of bone-marrow-derived cells with Purkinje neurons, cardiomyocytes and hepatocytes. Nature 425:968–973
55. Wislet-Gendebien S, Hans G, Leprince P, Rigo JM, Moonen G, Rogister B (2005) Plasticity of cultured mesenchymal stem cells: switch from nestin-positive to excitable neuron-like phenotype. Stem Cells 23:392–402
56. Corti S, Locatelli F, Strazzer S, Guglieri M, Comi GP (2003) Neuronal generation from somatic stem cells: current knowledge and perspectives on the treatment of acquired and degenerative central nervous system disorders. Curr Gene Ther 3:247–272
57. Noiseux N, Gnecchi M, Lopez-Ilasaca M, Zhang L, Solomon SD, Deb A, Dzau VJ, Pratt RE (2006) Mesenchymal stem cells overexpressing akt dramatically repair infarcted myocardium and improve cardiac function despite infrequent cellular fusion or differentiation. Mol Ther 14:840–850
58. Kajstura J, Rota M, Whang B, Cascapera S, Hosoda T, Bearzi C, Nurzynska D, Kasahara H, Zias E, Bonafe M, Nadal-Ginard B, Torella D, Nascimbene A, Quaini F, Urbanek K, Leri A, Anversa P (2005) Bone marrow cells differentiate in cardiac cell lineages after infarction independently of cell fusion. Circ Res 96:127–137
59. Harris MT, Butler DL, Boivin GP, Florer JB, Schantz EJ, Wenstrup RJ (2004) Mesenchymal stem cells used for rabbit tendon repair can form ectopic bone and express alkaline phosphatase activity in constructs. J Orthop Res 22:998–1003
60. Pochampally RR, Neville BT, Schwarz EJ, Li MM, Prockop DJ (2004) Rat adult stem cells (marrow stromal cells) engraft and differentiate in chick embryos without evidence of cell fusion. PNAS 101:9282–9285
61. Jiang Y, Vaessen B, Lenvik T, Blackstad M, Reyes M, Verfaillie CM (2002) Multipotent progenitor cells can be isolated from postnatal murine bone marrow, muscle, and brain. Exp Hematol 30:896–904
62. Lakshmipathy U, Verfaillie C (2005) Stem cell plasticity. Blood Rev 19:29–38
63. Hofstetter CP, Schwarz EJ, Hess D, Widenfalk J, El Manira A, Prockop DJ, Olson L (2002) Marrow stromal cells form guiding strands in the injured spinal cord and promote recovery. Proc Natl Acad Sci U S A 99:2199–2204
64. Deng J, Petersen BE, Steindler DA, Jorgensen ML, Laywell ED (2006) Mesenchymal stem cells spontaneously express neural proteins in culture and are neurogenic after transplantation. Stem Cells 24:1054–1064
65. Shi YY, Nacamuli RP, Salim A, Longaker, MT (2005) Plast Reconstr Surg 116:1686–1696
66. Min JY, Sullivan MF, Yang Y, Zhang JP, Converso KL, Morgan JP, Xiao YF (2002) Significant improvement of heart function by cotransplantation of human mesenchymal stem cells and fetal cardiomyocytes in postinfarcted pigs. Ann Thorac Surg 74:1568–1575
67. Shake JG, Gruber PJ, Baumgartner WA, Senechal G, Meyers J, Redmond JM, Pittenger MF, Martin BJ (2002) Mesenchymal stem cell implantation in a swine myocardial infarct model: engraftment and functional effects. Ann Thorac Surg 73:1919–1925; discussion 26
68. Wang Z, Cohen K, Shao Y, Mole P, Dombkowski D, Scadden DT (2004) Ephrin receptor, EphB4, regulates ES cell differentiation of primitive mammalian hemangioblasts, blood, cardiomyocytes, and blood vessels. Blood 103:100–109
69. Kawada H, Fujita J, Kinjo K, Matsuzaki Y, Tsuma M, Miyatake H, Muguruma Y, Tsuboi K, Itabashi Y, Ikeda Y, Ogawa S, Okano H, Hotta T, Ando K, Fukuda K (2004) Non-hematopoietic mesenchymal stem cells can be mobilized and differentiate into cardiomyocytes after myocardial infarction. Blood 104:3581–3587
70. Nagaya N, Fujii T, Iwase T, Ohgushi H, Itoh T, Uematsu M, Yamagishi M, Mori H, Kangawa K, Kitamura S (2004) Intravenous administration of mesenchymal stem cells improves cardiac function in rats with acute myocardial infarction through angiogenesis and myogenesis. Am J Physiol Heart Circ Physiol 287:H2670–2676
71. Gnecchi M, He H, Liang OD, Melo LG, Morello F, Mu H, Noiseux N, Zhang L, Pratt RE, Ingwall JS, Dzau VJ (2005) Paracrine action accounts for marked protection of ischemic heart by Akt-modified mesenchymal stem cells. Nat Med 11:367–368

72. Munoz JR, Stoutenger BR, Robinson AP, Spees JL, Prockop DJ (2005) Human stem/progenitor cells from bone marrow promote neurogenesis of endogenous neural stem cells in the hippocampus of mice. Proc Natl Acad Sci U S A 102:18171–18176
73. Caplan AI, Dennis JE (2006) Mesenchymal stem cells as trophic mediators. J Cell Biochem 98:1076–1084
74. Bartholomew A, Sturgeon C, Siatskas M, Ferrer K, McIntosh K, Patil S, Hardy W, Devine S, Ucker D, Deans R, Moseley A, Hoffman R (2002) Mesenchymal stem cells suppress lymphocyte proliferation in vitro and prolong skin graft survival in vivo. Exp Hematol 30:42–48
75. Krampera M, Glennie S, Dyson J, Scott D, Laylor R, Simpson E, Dazzi F (2003) Induction of neural-like differentiation in human mesenchymal stem cells derived from bone marrow, fat, spleen and thymus. Blood 101:3722–3729
76. Le Blanc K, Tammik C, Rosendahl K, Zetterberg E, Ringden O (2003) HLA expression and immunologic propertiesof differentiated and undifferentiated mesenchymal stem cells. Exp Hematol 31:890–896
77. Maitra B, Szekely E, Gjini K, Laughlin MJ, Dennis J, Haynesworth SE, Koc ON (2004) Human mesenchymal stem cells support unrelated donor hematopoietic stem cells and suppress T-cell activation. Bone Marrow Transplant 33:597–604
78. Yanez R, Lamana ML, Garcia-Castro J, Colmenero I, Ramirez M, Bueren JA (2006) Adipose tissue-derived mesenchymal stem cells have in vivo immunosuppressive properties applicable for the control of the graft-versus-host disease. Stem Cells 24: 2582–2591
79. Le Blanc K, Ringden O (2006) Mesenchymal stem cells: properties and role in clinical bone marrow transplantation. Curr Opin Immunol 18:586–591
80. Rasmusson I, Ringden O, Sundberg B, Le Blanc K (2005) Mesenchymal stem cells inhibit lymphocyte proliferation by mitogens and alloantigens by different mechanisms. Exp Cell Res 305:33–41
81. Le Blanc K, Rasmusson I, Sundberg B, Gotherstrom C, Hassan M, Uzunel M, Ringden O (2004) Treatment of severe acute graft-versus-host disease with third party haploidentical mesenchymal stem cells. Lancet 363:1439–1441
82. Ringden O, Uzunel M, Rasmusson I, Remberger M, Sundberg B, Lonnies H, Marschall HU, Dlugosz A, Szakos A, Hassan Z, Omazic B, Aschan J, Barkholt L, Le Blanc K (2006) Mesenchymal stem cells for treatment of therapy-resistant graft-versus-host disease. Transplantation 81:1390–1397
83. Estes BT, Wu AW, Storms RW, Guilak F (2006) Extended passaging, but not aldehyde dehydrogenase activity, increases the chondrogenic potential of human adipose-derived adult stem cells. J Cell Physiol 209:987–995
84. Ortiz LA, Dutreil M, Fattman C, Pandey AC, Torres G, Go K, Phinney DG (2007) Interleukin 1 receptor antagonist mediates the antiinflammatory and antifibrotic effect of mesenchymal stem cells during lung injury. Proc Natl Acad Sci U S A 104:11002–11007
85. Peister A, Mellad JA, Larson BL, Hall BM, Gibson LF, Prockop DJ (2003) Adult stem cells from bone marrow (MSCs) isolated from different strains of inbred mice vary in surface epitopes, rates of proliferation, and differentiation potential. Blood, 103: 1662–1668
86. Hermann A, Gastl R, Liebau S, Popa MO, Fiedler J, Boehm BO, Maisel M, Lerche H, Schwarz J, Brenner R, Storch A (2004) Efficient generation of neural stem cell-like cells from adult human bone marrow stromal cells. J Cell Sci 117:4411–4422
87. Abeyta MJ, Clark AT, Rodriguez RT, Bodnar MS, Pera RA, Firpo MT (2004) Unique gene expression signatures of independently-derived human embryonic stem cell lines. Hum Mol Genet 13:601–608
88. Jori FP, Galderisi U, Piegari E, Cipollaro M, Cascino A, Peluso G, Cotrufo R, Giordano A, Melone MA (2003) EGF-responsive rat neural stem cells: molecular follow-up of neuron and astrocyte differentiation in vitro. J Cell Physiol 195:220–233

89. Jori FP, Napolitano MA, Melone MA, Cipollaro M, Cascino A, Altucci L, Peluso G, Giordano A, Galderisi U (2005) Molecular pathways involved in neural in vitro differentiation of marrow stromal stem cells. J Cell Biochem 94:645–655
90. Bianco P, Gehron Robey P (2000) Marrow stromal stem cells. J Clin Invest 105:1663–1668
91. Beyer Nardi N, da Silva Meirelles L (2006) Mesenchymal stem cells: isolation, in vitro expansion and characterization. Handb Exp Pharmacol, 249–282
92. Hermann A, Liebau S, Gastl R, Fickert S, Habisch HJ, Fiedler J, Schwarz J, Brenner R, Storch A (2006) Comparative analysis of neuroectodermal differentiation capacity of human bone marrow stromal cells using various conversion protocols. J Neurosci Res 83:1502–1514
93. Kang SK, Putnam LA, Ylostalo J, Popescu IR, Dufour J, Belousov A, Bunnell BA (2004) Neurogenesis of Rhesus adipose stromal cells. J Cell Sci, jcs.01264
94. Kim S, Honmou O, Kato K, Nonaka T, Houkin K, Hamada H, Kocsis JD (2006) Neural differentiation potential of peripheral blood- and bone-marrow-derived precursor cells. Brain Res 1123:27–33
95. Krampera M, Marconi S, Pasini A, Galie M, Rigotti G, Mosna F, Tinelli M, Lovato L, Anghileri E, Andreini A, Pizzolo G, Sbarbati A, Bonetti B (2007) Bone marrow mesenchymal stem cells inhibit the response of naive and memory antigen-specific T cells to their cognate peptide. Bone 40:382–390
96. Mareschi K, Novara M, Rustichelli D, Ferrero I, Guido D, Carbone E, Medico E, Madon E, Vercelli A, Fagioli F (2006) Neural differentiation of human mesenchymal stem cells: Evidence for expression of neural markers and eag K+ channel types. Exp Hematol 34:1563–1572
97. Suon S, Jin H, Donaldson AE, Caterson EJ, Tuan RS, Deschennes G, Marshall C, Iacovitti L (2004) Transient differentiation of adult human bone marrow cells into neuron-like cells in culture: development of morphological and biochemical traits is mediated by different molecular mechanisms. Stem Cells Dev 13:625–635
98. Chopp M, Li Y (2002) Treatment of neural injury with marrow stromal cells. Lancet Neurol 1:92–100
99. Mahmood A, Lu D, Chopp M (2004) Intravenous administration of marrow stromal cells (MSCs) increases the expression of growth factors in rat brain after traumatic brain injury. J Neurotrauma 21:33–39
100. Mahmood A, Lu D, Qu C, Goussev A, Chopp M (2006) Long-term recovery after bone marrow stromal cell treatment of traumatic brain injury in rats. J Neurosurg 104:272–277
101. Mahmood A, Lu D, Wang L, Chopp M (2002) Intracerebral transplantation of marrow stromal cells cultured with neurotrophic factors promotes functional recovery in adult rats subjected to traumatic brain injury. J Neurotrauma 19:1609–1617
102. Chen X, Katakowski M, Li Y, Lu D, Wang L, Zhang L, Chen J, Xu Y, Gautam S, Mahmood A, Chopp M (2002) Intravenous bone marrow stromal cell therapy reduces apoptosis and promotes endogenous cell proliferation after stroke in female rat. J Neurosci Res 69:687–691
103. Mahmood A, Lu D, Lu M, Chopp M (2003) Treatment of traumatic brain injury in adult rats with intravenous administration of human bone marrow stromal cells. Neurosurgery 53:697–702; discussion 02–03
104. Mahmood A, Lu D, Qu C, Goussev A, Chopp M (2005) Human marrow stromal cell treatment provides long-lasting benefit after traumatic brain injury in rats. Neurosurgery 57:1026–1031; discussion 26–31
105. Borlongan CV, Lind JG, Dillon-Carter O, Yu G, Hadman M, Cheng C, Carroll J, Hess DC (2004) Bone marrow grafts restore cerebral blood flow and blood brain barrier in stroke rats. Brain Res 1010:108–116
106. Chen J, Li Y, Katakowski M, Chen X, Wang L, Lu D, Lu M, Gautam SC, Chopp M (2003) Human bone marrow stromal cell cultures conditioned by traumatic brain tissue extracts: growth factor production. J Neurosci Res 73:778–786
107. Ikeda N, Nonoguchi N, Zhao MZ, Watanabe T, Kajimoto Y, Furutama D, Kimura F, Dezawa M, Coffin RS, Otsuki Y, Kuroiwa T, Miyatake S (2005) Bone marrow stromal cells

that enhanced fibroblast growth factor-2 secretion by herpes simplex virus vector improve neurological outcome after transient focal cerebral ischemia in rats. Stroke 36:2725–2730
108. Li Y, Chen J, Zhang CL, Wang L, Lu D, Katakowski M, Gao Q, Shen LH, Zhang J, Lu M, Chopp M (2005) Gliosis and brain remodeling after treatment of stroke in rats with marrow stromal cells. Glia 49:407–417
109. Akiyama H, Barger S, Barnum S, Bradt B, Bauer J, Cole GM, Cooper NR, Eikelenboom P, Emmerling M, Fiebich BL, Finch CE, Frautschy S, Griffin WS, Hampel H, Hull M, Landreth G, Lue L, Mrak R, Mackenzie IR, McGeer PL, O'Banion MK, Pachter J, Pasinetti G, Plata-Salaman C, Rogers J, Rydel R, Shen Y, Streit W, Strohmeyer R, Tooyoma I, Van Muiswinkel FL, Veerhuis R, Walker D, Webster S, Wegrzyniak B, Wenk G, Wyss-Coray T (2000) Inflammation and alzheimer's disease. Neurobiol Aging 21:383–421
110. Shen LH, Li Y, Chen J, Zacharek A, Gao Q, Kapke A, Lu M, Raginski K, Vanguri P, Smith A, Chopp M (2007) Therapeutic benefit of bone marrow stromal cells administered 1 month after stroke. J Cereb Blood Flow Metab 27:6–13
111. Shen L H, Li Y, Chen J, Zhang J, Vanguri P, Borneman J, Chopp M (2006) Intracarotid transplantation of bone marrow stromal cells increases axon-myelin remodeling after stroke. Neuroscience 137, 393–399
112. Yano S, Kuroda S, Shichinohe H, Hida K, Iwasaki Y (2005) Do bone marrow stromal cells proliferate after transplantation into mice cerebral infarct?--a double labeling study. Brain Res 1065:60–67
113. Lee JA, Parrett BM, Conejero JA, Laser J, Chen J, Kogon AJ, Nanda D, Grant RT, Breitbart AS (2003) Biological alchemy: engineering bone and fat from fat-derived stem cells. Ann Plast Surg 50 610–617
114. Zurita M, Vaquero J (2004) Functional recovery in chronic paraplegia after bone marrow stromal cells transplantation. Neuroreport 15 1105–1108
115. Kinnaird T, Stabile E, Burnett MS, Shou M, Lee CW, Barr S, Fuchs S, Epstein SE (2004) Local delivery of marrow-derived stromal cells augments collateral perfusion through paracrine mechanisms. Circulation 109:1543–1549
116. Rehman J, Traktuev D, Li J, Merfeld-Clauss S, Temm-Grove CJ, Bovenkerk JE, Pell CL, Johnstone BH, Considine RV, March KL (2004) Secretion of angiogenic and antiapoptotic factors by human adipose stromal cells. Circulation 109:1292–1298
117. Nakagami H, Maeda K, Morishita R, Iguchi S, Nishikawa T, Takami Y, Kikuchi Y, Saito Y, Tamai K, Ogihara T, Kaneda Y (2005) Novel autologous cell therapy in ischemic limb disease through growth factor secretion by cultured adipose tissue-derived stromal cells. Arterioscler Thromb Vasc Biol 25:2542–2547
119. Schneider ML, Moore CF (2006) Neurobehavioral assessments of, nonhuman primate neonates. In: Sackett GP, Elias, RG, K, e. N. R. o. N. P. i. t. s. C. N. Y., and Springer (eds) (2006) Neurobehavioral assessments of nonhuman primate neonates, Springer, New York, pp 215–247
119. Mahmood A, Lu D, Chopp M (2004) Marrow stromal cell transplantation after traumatic brain injury promotes cellular proliferation within the brain. Neurosurgery 55:1185–1193
120. Hu Z, Zhou LF, Zhu J, Mao Y, Wu XH Upregulated gene expression of local brain-derived neurotrophic factor and nerve growth factor after intracisternal administration of marrow stromal cells in rats with traumatic brain injury. (2005) Chin. J. Traumatol. 8:23–26
121. Beltrami AP, Barlucchi L, Torella, D, Baker, M, Limana F, Chimenti S, Kasahara H, Rota M, Musso E, Urbanek K, Leri A, Kajstura J, Nadal-Ginard B, Anversa P (2003) Adult cardiac stem cells are multipotent and support myocardial regeneration. Cell 114, 763–776
122. Orlic D, Kajstura J, Chimenti S, Bodine DM, Leri A, Anversa P Mobilized bone marrow cells repair the infarcted heart, improving function and survival. (2001) Ann N Y Acad Sci 938:221–9; discussion 29–30
123. Orlic D, Kajstura J, Chimenti S, Jakoniuk I, Anderson SM, Li B, Pickel J, McKay R, Nadal-Ginard B, Bodine DM, Leri A, Anversa P(2001) Bone marrow cells regenerate infarcted myocardium. Nature 410:701–705

124. Orlic D, Kajstura J, Chimenti S, Limana F, Jakoniuk I, Quaini F, Nadal-Ginard B, Bodine DM, Leri A, Anversa P (2001) Mobilized bone marrow cells repair the infarcted heart, improving function and survival. Proc Natl Acad Sci U S A 98:10344–10349
125. Prockop DJ, Olson SD (2007) Clinical trials with adult stem/progenitor cells for tissue repair: let's not overlook some essential precautions. Blood 109:3147–3151
126. Fukuda K, Fujita J (2005) Mesenchymal, but not hematopoietic, stem cells can be mobilized and differentiate into cardiomyocytes after myocardial infarction in mice. Kidney Int 68:1940–1943
127. Makino S, Fukuda K, Miyoshi S, Konishi F, Kodama H, Pan J, Sano M, Takahashi T, Hori S, Abe H, Hata J, Umezawa A, Ogawa S (1999) Cardiomyocytes can be generated from marrow stromal cells in vitro. J Clin Invest 103:697–705
128. Strem BM, Zhu M, Alfonso Z, Daniels EJ, Schreiber R, Beygui R, MacLellan WR, Hedrick MH, Fraser JK (2005) Expression of cardiomyocytic markers on adipose tissue-derived cells in a murine model of acute myocardial injury. Cytotherapy 7:282–291
129. Wang T, Xu Z, Jiang W, Ma A (2006) Cell-to-cell contact induces mesenchymal stem cell to differentiate into cardiomyocyte and smooth muscle cell. Int J Cardiol 109:74–81
130. Xu W, Zhang X, Qian H, Zhu W, Sun X, Hu J, Zhou H, Chen Y (2004) Mesenchymal stem cells from adult human bone marrow differentiate into a cardiomyocyte phenotype in vitro. Exp Biol Med (Maywood) 229:623–631
131. Fukuhara S, Tomita S, Yamashiro S, Morisaki T, Yutani C, Kitamura S, Nakatani T (2003) Direct cell-cell interaction of cardiomyocytes is key for bone marrow stromal cells to go into cardiac lineage in vitro. J Thorac Cardiovasc Surg 125:1470–1480
132. Rangappa S, Entwistle JW, Wechsler AS, Kresh JY (2003) Cardiomyocyte-mediated contact programs human mesenchymal stem cells to express cardiogenic phenotype. J Thorac Cardiovasc Surg 126:124–132
133. Wakitani S, Saito T, Caplan AI (1995) Myogenic cells derived from rat bone marrow mesenchymal stem cells exposed to 5-azacytidine. Muscle Nerve 18:1417–1426
134. Wang M, Crisostomo PR, Herring C, Meldrum KK, Meldrum D R (2006) Human progenitor cells from bone marrow or adipose tissue produce VEGF, HGF, and IGF-I in response to TNF by a p38 MAPK-dependent mechanism. Am J Physiol Regul Integr Comp Physiol 291:R880–884
135. Ding DC, Shyu WC, Chiang MF, Lin SZ, Chang YC, Wang H J, Su CY, Li H (2007) Enhancement of neuroplasticity through upregulation of beta1-integrin in human umbilical cord-derived stromal cell implanted stroke model. Neurobiol Dis 27:339–353
136. Tang K, Rossiter HB, Wagner PD, Breen EC (2004) Lung-targeted VEGF inactivation leads to an emphysema phenotype in mice. J Appl Physiol 97:1559–1566; discussion 49
137. Chen SL, Fang WW, Ye F, Liu YH, Qian J, Shan SJ, Zhang J J, Chunhua RZ, Liao L, Lin S, Sun JP (2004) Effect on left ventricular function of intracoronary transplantation of autologous bone marrow mesenchymal stem cell in patients with acute myocardial infarction. Am J Cardiol 94:92–95
138. Katritsis DG, Sotiropoulou PA, Karvouni E, Karabinos I, Korovesis S, Perez SA, Voridis EM, Papamichail M (2005) Transcoronary transplantation of autologous mesenchymal stem cells and endothelial progenitors into infarcted human myocardium. Catheter Cardiovasc Interv 65:321–329
139. Katritsis DG, Sotiropoulou P, Giazitzoglou E, Karvouni E, Papamichail M (2007) Electrophysiological effects of intracoronary transplantation of autologous mesenchymal and endothelial progenitor cells. Europace 9:167–171
140. Zimmet JM, Hare JM (2005) Emerging role for bone marrow derived mesenchymal stem cells in myocardial regenerative therapy. Basic Res Cardiol 100:471–481

Part III
Cell Cycle and Differentiation in the Nervous System

Chapter 8
Neurogenesis in the Central Nervous System: Cell Cycle Progression/Exit and Differentiation of Neuronal Progenitors

Dimitra Thomaidou, Panagiotis K. Politis, and Rebecca Matsas

Abstract This chapter focuses on recent developments shedding light on the basic mechanisms occurring during the early stages of embryonic central nervous system (CNS) development leading to neuroepithelial cell proliferation and subsequent differentiation, as well as on the molecular determinants and signaling pathways regulating cell cycle progression/exit and cell fate determination. The morphogeneses of the embryonic spinal cord and cerebral cortex, which are two of the best characterized systems of CNS development, are used as complementary paradigms to describe the regulation of cell cycle progression/exit and differentiation, the types of cells participating in these mechanisms, as well as their divisional mode and differentiation potential. Particular emphasis is also given in describing the role and neurogenic potential of a revisited precursor cell type, namely the radial glia, as well as the contribution of neural stem cells (NCS) in adult neurogenesis. These two novel aspects of neurogenesis have been characterized during the last decade and have revolutionized the scenario concerning the timing of neuronal production in mammals and the cell types participating in this process. Most important, through progress in these fields it has been discovered that precursor cells residing in discrete regions of the adult brain are able to generate specific sub-classes of neurons throughout life under physiological conditions, but have also the initial intrinsic potential to contribute to neuronal regeneration under pathological conditions if appropriately stimulated. The intimate link between cell cycle control and neurogenesis during CNS development and the characterization of dual function molecules both instructing cell cycle exit and differentiation towards the neuronal lineage will be also discussed. The above link has received increasing scientific attention lately, as it is a key mechanism regulating the spatiotemporal networks that coordinate the size of different CNS regions. Finally, the developmental cell

D. Thomaidou and R. Matsas (✉)
Laboratory of Cellular and Molecular Neurobiology, Hellenic Pasteur Institute, 127. Vas. Sofias Avenue, Athens 11521, Greece
e-mail: thomaidou@mail.pasteur.gr, rmatsa@pasteur.gr

A. Giordano, U. Galderisi (eds.), *Cell Cycle Regulation and Differentiation in Cardiovascular and Neural Systems*, DOI 10.1007/978-1-60327-153-0_8,

death is also discussed as an alternative cell number control mechanism also contributing to the right number of differentiated neuronal and glial cells at the correct CNS areas.

Keywords Neural stem cells · Radial glia · Symmetric/asymmetric divisions · Spinal cord · Dorso-ventral patterning · RA · FGF signaling · Adult neurogenesis, Subventricular zone

The formation of the nervous system is governed by a delicate balance between cell proliferation, subsequent cell cycle withdrawal and differentiation to distinctive neuronal and glial phenotypes [1, 2]. During embryonic development multipotential progenitors generate gradually more restricted precursors that will finally produce neuronal or glial progeny [3–5]. Current observations have highlighted the existence of mechanisms coupling cell cycle exit and differentiation as well as functional cross-talk between intrinsic factors controlling these two mechanisms. A number of key factors regulating cell cycle progression have been implicated in cell fate determination and differentiation of neuronal precursors, while specification- and/or differentiation-inducing molecules are beginning to emerge as cell cycle regulators [6–11].

1 Pathways and Mechanisms of Differentiation in the Forebrain

The generation of new neurons in the developing cerebral cortex requires the tight coordination of specific cellular activities including cell cycle exit of precursor cells and their subsequent neuronal differentiation. Although the mechanisms that integrate these different cellular events into a coherent program of neurogenesis are not yet completely understood, genetic studies have led to the identification of several steps of this process and a number of molecules participating in it. Here, we will provide an overview of recent progress in the field of cortical neurogenesis regulation, emphasising to the cell types participating in it, their divisional mode, as well as molecules implicated in the cell cycle progression/exit and differentiation control.

1.1 The Role of Neural Stem Cells and Radial Glia During Embryonic Cortical Development

The mammalian cerebral cortex is composed of an enormous number of neurons and glia organized into cytologically and functionally distinct areas. All areas share a common basic structure with neurons arranged in six layers. The majority of cortical neurons are pyramidal cells found in all layers except layer I. These are the projection cells of the cortex that utilize the excitatory amino acid L-glutamate as a neurotransmitter [12]. The remaining neurons, scattered in all layers, are the nonpyramidal cells. They are the cortical interneurons that contain the inhibitory neurotransmitter GABA [12]. Recent evidence suggests that the two neuronal types are generated in distinct proliferative zones. Pyramidal cells are derived from the dorsal ventricular zone that lines the telencephalic ventricles [13, 14]. While few nonpyramidal cells are generated in the dorsal ventricular zone [15], their vast majority is derived from the ganglionic eminence of the ventral telencephalon [16, 17]. These ventrally derived neurons migrate tangentially in chains from the ventral telencephalon to their final positions in the developing cortex giving rise to approximately 25% of neurons of the dorsal telencephalon. Here we will use as a paradigm of cell cycle progression/exit and differentiation control the mode of division and properties of dividing cells in the dorsal VZ that migrate radially to reach their final positions in the cortical plate, where they differentiate to pyramidal projection neurons.

1.1.1 Neuroepithelial Cells

Going back to the first stages of forebrain morphogenesis, the cortical plate derives from a single layer of primary progenitors, known as neuroepithelial cells that form the neural plate. Neuroepithelial cells are highly polarized in the apical–basal axis, radially elongated cells contacting both the apical (ventricular) and basal (pial) surfaces. At this stage the neuroepithelium has a pseudostratified appearance, due to the interkinetic migration of the nuclei of neuroepithelial cells that lie at the ventricular surface during mitosis, but translocate towards the pial surface during interphase. These cells, which can be considered neural stem cells (NSCs), first undergo symmetric proliferative divisions, resulting in the generation of two daughter stem cells, thus increasing the pool of stem cells. Later in embryonic development, these divisions are followed by a large number of asymmetric, self-renewing divisions generating a daughter neural stem cell and a more differentiated cell, such as a neural progenitor or a neuron (Fig. 8.1a). This transition from a symmetric to an asymmetric pattern of divisions both marks the onset of neurogenesis and specifies a pool of neurogenic precursors in vivo, which keep on producing post-mitotic neurons [18]. There is a wealth of evidence supporting that repeated asymmetric divisions of cortical progenitor cells occurring at this developmental stage are the main reason for the slow increase in cortical cell [19, 14]. Additionally, clonal analysis in vitro and in vivo has demonstrated that most cortical progenitors between embryonic day 12 (E12) and E18 give rise mainly to cells of the neuronal lineage [20–22]. During the later stages of embryonic development neural progenitors undergo terminal symmetric divisions resulting in the production of two post-mitotic daughter neurons. At this time the neurogenic potential of precursors decreases while their gliogenic potential increases [23, 24].

Most known genes specifying neuronal fate at this early embryonic stage are transcription factors, such as proneural basic helix-loop-helix (bHLH) genes participating in the maintenance and differentiation of neural stem cells during forebrain development [25, 26] or patterning genes [27, 16] expressed transiently by proliferating precursors. In the embryonic mammalian telencephalon the expression of the proneural genes neurogenin1 (Ngn1), neurogenin2 (Ngn2) [28] and Mash1 [29] is

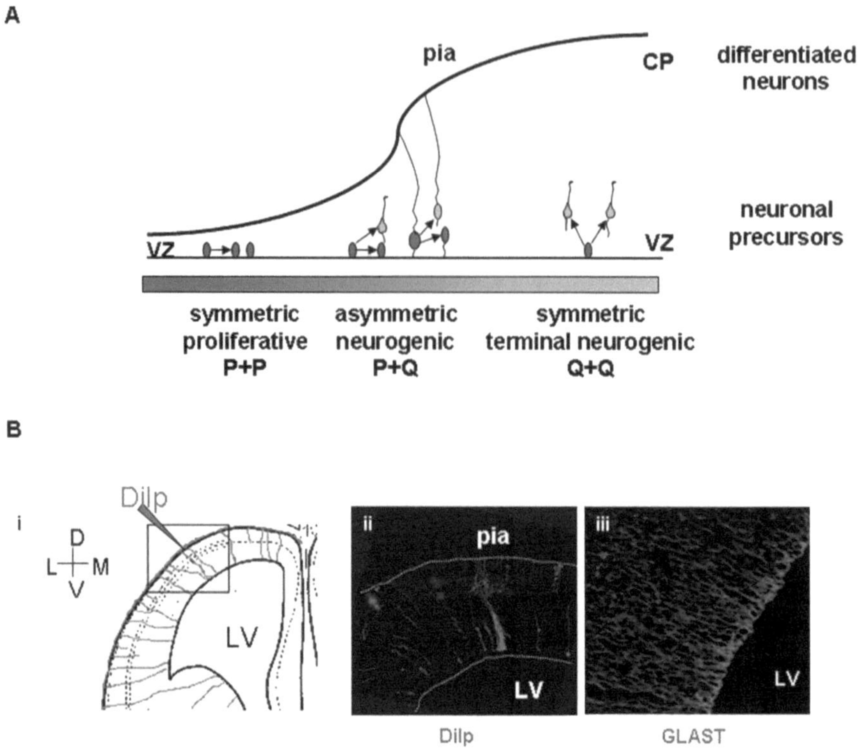

Fig. 8.1 Embryonic cortical development. (**a**) Modes of division during embryonic cortical development. Schematic drawing, based on the diagram on cortical neurogenesis *by Takahashi et al. (1996),* modified to illustrate the switch from symmetric P+P to asymmetric P+Q and final symmetric Q+Q divisions during cortical neurogenesis. P, proliferative; Q, quiescent. (**b**) Radial glial cells in the developing cortex. (i) Schematic drawing depicting DiIp back-tracing of precursors with radial processes (radial glia cells) from the pial surface. Cortical radial glia was traced by applying the fluorescent dye DiI as small crystals onto the pial surface of the cerebral cortex. (ii) Coronal vibratome section of E16 rat brain following DiI labelling of radial glial cells spanning the whole thickness of the cortex. (iii) Immunofluorescence labelling of E16 mouse cortex with antibodies to the radial glial cells marker GLAST (modified from Koutmani et al., 2004). LV, lateral ventricle

a prerequisite for the initiation of the differentiation of neuroepithelial cells towards the neuronal lineage. In particular Ngn1 and Ngn2 are implicated in neurogenesis of the dorsal telencephalon, whereas Mash1 in neurogenesis of the ventral telencephalon. However, studies with knockout mice lacking the above-mentioned genes imply that other genes are also involved in the neurogenic cell fate decision of telencephalic neuroepithelial cells. On the other hand, the Notch-signalling pathway and its downstream effectors Hes1 and Hes5 – also belonging to the bHLH gene family – antagonize differentiation-inducing genes, such as Mash1 thus regulating NSC maintenance. Consequently, in the absence of Hes1 genes NSCs are prematurely differentiated into neurons and depleted without generating the later-born cell types of the glial lineage [30]. In a negative feedback loop, Hes genes have been also shown to negatively regulate Notch1 expression, which is spatiotemporally correlated with asymmetric neurogenic cell divisions [31]. These data suggest a closely regulated relationship between cell cycle exit and neuronal differentiation.

The expression of proneural genes in neuroepithelial cells triggers a cascade of related events, such as up-regulation of genes inducing the neuronal phenotype [32], down-regulation of genes promoting the glial phenotype [33, 34] and the inhibition of differentiation of neighbouring cells through a mechanism known as "lateral inhibition", which results in the controlled differentiation of certain numbers of cells from the neuroepithelial pool [35]. In this signalling pathway, proneural genes induce the expression of the plasma membrane proteins Delta and Serrate/Jagged which are ligands for the Notch receptors expressed by neighbouring cells. Thus, expression of Notch in a cell keeps it in an undifferentiated neuroepithelial state through an autocrine loop that prohibits proneural gene expression in this cell.

1.1.2 Radial Glial Cells

With the generation of the first post-mitotic neurons the neuroepithelium looses its apical–basal polarity and transforms to a tissue with multiple cell layers, the one lining the ventricle referred to as ventricular zone. By the time point that neurogenesis commences, which occurs around E10 in mice, embryonic neural stem cells give rise to a distinct but highly related cell type, the radial glia cells, exhibiting both neuroepithelial and astroglial characteristics (Fig. 8.1b). More specifically, radial glial cells maintain many neuroepithelial properties, which include apical–basal polarity and expression of neuroepithelial markers such as the intermediate-filament protein nestin and its post-translational modifications labelled by the RC1 and RC2 antibodies [4]. In parallel, by contrast to neuroepithelial cells, radial glial cells exhibit certain astroglial traits. These include the presence of glycogen granules in their cytoplasm, as well as the expression of several astroglial markers such as the glial fibrillary acidic protein (GFAP), the astrocyte-specific glutamate transporter (GLAST) and the Ca^{++}-binding protein S100β. Radial glial cells gradually replace neuroepithelial cells in the middle embryonic period and mostly divide asymmetrically to generate a neuron and a radial glial progenitor [36, 37]. In terms of multipotency radial glial cells seem to have a more restricted cell fate as compared to neuroepithelial cells. Indeed, in vivo fate mapping experiments using Cre-recombinase under various cell type-specific promoters that specifically drive its expression either in radial glia or in neuroepithelial cells at early embryonic ages have demonstrated that by contrast to neuroepithelial cells' progeny, the progeny that inherited the recombination from radial glial cells was more restricted in terms of cell identity. More specifically, in transgenic mice where part of the nestin promoter was used to drive expression of Cre-recombinase at embryonic day 10 (E10), when radial glial cells have not appeared yet, the recombined genes were found in all CNS cell types [38]. By contrast, when Cre-recombinase was placed under the control of the human glial fibrillary acidic protein (GFAP) promoter, which is only active at the time of radial glial cells differentiation [39], the progeny of recombined precursors in most cases gave rise to not only single cells type, predominantly neurons, but also glial cells, depending on the age and the telencephalic area. For example the progeny of radial glial cells from the dorsal telencephalon gave rise to cortical projection neurons [40], whereas ventral telencephalic radial glia generated predominantly glial cells [39].

1.2 Symmetric Versus Asymmetric Cell Divisions During Neurogenesis

The apical–basal polarity of neuroepithelial and radial glial cells plays an important role for the balance between their symmetric versus asymmetric divisions, as judged by the distribution of cellular components in their progeny. In support, it has been demonstrated that loss of the mammalian homologue of the *Drosophila* lethal giant larvae gene *Lgl1* results in disruption of neuroepithelial cell polarity and hyperproliferation of neuroepithelial and radial glial cells of the mouse brain [41]. The mechanism that has been initially proposed to explain the relation between the cells' polarity and mode of division was that vertical cleavage planes result in symmetric divisions, whereas horizontal (parallel to the ventricular zone surface) in asymmetric divisions. However, it was later observed that the majority of cells divide in the vertical orientation and that horizontal cleavage planes rarely occur. Moreover, it was recently shown, that the shape of elongated neuroepithelial and radial glial cells is such that their apical plasma membrane and adjacent adherens junctions constitute only a 1–2% fraction of their whole plasma membrane [42]. Hence vertical cleavage planes can occur in such an angle that the apical plasma membrane and surrounding junctional complexes are either bisected or bypassed and thus inherited to either both or only one of the daughter cells resulting in symmetric or asymmetric divisions, respectively. Genetic studies in Drosophila have identified a number of genes acting to specify asymmetric divisions [43–45]. Accordingly in vertebrates, Numb [46–50], the anti-proliferative genes PC3/Tis21 [51] and pRb [52] and the transcription factor Pax6 [53, 54] seem to affect the progression from symmetric/proliferative to asymmetric/neuron-generating divisions. The anti-proliferative protein Tis21 in particular has been shown not to be expressed by symmetrically dividing neuroepithelial cells and to be present only in one of the two daughter cells generated by asymmetric divisions, which in 90% of cases examined was the cell that would become a post-mitotic neuron [42]. Also, the expression levels of the protein BM88/Cend1 have been correlated with the asymmetric neurogenic divisions of neuroepithelial precursors during embryonic cortical development [55]. By contrast the transcription factor Emx2 promotes not only a vertical cleavage plane orientation but also symmetric, proliferative cell divisions [7].

Several studies indicate that distinct cell fates may be determined during the final cell cycle of progenitors [1, 56–58] and therefore determination of the involvement of genes that act during this stage is of major importance. For example, transcription of proneural genes is up-regulated in the last cell cycle, when transcription of cell cycle activators has been down-regulated. Moreover, homeodomain proteins specifying neuronal fates such as Prospero in Drosophila [59], its mammalian homolog Prox-1 [60] and Phox2b in vertebrates [8] induce exit from the cell cycle. Conversely, inhibitors of cell cycle progression, such as p27Xic1 in Xenopus and $p27^{KIP}$ in mammals and pRb, are implicated in neuronal differentiation [10, 61, 62].

1.3 Brain Size Control

Switching from a pattern of symmetric divisions, in which both cells remain in proliferative status, to asymmetric divisions, in which one daughter cell exits from the cell cycle, is a very efficient way to regulate neuronal numbers [63]. This kind of control mechanism explains why transgenic mice expressing constitutively active β-catenin, a protein of the wnt pathway implicated in the control of the polarity leading to asymmetric divisions under the control of a neuron-specific promoter, have bigger brains [64]. In agreement, β-catenin transgenic mice exhibit a twofold increase in the number of precursors undergoing symmetrical proliferative divisions [18, 65].

Interestingly, several findings suggest that lengthening of the G1 phase of the cell cycle is responsible for the onset of asymmetric/differentiative final divisions of cortical progenitors [66]. The question that arises from these observations is how does lengthening the neuroepithelial cell cycle and more specifically the G1 phase could trigger the switch to neurogenesis? According to a model referred to as the "cell cycle length hypothesis", an unequal inheritance of a cell fate determinant by the daughter cells upon progenitor cell division may or may not lead to asymmetric daughter cell fate, depending on the length of time it is allowed to function [67]. This hypothesis, which is based on both in vitro and in vivo observations, could also explain the symmetric versus asymmetric fate of dividing cells. If the relevant G1 phase of the cell cycle is too short for the cell fate determinant to induce differentiation then both daughter cells will continue to proliferate. Alternatively if the cell cycle is longer, the cell fate determinant will promote the differentiation programme of one of the two daughter cells and in an even longer cell division, differentiation will be induced in both daughter cells. The above hypothesis has been confirmed by the observation that the protein PC3/Tis1, that blocks the G1-to-S-phase progression by lengthening the G1 phase of the cell cycle, is sufficient to increase neurogenesis and inhibits neuroepithelial cell proliferation at the same time [68, 69]. These observations could explain the progressive lengthening of the cell cycle from symmetric proliferative to asymmetric neurogenic and finally to symmetric terminal neurogenic divisions and further suggest the involvement of cell cycle mediators in coupling the mode of progenitor cell division with differentiation. In addition, three recent studies in mammalian embryos also support the cell cycle length hypothesis. First, shortening the G1 phase of the cell cycle by administration of insulin growth factor-1 (IGF1) in mouse neuroepithelial or radial glial cells increases the probability that their progeny will re-enter the cell cycle, indicating a shift towards symmetric divisions [70]. Second, in the primate cortex, progenitors of the area 17 of the subventricular zone (SVZ) that have a shorter G1 phase than those of area 18 exhibit a greater probability of cell cycle re-entry, a process regulated by cyclin E and p27 [66]. Third, at any given neurogenesis stage and brain region, progenitors undergoing neurogenic divisions have a longer cell cycle than those undergoing proliferative divisions [71]. It is also possible that during an elongated G1-phase neurogenic and differentiation-promoting genes are up-regulated thus forcing one of the daughter cells towards neuronal differentiation. This is the case for the dual function neurogenic protein BM88/Cend1, whose up-regulation leads to elongation of the G1 phase of the cell cycle and subsequent cell cycle exit of a sub-population of proliferating cells [72].

Cell size is another factor determining brain size. In related invertebrate species with similar numbers of neurons, for example, there is much variation in the size of these neurons. A way that this size difference is accomplished is through the process of endoreplication, where the cell cycle is completed without cytokinesis. This process takes place both in *Drosophila* and in mice [73] and in both cases cyclin E plays a regulatory role. As a result natural giant neurons are often polyploid and it has been shown that unnaturally large neurons can be produced by interfering in the cytokinesis pathway [74]. Another way for cell size control is the control of cell growth during the G (G1 and G2) phases of the cell cycle. The insulin receptor PI3K pathway has been shown to be implicated in this mechanism, as transgenic mice lacking PTEN, a component of this pathway, show increased neuron size [75]. In support, in *Drosophila* mutations of almost all components of this pathway result in production of cells with abnormal size [76, 77]. Both mechanisms described above link developmental pathways of neurogenesis to components of the cell cycle.

1.4 Cell Cycle Exit Signalling in the CNS

The overall size of the brain is governed by the cell cycle machinery. This is clearly demonstrated by enlarged brain of mice lacking the cell cycle inhibitor p27kip1 [78, 79]. But as some parts of the CNS, such as the forebrain, are larger than others, such as the spinal cord, cell cycle components are expressed or activated at certain time points in specific areas so that each part of the CNS ends up being the correct size. This process is to a great extent coordinated by the control of cell cycle progression and the precise timing of cell cycle exit [80, 81]. Emerging evidence suggests that progression of progenitors towards neuronal differentiation is tightly linked with cell cycle control and that the two events may be coordinately regulated. This control of cell cycle progression plays an essential role in the generation of the appropriate number of functional neurons at certain brain areas. When a neuronal progenitor is committed to undergo differentiation, it exits from the G1 phase of the cell cycle and enters into an irreversible quiescent state referred to as G0. The tumour suppressor proteins p53 and pRb are central regulators of this progression [82]. Having a short half-life, p53 is normally maintained at low levels in unstressed mammalian cells by continuous ubiquitination and subsequent degradation [83]. When the cell is confronted with stress or a developmental cell cycle arrest signal, p53 ubiquitination is suppressed and p53 protein becomes stabilized, accumulates in the nucleus and activates or represses the transcription of specific target genes [84]. Activation by growth arrest stimuli of functional p53 or related protein family members, such as p63 and p73, causes G1 arrest at the G0 restriction point through induction of a number of transcription factors, including $p21^{CIP1/WAF}$, $p27^{kip1}$ and PC3/Tis21, which in turn inhibit cyclin/cdk action [9, 62, 85, 86]. D-type cyclins and related cyclin-dependent kinases (CDKs) are responsible for pRb phosphorylation [87–89] leading to cell cycle progression and thus loss of their activity induces G1 arrest through inhibition of pRb phosphorylation [82]. pRb is therefore another key molecule responsible for growth arrest and accumulation of cells in G1 in response to anti-proliferative signals. Under these conditions hypo-phosphorylated pRb associates with the E2F family of transcription factors and impairs their ability to transactivate genes required for cell cycle progression [90]. As a consequence cells do not progress through the G1-to-S-phase transition. Increasing evidence suggests that transcription factors of the Sp1 family are critical for the cellular responses to p53, including activation of its downstream growth arrest effector $p21^{WAF/Cip1}$, both in neuronal [91, 92] and non-neuronal cells [93].

Several studies have indicated that a major event associated with cell cycle withdrawal and differentiation both in neuronal and non-neuronal cells is the cellular compartmentalization of cyclin D1, which shifts from a predominantly nuclear localization to cytoplasmic sequestration [94, 95]. In particular, Sumrejkanchanakij et al. [94] have shown that cyclin D1 becomes predominantly cytoplasmic as primary cortical progenitor cells undergo cell cycle withdrawal and terminal differentiation. In the same study it was also shown that exogenously expressed cyclin D1 sequesters in the cytoplasm of post-mitotic neurons by a mechanism inhibiting its nuclear import, whereas it efficiently enters the nucleus of proliferating progenitor cells. Furthermore, forced cyclin D1 expression in the nucleus of differentiated neurons resulted in apoptotic induction.

As mentioned above, the G1 restriction point located at the end of G1 phase is a key proliferation check point, as if cells pass this point, they will almost invariably complete the cell cycle, otherwise they will become post-mitotic [96, 97]. In these lines increasing evidence suggests that while select G1-phase components affect cell fate determination at this stage, the opposite is also true. Thus a number of neuronal determinants affect cells lying at G1, allowing them to take the G0 branch [98–100]. Characterization of dual function molecules, such as the cyclin-dependent kinase inhibitor p27Kip1 or its Xenopous homologue p27Xic1 [57], Geminin [101–103], the neural proliferation and differentiation control protein NPDC-1 [104] or the Hu family of neuronal RNA-binding proteins [105], has yielded intriguing insights into the functional link between the control of cell cycle progression and neuronal commitment/differentiation (Fig. 8.2). In support, the role of the BM88/Cend1 protein, which has been previously related with asymmetric neurogenic divisions during embryonic cortical development [55], in the control of cell cycle exit via cyclin D1 down-regulation and pRb hypophosphorylation has been identified [72]. Additionally, it should be noted that during embryonic CNS

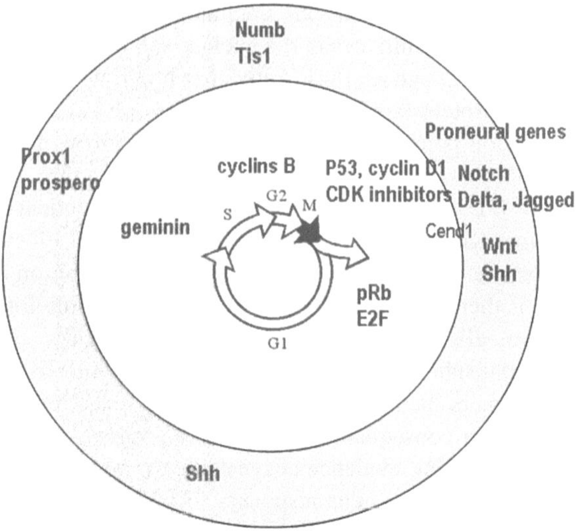

Fig. 8.2 Cross-talk between cell cycle and cell fate determinants. The cell cycle factors also influencing neurogenesis are shown in the white circle, whereas the neuronal fate determinants also participating in cell cycle control in the *grey circle*. Both sets of factors are positioned in their corresponding time of action during the different phases of the cell cycle

development withdrawal from the cell cycle is accompanied by acquisition and maintenance of the neuronal phenotype, which involves a large number of genes that are expressed in a predetermined and coordinated manner. These include proneural genes, transcription factors involved in patterning and members of the bHLH family [25, 106–108]. Along these lines it has been demonstrated that overexpression of the proneural genes NeuroD, Mash 1 and neurogenin1 induces P19 embryocarcinoma cells to differentiate towards a neuronal phenotype through a mechanism that involves expression of the cyclin D1 inhibitor p27kip1 and subsequent cell cycle arrest at G1 [32]). In addition, external neuronal determination signals, such as Wnt, sonic hedgehog (Shh) and retinoic acid (RA), have been shown to regulate cell cycle exit through modulation of cyclin D1, cyclin D2 and N-myc transcription [109, 110–112].

1.5 Proliferation and Cell Death During CNS Development

Apoptosis has been recognized as a prominent event during the development of several regions of the vertebrate nervous system, as well as during disease and trauma. During embryogenesis, cell death has a morphogenetic function at various stages of the formation of the central nervous system (CNS): during the closure of the neural tube [113]; during the development of the mesencephalic region and in the process of negative selection of certain progenitor cells from inappropriate regions of the CNS [114]. Later in development, neurons generated in some areas of the nervous system may die as a result of limited availability of trophic factors or lack of synaptic inputs necessary to suppress the endogenous genetic death program [115, 116]. Based on the time of occurrence of cell death, it is possible to hypothesize the existence of two functionally distinct types of death in the nervous system of developing mammals. They may share morphological (apoptosis) and/or biochemical (activation of cell cycle genes) similarities, but differ substantially in the type of cells involved: the "proliferative" type of cell death involves actively cycling cells whereas "target-related" cell death involves post-mitotic neurons. The simultaneous occurrence of both proliferation and apoptosis in some cell populations has suggested that the two processes may be related. Indeed, besides several morphological similarities between dying and dividing cells a wealth of studies have shown that molecules acting during cell cycle progression are required for apoptosis. These include mitotic kinases [117]; the tumour suppressor gene p53 [118, 119]; and cyclin D1 [120]. Similarly, molecules which act as checkpoints during the progression through the cell cycle, such as pRb, have been shown to prevent apoptosis [121]. More specifically, following DNA damage, pRb acts to prevent entry into S phase, which would otherwise trigger apoptosis [122]. In contrast, bcl-2 acts independent of the cell cycle to directly suppress apoptosis [123]. Alike Bcl-2, BM88/Cend1 which shares many similarities in terms of localization and function with Bcl-2, has also been shown to protect cells from C2-ceramide-induced apoptosis via a mechanism involving calcium dynamics (Masgrau, Thomaidou and Matsas, unpublished data).

The occurrence of developmental cell death in the cerebral cortex of embryos has initially been regarded as a rare event [124–126]. However, more recent studies using sensitive techniques to visualize apoptotic cells over a longer time period before their clearance by macrophages suggest that CNS progenitor cells undergo "proliferative" apoptosis, similar to many other proliferating tissues, but this event takes place so quickly that only a few apoptotic nuclei may be detected in a histological section at any time point [127]. In support of this comes evidence showing that the histologically visible stages of the apoptotic process are very short, lasting from a few minutes to a maximum of 3 h [128]. Lately there has been evidence defining a molecular pathway that includes activation of cyclin-dependent kinase-4 (Cdk-4) in neurons which leads to hyper-phosphorylation of the Rb family member p130 and subsequent dissociation of the protein complex consisting of the chromatin modifiers Suv39H1, HDAC1, the transcription factor E2F and p130, finally resulting in the induction of the pro-apoptic Bh3-only protein Bim. Bim then interacts with the core cellular apoptotic machinery, leading to caspase activation and apoptotic death [129]. This pathway is supported by a variety of observations indicating it as a required element for neuronal loss during normal development and in many nervous system injuries and disorders. Thus the components of this signalling pathway may represent potential therapeutic targets for prevention of neurodegeneration-associated neuronal death.

2 Pathways and Mechanisms of Differentiation in the Spinal Cord

During embryonic development competent ectodermal cells are committed to neural fate to form the neural plate and later the neural tube. The early neural plate is initially rostral in character (forebrain like) and more caudal regions, such as spinal cord, form as a result of the caudal regression of the organizer. The caudal character of neural cells emerges soon after neural induction, through the reprogramming of cell fates by a series of extrinsic signals. Many different classes of secreted factors have been implicated in this rostrocaudal extension of the neural plate: fibroblast growth factors (FGFs), retinoids, bone morphogenetic proteins (BMPs), Wnts and a caudalizing activity from the paraxial mesoderm [130–133].

The spinal cord develops from a small number of highly plastic NSCs that proliferate, acquire regional identities and generate a progressively restricted repertoire of cell types, first neurons and later oligodendrocytes and astrocytes. The induction of neural fate does not affect the proliferative capacity of cells [134]. It is only later that committed precursors are instructed to become post-mitotic, as progenitors exit the cell cycle, cease to proliferate and differentiate into neurons and glial cells. Similarly to cortical development, both neural induction and initiation of differentiation pathways in the spinal cord, either neurogenic or gliogenic, appear to be connected to cell cycle control systems that regulate whether stem cells will maintain their proliferative ability or differentiate into the appropriate neural cell type [2]. This regulation is essential for the generation of appropriate number of neurons, neuronal subtypes and proper wiring of neuronal circuits in the spinal cord.

2.1 Dorso-ventral Patterning in the Spinal Cord

As soon as the early neural tube is formed, it already acquires dorso-ventral organization. Neural precursors in the dorsal part show higher proliferation rates than ventral precursors, whereas the opposite is true for the differentiation rates [135]. In this early neural tube, the region between roof and floor plate is densely filled with neural precursors. These precursors initially undergo rapid proliferation and subsequently give rise to post-mitotic cells that differentiate into distinct neuronal and glia phenotypes. The early neural tube is patterned by extrinsic signals that activate hierarchies of transcription factors expressed in a region- and cell-specific manner [136, 137]. These transcription factors act to subdivide the ventricular zone (VZ) into defined progenitor domains with restricted developmental potential, and subsequently, to establish distinct differentiation programmes in the neurons that emerge from each domain. Current evidence suggests that progression of progenitor cells towards neuronal differentiation is tightly linked with cell cycle control and that the two events are coordinately regulated [2]. Therefore, upon progression of development, VZ precursors exit the cell cycle and differentiate first into neurons (early neurogenic phase) and then glia (late gliogenic phase). Newly born neurons migrate laterally out of the ventricular zone into their final positions in the periphery of the spinal cord (mantle zone), where they become incorporated into the local neural circuitry.

A large number of molecular markers characterizing different types of neurons in the spinal cord have been used to identify distinct domains of dorsal and ventral neuronal populations, as well as the corresponding progenitor domains that generate these cell populations. Accordingly, non-overlapping expression domains of proneural genes, which encode basic helix-loop-helix (bHLH) transcription factors and homeodomain transcription factors, define six distinct progenitor cell types in the early dorsal neural tube, designated dp1–dp6. These cells differentiate to give the six types of dorsal interneurons (dI), dI1–dI6, which can be distinguished on the basis of the homeodomain and bHLH transcription factors that they express (Fig. 8.3a) [138]. In a similar manner the ventral spinal cord is subdivided in five progenitor domains: vp0, vp1, vp2, pMN and vp3, which generate five distinct groups of neuronal populations: v0, v1, v2, motor neurons (MNs) and v3 (Fig. 8.3a) [136, 137].

2.1.1 Down-Regulation of FGF Signalling Induces the Onset of Neuronal Differentiation and Establishment of the Patterning System in the Spinal Cord

The characteristic dorso-ventral patterning of the spinal cord is initially generated by the action of four extracellular signalling molecules, which are distributed in a gradient fashion, namely fibroblast growth factors (FGFs), retinoic acid (RA), sonic hedgehog (Shh) and bone morphogenetic proteins (BMPs).

The cells of the caudal neural plate, which regress alongside the primitive streak, constitute the caudal stem zone and cells from this region give rise to neural progenitors, which are left behind by the zone and form the spinal cord (Fig. 8.4). These ancestral cells subsequently undergo differentiation and patterning in the extending body axis [139]. FGF inhibits differentiation in this caudal stem zone, while RA provided rostrally by somitic mesoderm is required for neuronal differentiation and the establishment of ventral neural patterning. FGF signalling is a major pathway involved in the maintenance of caudal progenitors in this stem zone, whereas inhibition of this signalling is necessary for the progression of neuronal differentiation. In particular, it was demonstrated that removal of presomitic mesoderm results in the precocious onset of the ventral patterning gene Pax6 [35]. The signal responsible was identified as FGF, which is produced by presomitic mesoderm cells and is able to repress Pax6. Moreover, recent findings have shown that attenuation of FGF signalling is also necessary for neuronal differentiation [140]. Repression by presomitic mesoderm and in particular by FGF appears to be a general mechanism that represses markers for generic neuronal differentiation, such as NeuroM, and ventral patterning genes, such as Pax6 and Irx3, and consequently restrains differentiation at the caudal end of the developing spinal cord [141, 142].

2.1.2 Ventral Patterning and Shh

Extensive experimental data, based on both gain and loss-of-function approaches performed in all the main vertebrate model systems, namely fish, frog, chick and mouse, have shown that hedgehog (HH) signalling molecules are crucial for the specification of ventral cell fates in the spinal cord [137, 143]. The *Shh* member of the HH family of secreted proteins begins to be expressed in the notochord as soon

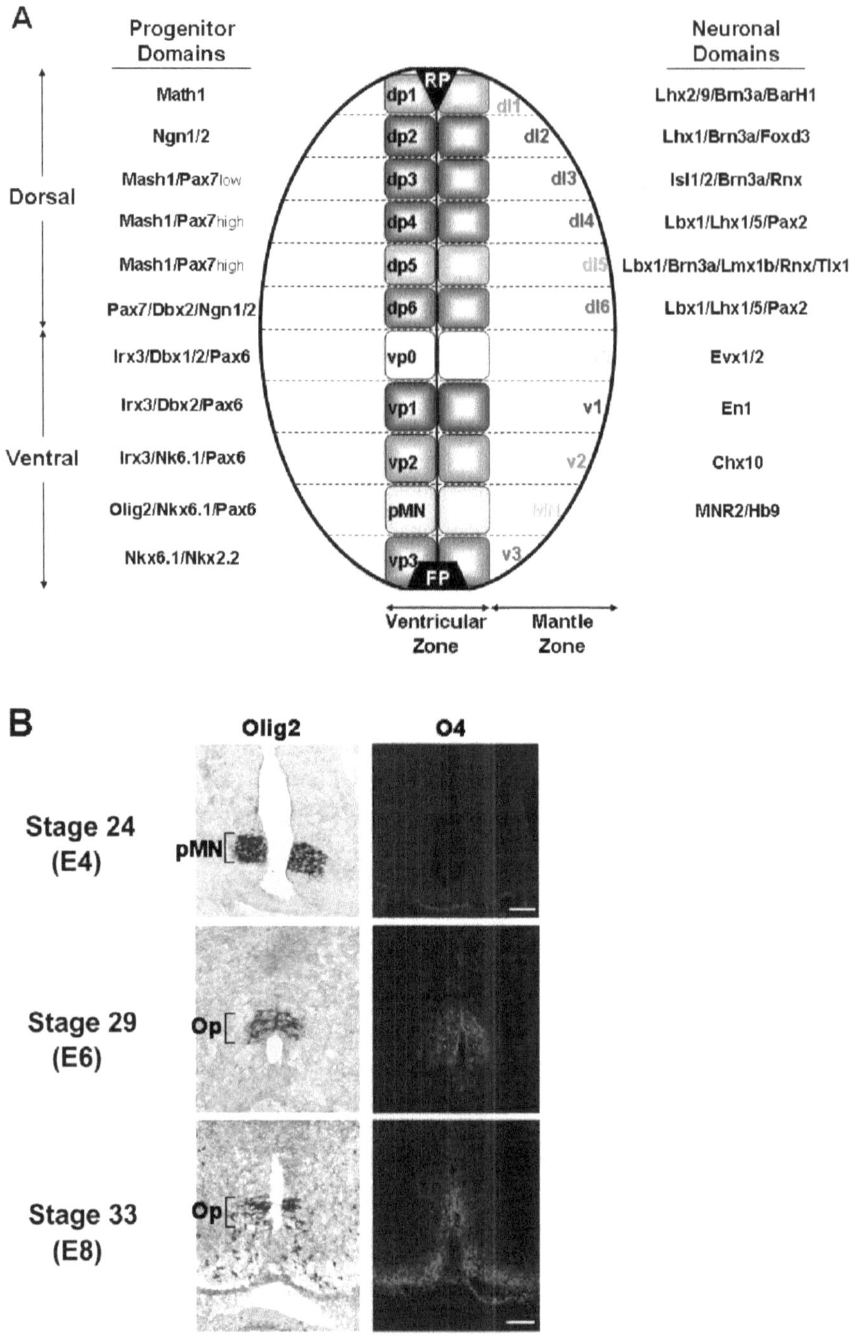

Fig. 8.3 A combinatorial code of transcription factors specifies the domains for cell types in ventricular (VZ) and mantle zone (MZ) in the developing spinal cord. (**a**) Schematic of the regionalization patterns of the six types of dorsal neurons (dI1–dI6) and the five types of ventral

as the cells, which are destined to give this structure, have left the regressing node. *Shh* is, like FGF, an extracellularly secreted molecule which acts in a concentration-dependent fashion. In the neural tube, the ventral-most cells, comprising the floor plate, constitute a prominent source of this molecule, leading to the establishment of dorso-ventral gradient of Shh activity, which decreases from the ventral-to-dorsal part, thereby providing a means for generating molecular differences in the VZ.

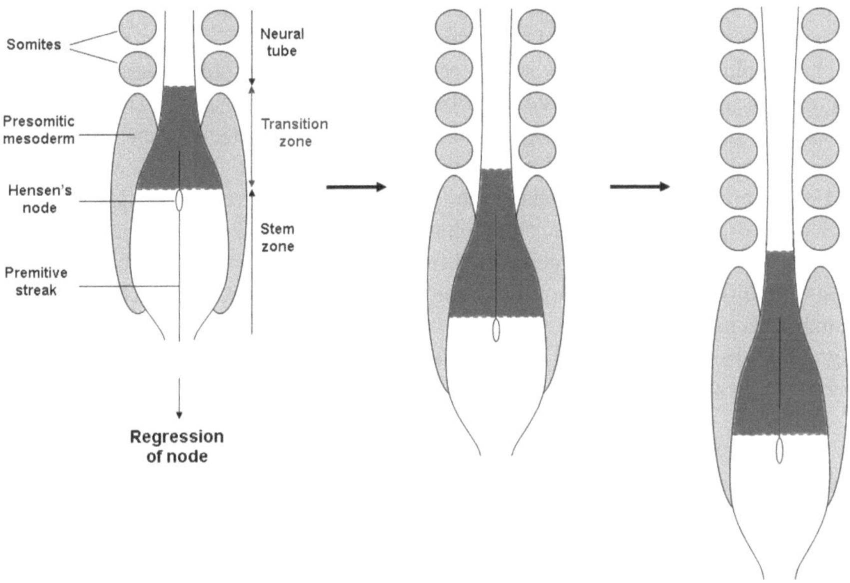

Fig. 8.4 Caudal stem zone progressively generates early spinal cord. The stem zone in the caudal part of the neural plate is constituted of neural precursor cells which regress alongside the primitive streak and progressively give rise to neural progenitors, which are left behind to generate the spinal cord. In particular, the cells from the stem zone divide and progressively become neural progenitors in the transition zone. The cells that enter the transition zone acquire a fixed rostrocaudal position and when somites form adjacent to the neural tube produce signals that induce the differentiation of progenitors in the transition zone to generate the early spinal cord

Fig. 8.3 (continued) neurons (v0, v1, v2, MN and v3) in the developing neural tube. On the *left* are the transcription factors which are used to identify the progenitor domains of the different dorso-ventral domains for post-mitotic neuronal populations. On the *right* are the transcription factors which are used to identify the neuronal types in the MZ, where differentiated neurons migrate. FP, floor plate; RP, roof plate. (**b**) Comparison of *Olig2* and *O4* expression in the ventral chick spinal cord during transition of neuroepithelial precursors from neurogenic (E4) to oligodendrogenic (E6 and E8) phase. Adjacent transverse sections of Hambourger–Hamilton (HH) stage 24 (E4), HH stage 29 (E6) and HH stage 33 (E8) were subjected to *cOlig2* in situ hybridization (*left panel*) or immunofluorescence labelling with anti-O4 antibody (*right panel*). At HH stage 24 the Olig2+ neuroepithelial precursors of the pMN domain generate motor neurons and thereby they do not express the early oligodendrocyte precursor marker O4. At later stages of embryonic development (HH 29 and 33), the Olig2+ domain (Op, Oligodendrocyte progenitors) becomes destined to generate O4+ oligodendrocytes (*right panel*) (modified from Politis et al., 2007)

Gain and loss-of-function and studies indicate that *Shh* is necessary and sufficient to induce the floor plate and all of the ventral progenitor domains, with the possible exception of the p0 domain and zebrafish medial floor plate. Thus, overexpression of *Shh* in the neural tube in vivo ventralizes the spinal cord, and incubation of spinal cord explants in increasing concentrations of purified Shh protein in vitro specifies progenitor domains of progressively more ventral character [136, 137, 144–146]. Conversely, incubation of spinal cord explants in anti-Shh antibodies prevents ventral specification, and mouse embryos with a targeted knockout of *Shh* lack floor plate, p3, pMN and p2 domains, whereas p1 and p0 cells are displaced to the most ventral part of the spinal cord [137, 146–148]. In the *Shh−/−* mutant mouse, however, V0 and V1 interneurons still develop, albeit in reduced numbers [144], and thereby these neuronal classes cannot be totally dependent on Shh. This raises the possibility that alternative member(s) of HH signalling molecules specify V1 and V0 neurons in the *Shh* knockout mice. Consistently, cells that lack *smoothened* (*SMO*), a component of the HH signalling pathway required for signalling by all HH proteins, do not form the floor plate, MNs or V3, V2 or V1 neurons [149]. Although some V0 neurons are formed from Smo mutant cells, they are found only at abnormally ventral positions [149], suggesting that fate specification of all ventral spinal cord neurons, with the possible exception of some V0 neurons, requires HH signalling. In fact, V0 and V1 interneurons are also dependent on RA signalling for their specification [150, 151].

Although these observations support the hypothesis of a Shh concentration gradient, this Shh gradient has never actually been detected in the spinal cord. Immunocytochemical and in situ hybridization studies only show the protein and mRNA, respectively, to be present in the notochord and floor plate. However, expression of the HH target gene patched 1 (*Ptc1*), considered to be a direct readout of HH activity, is detected with a dorso-ventral gradient (low to high) throughout the ventral half of the spinal cord [152–154]. This suggests that all six ventral domains could be specified by distinct thresholds of a ventral-to-dorsal morphogenetic gradient of HH activity. In agreement, overexpression of a constitutively active form of SMO ventralizes the spinal cord in a cell-autonomous manner [155]. Similar results are obtained with active forms of *Gli* genes, which are downstream effectors of the Shh and encode zinc finger transcription factors acting in the nucleus in response to Shh signals [156]. *Gli* constructs endowed with increasing degrees of activity cause transfected cells to acquire progressively more ventral character [157]. Conversely, overexpression of a constitutively active form of *Ptc*, which represses HH signalling, cell-autonomously inhibits V3–V0 and MN fates within their normal domains and induces ectopic V2, V1 and V0 neurons in the ventral spinal cord [158]. These experimental results indicate that HH signals act in a concentration-dependent manner as long-range diffusible factors in the ventral spinal cord, directly inducing different cell types at increasing distances from their source, and this gradient is translated into cell-autonomous specification of different ventral fates by graded *Gli* activation inside target cells [157].

Although these findings support the hypothesis that the position of a progenitor cell within a ventral-to-dorsal gradient of Shh specifies its differentiation into specific neuronal subtypes, they still leave open the question of how an inductive gradient is converted into well-defined progenitor domains. Experiments in chick and mouse embryos provided evidence that a group of homeodomain proteins expressed by ventral progenitor cells act as intermediary factors in the interpretation of graded Shh signalling [151, 159, 160]. These homeodomain proteins can be divided into two major categories on the basis of their pattern of expression and mode of regulation by Shh [159]: class I proteins, which are repressed by various concentrations of Shh and class II proteins, which are induced by Shh activity. Pairs of class I and class II factors have been shown to negatively regulate each other via mutual transcriptional repression [161]. These cross-repressive interactions are thought to result in a delineation of cells expressing either one factor or the other, ultimately leading to sharpening and stabilization of the boundaries between the initially coarsely separated progenitor domains [137, 159]. Thus, for example, if *Pax6* is ectopically expressed in ventral regions of the chick neural tube, then *Nkx2.2* is repressed. Conversely, if *Nkx2.2* is ectopically expressed in more dorsal regions, then *Pax6* is repressed [159]. Thereby, the combinatorial expression profile of these two classes of homeodomain proteins defines five progenitor cell domains within the ventral neural tube. For example, the combinatorial actions of three homeodomain proteins, Nkx6.1, Nkx2.2 and Irx3, restrict the generation of motor neurons to a single progenitor domain. Within this domain, Nkx6.1 activity directs the domain-restricted expression of downstream factors, such as the homeodomain protein MNR2 [162]. MNR2 is first expressed during the final division cycle of motor neuron progenitors and functions as a dedicated determinant of motor neuron identity. Once induced, MNR2 positively regulates its own expression [162], further inducing the progression of progenitor cells to a motor neuron fate.

2.1.3 RA Participates in Defining the Patterning in the Spinal Cord

Retinoic acid, which is initially produced by somitic mesoderm, is required for neuronal differentiation and establishment of ventral neural patterning. Impairment of signalling between somitic tissue and neural tube results in a decrease in Pax6, Irx3 and NeuroM [35, 141], indicating that a signal from the somite normally activates their expression. This is further confirmed by the ability of somitic tissue to induce Pax6 and NeuroM in stem zone explants [141, 163]. The somite-derived activator appears to be retinoic acid, which is produced by Raldh2, an enzyme present at somitic stages in rostral presomitic mesoderm and somites [164, 165] but absent in more caudal regions. It has been shown that somites synthesize several active retinoids [165], and from gastrulation onwards, paraxial mesoderm expresses high levels of Raldh2 [164, 166–168]. Upon treatment of stem zone explants with RA or a retinoic acid receptor (RAR) agonist, the expression of the neuronal marker NeuroM is increased, whereas interference with the retinoid pathway blocks the ability of somites to promote neuronal differentiation [142]. Moreover the requirement for RA in spinal cord differentiation and patterning is supported by many different experiments in diverse species, where the retinoid pathway has been attenuated. These embryos showed dramatically abnormal development of the spinal cord as indicated by reduced neural tube size, neuron number, expression of proneural genes, such as neurogenins1 and -2, and ventral patterning transcription factors, such as Olig2, Pax6, Irx3 and Nkx6.2 [142, 169, 170]. For example, electroporation of a dominant-negative RA receptor into the neural tube to inhibit RA signalling reduces the expression of *Pax6*, *Irx3*, *Dbx1* and *Dbx2* [170]. Conversely, when RA is added to neural plate explants, then these ventral patterning genes (Fig. 8.3a) and others such as *Evx1/2* and *En-1* are induced [151]. These genes are involved in the specification of ventral interneurons and motor neuron subtypes in the ventral spinal cord, and therefore changes in their expression have dramatic effects in spinal cord development.

Furthermore, RA is also required for subsequent steps leading to motor neuron differentiation and specification of the columnar identity of these neurons [170–172]. Thus, sets of motor neurons cell bodies are organized in arrays along the rostrocaudal axis of spinal cord generating longitudinal columns that project axons to distinct regions in the periphery. Lateral motor column (LMC) neurons are generated only at limb levels and project axons into the limb mesenchyme. The median motor column (MMC) is divided into a medial group (mMMC), which is found at all rostrocaudal levels and projects to axial muscles, and a lateral group (lMMC) found only at thoracic levels and projects to the muscles of body wall. Signals provided by early-born motor neurons of the lateral motor column (LMC) help to specify the fate of later-born lateral LMC neurons. A number of studies have implicated retinoid signalling in this aspect of motor neuron subtype specification, since LMC neurons selectively express Raldh2 and synthesize biologically active retinoids [173–175]. Therefore, retinoids provided by early LMC neurons can induce the expression of the Lim1 homeodomain gene, which is a defining marker for lateral LMC identity.

2.1.4 Dorsal Patterning and BMPs

Neuronal patterning in the dorsal half of the spinal cord proceeds analogously to ventral patterning, and instead of Shh and floor plate, it requires the inductive activities of BMPs produced in the overlying ectoderm and roof plate [176–178]. Initial signals from the surface ectoderm specify the roof-plate cells [177, 179], which then provide the signals to specify the dorsal cell types of the spinal cord. BMPs are expressed in the surface ectoderm at the time of roof-plate generation, and naïve explants of neural tissue from chick embryos are induced to express roof plate or dorsal neural markers upon treatment with BMPs [177, 179, 180]. The roof plate is the dorsal equivalent of the notochord, because genetic ablation experiments demonstrated that the roof plate is essential for the specification of some dorsal neural cell types. In particular, ablation of roof plate leads to reduction of the dorsal *Pax7* domain, whereas the ventral *Pax6* domain expands, *Math1*+ and *Ngn*+ cells are missing, as are the respective neuronal population they produce [181]. The roof plate expresses *Bmp4*, *Bmp5* and *Bmp7* and these proteins are sufficient to induce dorsal markers such as *Pax3* and *Msx* and dorsal neuronal subtypes in intermediate neural plate cultures [176, 177]. Gain-of-function experiments in the chick neural tube have shown that many different BMPs can promote dorsal neural cell fates [176, 182, 183]. In addition, electroporation or viral infection of constitutively active BMP receptor in the same system resulted in a number of concentration-dependent dorsalization effects [183]: *Pax7* was ectopically expressed more ventrally, *Pax6* was repressed at high levels of the electroporated gene, while at lower levels was shifted, *Msx1* and *-2* were induced, and *Dbx1* and *-2* were repressed. Of the neuronal specification genes, *Cath1* was up-regulated whereas *Cash1*, *Ngn1* and *Ngn2* were repressed. With regard to neuronal subtypes, Evx1 and En-1 interneurons were reduced, LH2A and B neurons were induced, Lim1/2 neurons were reduced, and dorsal interneurons expressing Islet1 were reduced or absent. In support, zebrafish mutants with compromised BMP signalling activity showed abnormalities in the dorso-ventral patterning such as loss of dorsal sensory neurons and expansion of interneuron domains [184, 185]. Conversely, double deletion of both receptors for BMPs, *Bmpr1a* and *Bmpr1b* results in loss of dl1, *Math1* sensory interneurons and a reduction and dorsal shift in dl2 neurons [186]. These findings suggest that BMPs provide positional cues in dorsal and intermediate regions by setting borders of expression of homeodomain target genes in a similar fashion to Shh ventrally.

Another family of candidate extracellular signalling molecules for transducing the effect of roof plate in the dorsal neural tube progenitors is the Wnts. *Wnt1* and *Wnt3a* are expressed in the roof plate as soon as the neural tube closes, and their expression continues throughout neurogenesis, making these factors likely candidates to contribute to the dorsal neural patterning. These molecules act as mitogens in a concentration-dependent fashion and are responsible for the enhanced rates of proliferation in the progenitors of dorsal neural tube [135]. In addition, there are data suggesting that they act downstream of Bmps to induce dorsal neuronal fates. Mouse double deletion mutants for Wnt1 and Wnt3a showed fewer dI1–dI3 interneurons, as indicated by the expression of Math1 and Ngn1 proneural genes in precursor cells and by the expression of Lh2 and Isl1 genes in the post-mitotic neurons. This decrease in dI1–dI3 populations was accompanied by an expansion in dI4 population, assayed by the expression of lim1/2 and Pax2, implying that the Wnts participate in a choice between dI4 and more dorsal fates [187]. The observation that in the Bmp receptors double mutant mouse the expression of Wnt1 and Wnt3a is down-regulated further suggests that Wnts may be downstream of the Bmp signalling.

2.2 Specification of Neuronal Subtype Identities

The dorso-ventral patterning gradient generated by the action of FGF, Shh, RA and BMP signalling acts coordinately to induce subsequent genes depending on the relative strength of each signalling input. The protein products of these downstream genes act as either transcriptional activators or repressors. The combinatorial actions of these proteins direct the generation of specific neuronal identities within restricted domains of the developing spinal cord. These modulators of transcription usually act as repressors of other transcriptional repressors that inhibit the generation of alternative fates [161, 188–190]. By this way, the generation of individual neuronal subtypes is achieved by the repression of alternative cell fates, in a de-repressive manner [161, 191]. The latter is particularly true for the specification of motor neurons, where most of the transcription factors involved in this specification function as repressors [189, 192]. Therefore, expression of *Olig2,* which marks MN progenitor state and is involved in MN specification, is dependent on the co-expression of *Nkx6* and *Pax6* which are required to prevent the expression of other transcription factors capable of repressing *Olig2* expression [159, 193–195]. *Olig2* itself then functions as a transcriptional repressor to direct the expression of downstream homeodomain regulators of MN identity via MNR2, Hb9 and LIM (Isl1/2 and Lim3) proteins, indicating a de-repression mechanism that functions during this later phase of MN specification [28, 194, 196–198].

Moreover, these signalling molecules are able to induce the expression of homeodomain genes that subsequently specify the identity of each of the classes of post-mitotic neurons that derive from individual progenitor domains (see class I and II genes in Shh). Thereby, gain-of-function experiments in chick neural tube have shown that individual homeodomain proteins are sufficient to change the fate and position at which individual types of neurons are generated, as predicted by the normal profile of homeodomain protein expression [159]. Additionally, there are predictable changes in progenitor domain identity and neuronal fate in mice in which individual homeodomain proteins have been knocked out by gene targeting [160, 193, 199, 200]. These observations indicate the involvement of homeodomain genes in the specification of neuronal subtype identities.

2.2.1 Common Spinal Cord Patterning Mechanisms Operate to Specify Neurons and Glial Cells

Early pulse chase labelling studies have indicated that, alike forebrain morphogenesis, development of neurons precedes that of glial subtypes during spinal cord development. However, the process of glial cell generation in spinal cord requires precise interplay between cell-intrinsic and regionally restricted extrinsic factors, which has much in common with the mechanisms that underlie the development of neurons. The first indications that neurons and glia might share common specification mechanisms came from the observation that oligodendrocyte progenitors emerge from a discrete region in the ventral neural tube [201], rather than from diffuse locations. In situ expression analysis of many markers for oligodendrocyte progenitors, such as PDGFRa and PLP/DM20 in mouse embryos [202, 203] and O4 in chick embryos [204], suggested that oligodendrocyte progenitor development is initiated in the same domain that generates motor neurons (pMN domain) (Fig. 8.3b). In mouse embryos, initially the pMN precursor domain, around embryonic day 9–10.5, gives rise in a first wave of motor neuron progenitors and in a later wave, around E12.5, to oligodendrocyte progenitors [205]. Subsequently these progenitors migrate out from the pMN to lateral and dorsal regions throughout the spinal cord. Finally, at post-natal stages, myelinating oligodendrocytes are found all over the spinal cord, most abundantly not only in white matter tracts but also in grey matter. Most importantly, Shh signalling is necessary and sufficient for oligodendrocyte development in the spinal cord, indicating further parallels with the development of motor neurons [206–210]. Several studies indicate that a prolonged period of Shh activity is necessary to ensure normal cell fate acquisition in pMN-oligodendrocyte progenitors, which is completed by about E12.5 in the mouse and stage 24 (E4) in the chick. Therefore, by using ventral chick spinal cord explants, isolated at various development stages, it was shown that neuroepithelial precursors are unable to generate oligodendrocytes in culture until E5 but become able to do so in an autonomous way from E5.5. These observations indicate that the induction of oligodendrocyte precursors is a late event that occurs between E5 and E5.5, precisely at the time when the ventral neuroepithelium stops producing motor neurons (Fig. 8.3b). Moreover, in the same experimental system Shh is sufficient to induce oligodendrocyte formation from ventral neuroepithelial explants dissected at E5 [208]. However, later stages of oligodendrocyte maturation are Shh independent [208, 210], consistent with the migration of these progenitors away from the original morphogenetic source of Shh.

Moreover, *Olig1* and *Olig2* genes, which are initially expressed in the pMN domain and genetic analysis showed that they lie downstream of Shh signalling [209, 211–213], are required for the specification of pMN domain [214]. Moreover, a null mutation of *Olig2* alone results in failure of development of pMN progeny [195, 215], including motor neurons and oligodendrocytes. These observations provided evidence for a single cell-intrinsic determinant that is essential for the specification of both neuronal and glial lineages.

3 Adult Neurogenesis

In adult mammals, neurogenesis occurs within two discrete brain regions, the dentate gyrus of the hippocampus and the subventricular zone (SVZ) lining the lateral wall of the lateral ventricles (LV) [216], a process sustained by the lifelong persistence of neural stem cells (NSCs) within these two areas (Fig. 8.5). However, the mechanisms determining how neurogenesis is restricted only to few regions in the adult, in contrast to its more widespread location during embryogenesis, largely depend on controlling the balance between precursor cell proliferation and differentiation. The SVZ is the richest source of NSCs in the adult CNS, providing a continuous supply of neuroblasts to the olfactory bulb (OB), one of the few structures remaining active in the adult forebrain [217, 218]. In this area, the source of permanent cell production consists of NSCs belonging to the astroglial lineage. Several observations have concluded indeed that a subset of astroglia acts as multipotent NSCs [219]. These, called type B cells, are slowly dividing astrocytes that

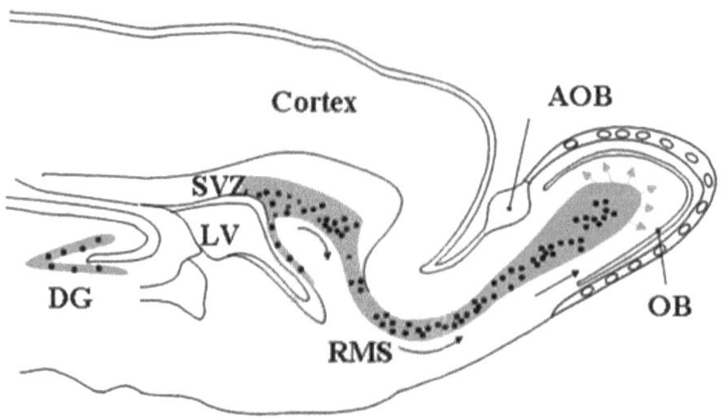

Fig. 8.5 Neurogenic areas in the adult brain. Schematic drawing of a sagittal section of the adult brain, where the two neurogenic regions – subventricular zone (SVZ) and dentate gyrus (DG) – are marked in *grey*. *Black dots* correspond to dividing cells. In the olfactory system *black arrows* reflect tangential migration through the rostral migratory stream (RMS) to the olfactory bulb (OB), whereas *grey arrows* reflect radial migration to the outer layers of the OB

generate committed neuronal precursors (type A cells), passing through an intermediate type of rapidly dividing cells (transit amplifying type C cells) [218, 220, 221]. Moreover, this subset of astrocytes has the capacity to restore adult neurogenesis after all rapidly proliferating cells have been eliminated [220]. An increasing number of molecules characterizing and/or participating in this B-C-A transition and cell fate restriction have been identified [4, 222]. These include basic fibroblast growth factor (bFGF), brain-derived neurotrophic factor (BDNF) and ciliary neurotrophic factor (CNTF) that regulate NSC maintenance and self-renewal within the SVZ niche [223, 224], while Notch1, acting downstream of CNTF, controls NSC numbers [225]. EGF receptors, on the other hand, are predominantly found in transit amplifying type C cells [218, 226, 227], controlling their fast proliferation rate in cooperation with sonic hedgehog (Shh) [228]. Numb/Numb-like proteins are present in all types of SVZ precursors and have been recently shown to participate in repair mechanisms of the subventricular niche [229]. Additionally, the transcription factor Mash1, which is transiently expressed in embryonic precursor cells, is also localized in transit amplifying type C cells, where it possibly controls the activation of downstream genes participating in cellular differentiation networks. Among factors also known to control neuronal progenitor proliferation and differentiation are unexpectedly neurotransmitters, such as dopamine [227] and GABA [230]. Although these findings have changed the scenario concerning adult NSC origin and identity, lack of specific markers for "stem-like" cells has hampered the identification and characterization of adult-derived NSCs. Recent findings suggest that during adult neurogenesis radial glial cells act as cellular intermediates between embryonic and adult neural stem cells [4]. So, one key question that arises is why all astrocytes of the mammalian adult brain cannot maintain this neurogenic potential like their ancestors, the radial glial cells? One major difference that may account for this is the fact that only the astrocytes that have an apical membrane contacting the ventricle undergo adult neurogenesis [218]. This finding might imply that, similarly to embryonic development, the polarity of adult NSC plays a crucial role in the control of adult neurogenesis.

3.1 Adult NSC: Possible Candidates for CNS Repair?

Due to their extreme plasticity, adult SVZ NSCs are very powerful candidates for cell replacement therapies of neurological disorders through their directed transplantation in damaged brain areas. Alternatively, stimulation of endogenous neurogenesis from SVZ neuronal precursors seems a suitable approach, given the fact that a self-repair mechanism exists in the SVZ "niche" in several animal models of neurodegeneration. Indeed, recent progress shows not only that neurons suitable for transplantation can be generated from adult NSCs maintained in culture but also that the adult brain itself produces new cells that differentiate into functional neurons in diseased areas [231, 232]. In support, a number of studies report initial migration of SVZ neuroblasts to distant areas triggered by brain injury, such as stroke [233] and endogenous activation of SVZ NSC in an animal model of inflammatory demyelination of the corpus callosum [231, 234]. In addition, striatal dopaminergic neuron degeneration in the 6-hydroxy-dopamine (6-OHDA) chemical model of Parkinson's disease (PD) in rodents impairs SVZ neural precursor generation most probably due to dopamine depletion [235], indicating that a feedback mechanism exists between neuronal production in the SVZ and neurodegeneration in neighbouring brain areas. More importantly, similar results to the ones reported in animal models have been also observed in post-mortem brains of patients with Huntington's disease [236] and multiple sclerosis [237] that exhibited a significant enhancement in cell density and proliferation of the SVZ area. By contrast, cell proliferation and neurogenesis are reduced in PD patients, as a result of dopamine depletion [226]. Although some extrinsic factors – mainly neurotransmitters and factors secreted by reactive astrocytes and macrophages involved in inflammation – controlling the response of the SVZ NSC "niche" in tissue damage have been identified, the intrinsic cascade of molecular events altering the precursor cell fate remains unknown. Therefore, study of the cell biology and identification of the intrinsic molecules and pathways stimulating or prohibiting proliferation and differentiation of endogenous precursors towards the neuronal lineage appears to be the first step for directing the controlled dispersal, functional integration and long-term survival of new neurons into host neural circuitries following brain damage.

Over the last century our concept of the adult central nervous system (CNS) as a rigid and inflexible structure has radically changed. It is now well recognized that the CNS is considerably more plastic than originally thought. In particular, the discovery that new neurons are continuously being generated throughout life has created a major breakthrough with important implications for the development of restorative approaches for CNS repair after damage. However, despite the new knowledge acquired and the progress made it is still a fact that the ability of the CNS for self-repair is very limited.

3.2 Conclusions and Future Perspectives

Although considerable progress has been made in the field NSC biology during the last decade, several fundamental questions that will create new knowledge on the basic mechanisms of neurogenesis remain to be answered. These can be summarized as follows:

1. Is adult neurogenesis a mere recapitulation of embryonic neurogenesis or a unique feature of the adult brain?
2. What is the physiological function of adult neurogenesis and why it is restricted to two specific regions of the adult brain as opposed to its widespread location during embryogenesis?
3. Which factors influence a NSC to adopt a neuronal versus a glial cell fate in the developing and adult brain and how their mechanisms of action are altered under pathological conditions?

To reply such questions new transgenic mice models are currently developed in which NSC and newborn neurons are in vivo ablated and essential changes in the molecular phenotype, cell fate and functional integration of remaining cells are monitored. As several groups have demonstrated that the adult SVZ is subjected to major modifications following neurological diseases, the mobilization of the neighbouring cell types to NSC ablation is a very suitable way to trace their flexibility and possible reprogramming in response to an environmental change. In these transgenic models it will also be very interesting to explore the role of molecules known to enhance the migration and differentiation properties of neuronal precursors, such as PSA-NCAM [238] and BM88/Cend1 [72], respectively, in promoting neuronal regeneration.

Apart from the subventricular zone and the hippocampal region, there is evidence, although still controversial, that adult NSCs also exist in other brain areas that are traditionally considered as non-neurogenic regions, e.g. the cortex, the substantia nigra and the spinal cord [239, 240]. Under the light of this new evidence originated with the use of more sensitive and reliable techniques, we should keep an open mind to the possibility of a more plastic and neurogenic adult brain than originally thought and try to explore the ways to enhance functional neurons production following neurodegeneration. Understanding the mechanisms of division and the physiological function of these cells will provide the means for appropriately manipulating them in order to use them as an alternative source of endogenous neural precursors for brain repair therapies in the future.

References

1. Cremisi F, Philpott A, Ohnuma S (2003) Cell cycle and cell fate interactions in neural development. Curr Opin Neurobiol 13(1):26–33
2. Ohnuma S, Harris WA (2003) Neurogenesis and the cell cycle. Neuron 40(2):199–208

3. Desai AR, McConnell SK (2000) Progressive restriction in fate potential by neural progenitors during cerebral cortical development. Development 127(13):2863–2872
4. Mori T, Buffo A, Gotz M (2005) The novel roles of glial cells revisited: the contribution of radial glia and astrocytes to neurogenesis. Curr Top Dev Biol 69:67–99
5. Temple S, Qian X (1995) bFGF, neurotrophins, and the control or cortical neurogenesis. Neuron 15(2):249–252
6. Dou CL, Li S, Lai E (1999) Dual role of brain factor-1 in regulating growth and patterning of the cerebral hemispheres. Cereb Cortex 9(6):543–550
7. Heins N, Cremisi F, Malatesta P et al (2001) Emx2 promotes symmetric cell divisions and a multipotential fate in precursors from the cerebral cortex. Mol Cell Neurosci 18(5):485–502
8. Dubreuil V, Hirsch MR, Pattyn A, Brunet JF, Goridis C (2000) The Phox2b transcription factor coordinately regulates neuronal cell cycle exit and identity. Development 127(23):5191–5201
9. Dyer MA, Cepko CL (2000) p57(Kip2) regulates progenitor cell proliferation and amacrine interneuron development in the mouse retina. Development 127(16):3593–3605
10. Nguyen L, Besson A, Roberts JM, Guillemot F (2006) Coupling cell cycle exit, neuronal differentiation and migration in cortical neurogenesis. Cell Cycle 5(20):2314–2318
11. Politis PK, Rohrer H, Matsas R (2007) Expression pattern of BM88 in the developing nervous system of the chick and mouse embryo. Gene Expr Patterns 7(1–2):165–177
12. Parnavelas JG, Papadopoulos GC (1989) The monoaminergic innervation of the cerebral cortex is not diffuse and nonspecific. Trends Neurosci 12(9):315–319
13. Tan SS, Kalloniatis M, Sturm K, Tam PP, Reese BE, Faulkner-Jones B (1998) Separate progenitors for radial and tangential cell dispersion during development of the cerebral neocortex. Neuron 21(2):295–304
14. Mione MC, Cavanagh JF, Harris B, Parnavelas JG (1997) Cell fate specification and symmetrical/asymmetrical divisions in the developing cerebral cortex. J Neurosci 17(6):2018–2029
15. Parnavelas JG (2000) The origin and migration of cortical neurones: new vistas. Trends Neurosci 23(3):126–131
16. Anderson SA, Marin O, Horn C, Jennings K, Rubenstein JL (2001) Distinct cortical migrations from the medial and lateral ganglionic eminences. Development 128(3):353–363
17. Lavdas AA, Grigoriou M, Pachnis V, Parnavelas JG (1999) The medial ganglionic eminence gives rise to a population of early neurons in the developing cerebral cortex. J Neurosci 19(18):7881–7888
18. Chenn A, McConnell SK (1995) Cleavage orientation and the asymmetric inheritance of Notch1 immunoreactivity in mammalian neurogenesis. Cell 82(4):631–641
19. Kornack DR, Rakic P (1995) Radial and horizontal deployment of clonally related cells in the primate neocortex: relationship to distinct mitotic lineages. Neuron 15(2):311–321
20. Gotz M, Williams BP, Bolz J, Price J (1995) The specification of neuronal fate: a common precursor for neurotransmitter subtypes in the rat cerebral cortex in vitro. Eur J Neurosci 7(5):889–898
21. Grove EA, Williams BP, Li DQ, Hajihosseini M, Friedrich A, Price J (1993) Multiple restricted lineages in the embryonic rat cerebral cortex. Development 117(2):553–561
22. Mione MC, Danevic C, Boardman P, Harris B, Parnavelas JG (1994) Lineage analysis reveals neurotransmitter (GABA or glutamate) but not calcium-binding protein homogeneity in clonally related cortical neurons. J Neurosci 14(1):107–123
23. Morrow T, Song MR, Ghosh A (2001) Sequential specification of neurons and glia by developmentally regulated extracellular factors. Development 128(18):3585–3594
24. Qian X, Shen Q, Goderie SK et al (2000) Timing of CNS cell generation: a programmed sequence of neuron and glial cell production from isolated murine cortical stem cells. Neuron 28(1):69–80
25. Nieto M, Schuurmans C, Britz O, Guillemot F (2001) Neural bHLH genes control the neuronal versus glial fate decision in cortical progenitors. Neuron 29(2):401–413

26. Kageyama R, Ohtsuka T, Hatakeyama J, Ohsawa R (2005) Roles of bHLH genes in neural stem cell differentiation. Exp Cell Res 306(2):343–348
27. Simeone A (1998) Otx1 and Otx2 in the development and evolution of the mammalian brain. Embo J 17(23):6790–6798
28. Scardigli R, Schuurmans C, Gradwohl G, Guillemot F (2001) Crossregulation between Neurogenin2 and pathways specifying neuronal identity in the spinal cord. Neuron 31(2):203–217
29. Fode C, Ma Q, Casarosa S, Ang SL, Anderson DJ, Guillemot F (2000) A role for neural determination genes in specifying the dorsoventral identity of telencephalic neurons. Genes Dev 14(1):67–80
30. Hatakeyama J, Bessho Y, Katoh K et al (2004) Hes genes regulate size, shape and histogenesis of the nervous system by control of the timing of neural stem cell differentiation. Development 131(22):5539–5550
31. Hatakeyama J, Kageyama R (2006) Notch1 expression is spatiotemporally correlated with neurogenesis and negatively regulated by Notch1-independent Hes genes in the developing nervous system. Cereb Cortex 16(Suppl 1):i132–37
32. Farah MH, Olson JM, Sucic HB, Hume RI, Tapscott SJ, Turner DL (2000) Generation of neurons by transient expression of neural bHLH proteins in mammalian cells. Development 127(4):693–702
33. Vetter M (2001) A turn of the helix: preventing the glial fate. Neuron 29(3):559–562
34. Inoue T, Hojo M, Bessho Y, Tano Y, Lee JE, Kageyama R (2002) Math3 and NeuroD regulate amacrine cell fate specification in the retina. Development 129(4):831–842
35. Bertrand N, Medevielle F, Pituello F (2000) FGF signalling controls the timing of Pax6 activation in the neural tube. Development 127(22):4837–4843
36. Miyata T, Kawaguchi A, Okano H, Ogawa M (2001) Asymmetric inheritance of radial glial fibers by cortical neurons. Neuron 31(5):727–741
37. Hall AC, Mira H, Wagner J, Arenas E (2003) Region-specific effects of glia on neuronal induction and differentiation with a focus on dopaminergic neurons. Glia 43(1): 47–51
38. Anthony TE, Klein C, Fishell G, Heintz N (2004) Radial glia serve as neuronal progenitors in all regions of the central nervous system. Neuron 41(6):881–890
39. Malatesta P, Hack MA, Hartfuss E et al (2003) Neuronal or glial progeny: regional differences in radial glia fate. Neuron 37(5):751–764
40. Noctor SC, Flint AC, Weissman TA, Wong WS, Clinton BK, Kriegstein AR (2002) Dividing precursor cells of the embryonic cortical ventricular zone have morphological and molecular characteristics of radial glia. J Neurosci 22(8):3161–3173
41. Klezovitch O, Fernandez TE, Tapscott SJ, Vasioukhin V (2004) Loss of cell polarity causes severe brain dysplasia in Lgl1 knockout mice. Genes Dev 18(5):559–571
42. Kosodo Y, Roper K, Haubensak W, Marzesco AM, Corbeil D, Huttner WB (2004) Asymmetric distribution of the apical plasma membrane during neurogenic divisions of mammalian neuroepithelial cells. Embo J 23(11):2314–2324
43. Frise E, Knoblich JA, Younger-Shepherd S, Jan LY, Jan YN (1996) The Drosophila Numb protein inhibits signaling of the Notch receptor during cell-cell interaction in sensory organ lineage. Proc Natl Acad Sci U S A 93(21):11925–11932
44. Rhyu MS, Jan LY, Jan YN (1994) Asymmetric distribution of numb protein during division of the sensory organ precursor cell confers distinct fates to daughter cells. Cell 76(3): 477–491
45. Arduengo PM, Appleberry OK, Chuang P, L'Hernault SW (1998) The presenilin protein family member SPE-4 localizes to an ER/Golgi derived organelle and is required for proper cytoplasmic partitioning during Caenorhabditis elegans spermatogenesis. J Cell Sci 111 (Pt 24):3645–3654
46. Wakamatsu Y, Maynard TM, Jones SU, Weston JA (1999) NUMB localizes in the basal cortex of mitotic avian neuroepithelial cells and modulates neuronal differentiation by binding to NOTCH-1. Neuron 23(1):71–81

47. Zhong W, Feder JN, Jiang MM, Jan LY, Jan YN (1996) Asymmetric localization of a mammalian numb homolog during mouse cortical neurogenesis. Neuron 17(1):43–53
48. Zhong W, Jiang MM, Weinmaster G, Jan LY, Jan YN (1997) Differential expression of mammalian Numb, Numblike and Notch1 suggests distinct roles during mouse cortical neurogenesis. Development 124(10):1887–1897
49. Cayouette M, Raff M (2002) Asymmetric segregation of Numb: a mechanism for neural specification from Drosophila to mammals. Nat Neurosci 5(12):1265–1269
50. Shen Q, Zhong W, Jan YN, Temple S (2002) Asymmetric Numb distribution is critical for asymmetric cell division of mouse cerebral cortical stem cells and neuroblasts. Development 129(20):4843–4853
51. Malatesta P, Gotz M, Barsacchi G, Price J, Zoncu R, Cremisi F (2000) PC3 overexpression affects the pattern of cell division of rat cortical precursors. Mech Dev 90(1):17–28
52. Ferguson KL, Vanderluit JL, Hebert JM et al (2002) Telencephalon-specific Rb knockouts reveal enhanced neurogenesis, survival and abnormal cortical development. Embo J 21(13):3337–3346
53. Estivill-Torrus G, Pearson H, van Heyningen V, Price DJ, Rashbass P (2002) Pax6 is required to regulate the cell cycle and the rate of progression from symmetrical to asymmetrical division in mammalian cortical progenitors. Development 129(2):455–466
54. Heins N, Malatesta P, Cecconi F et al (2002) Glial cells generate neurons: the role of the transcription factor Pax6. Nat Neurosci 5(4):308–315
55. Koutmani Y, Hurel C, Patsavoudi E et al (2004) BM88 is an early marker of proliferating precursor cells that will differentiate into the neuronal lineage. Eur J Neurosci 20(10): 2509–2523
56. Edlund T, Jessell TM (1999) Progression from extrinsic to intrinsic signaling in cell fate specification: a view from the nervous system. Cell 96(2):211–224
57. Ohnuma S, Hopper S, Wang KC, Philpott A, Harris WA (2002) Co-ordinating retinal histogenesis: early cell cycle exit enhances early cell fate determination in the Xenopus retina. Development 129(10):2435–2446
58. Isshiki T, Doe CQ (2004) Maintaining youth in Drosophila neural progenitors. Cell Cycle 3(3):296–299
59. Li L, Vaessin H (2000) Pan-neural Prospero terminates cell proliferation during Drosophila neurogenesis. Genes Dev 14(2):147–151
60. Torii M, Matsuzaki F, Osumi N et al (1999) Transcription factors Mash-1 and Prox-1 delineate early steps in differentiation of neural stem cells in the developing central nervous system. Development 126(3):443–456
61. Toma JG, El-Bizri H, Barnabe-Heider F, Aloyz R, Miller FD (2000) Evidence that helix-loop-helix proteins collaborate with retinoblastoma tumor suppressor protein to regulate cortical neurogenesis. J Neurosci 20(20):7648–7656
62. Dyer MA, Cepko CL (2001) p27Kip1 and p57Kip2 regulate proliferation in distinct retinal progenitor cell populations. J Neurosci 21(12):4259–4271
63. Takahashi T, Nowakowski RS, Caviness VS Jr (1996) The leaving or Q fraction of the murine cerebral proliferative epithelium: a general model of neocortical neuronogenesis. J Neurosci 16(19):6183–6196
64. Bellaiche Y, Radovic A, Woods DF et al (2001) The Partner of Inscuteable/Discs-large complex is required to establish planar polarity during asymmetric cell division in Drosophila. Cell 106(3):355–366
65. Chenn A, Walsh CA (2002) Regulation of cerebral cortical size by control of cell cycle exit in neural precursors. Science 297(5580):365–369
66. Lukaszewicz A, Savatier P, Cortay V, Kennedy H, Dehay C (2002) Contrasting effects of basic fibroblast growth factor and neurotrophin 3 on cell cycle kinetics of mouse cortical stem cells. J Neurosci 22(15):6610–6622
67. Calegari F, Huttner WB (2003) An inhibition of cyclin-dependent kinases that lengthens, but does not arrest, neuroepithelial cell cycle induces premature neurogenesis. J Cell Sci 116 (Pt 24):4947–4955

68. Malatesta P, Hartfuss E, Gotz M (2000) Isolation of radial glial cells by fluorescent-activated cell sorting reveals a neuronal lineage. Development 127(24):5253–5263
69. Canzoniere D, Farioli-Vecchioli S, Conti F et al (2004) Dual control of neurogenesis by PC3 through cell cycle inhibition and induction of Math1. J Neurosci 24(13):3355–3369
70. Hodge RD, D'Ercole AJ, O'Kusky JR (2004) Insulin-like growth factor-I accelerates the cell cycle by decreasing G1 phase length and increases cell cycle reentry in the embryonic cerebral cortex. J Neurosci 24(45):10201–10210
71. Calegari F, Haubensak W, Haffner C, Huttner WB (2005) Selective lengthening of the cell cycle in the neurogenic subpopulation of neural progenitor cells during mouse brain development. J Neurosci 25(28):6533–6538
72. Georgopoulou N, Hurel C, Politis PK, Gaitanou M, Matsas R, Thomaidou D (2006) BM88 is a dual function molecule inducing cell cycle exit and neuronal differentiation of neuroblastoma cells via cyclin D1 down-regulation and retinoblastoma protein hypophosphorylation. J Biol Chem 281(44):33606–33620
73. Geng Y, Yu Q, Sicinska E et al (2003) Cyclin E ablation in the mouse. Cell 114(4):431–443
74. Wu CF, Sakai K, Saito M, Hotta Y (1990) Giant Drosophila neurons differentiated from cytokinesis-arrested embryonic neuroblasts. J Neurobiol 21(3):499–507
75. Backman SA, Stambolic V, Suzuki A et al (2001) Deletion of Pten in mouse brain causes seizures, ataxia and defects in soma size resembling Lhermitte-Duclos disease. Nat Genet 29(4):396–403
76. Tapon N, Moberg KH, Hariharan IK (2001) The coupling of cell growth to the cell cycle. Curr Opin Cell Biol 13(6):731–737
77. Saucedo LJ, Gao X, Chiarelli DA, Li L, Pan D, Edgar BA (2003) Rheb promotes cell growth as a component of the insulin/TOR signalling network. Nat Cell Biol 5(6):566–571
78. Fero ML, Rivkin M, Tasch M et al (1996) A syndrome of multiorgan hyperplasia with features of gigantism, tumorigenesis, and female sterility in p27(Kip1)-deficient mice. Cell 85(5):733–744
79. Nakayama K, Ishida N, Shirane M et al (1996) Mice lacking p27(Kip1) display increased body size, multiple organ hyperplasia, retinal dysplasia, and pituitary tumors. Cell 85(5):707–720
80. Hu M, Easter SS (1999) Retinal neurogenesis: the formation of the initial central patch of postmitotic cells. Dev Biol 207(2):309–321
81. McConnell SK (1995) Constructing the cerebral cortex: neurogenesis and fate determination. Neuron 15(4):761–768
82. Mittnacht S (1998) Control of pRB phosphorylation. Curr Opin Genet Dev 8(1):21–27
83. Bode AM, Dong Z (2004) Post-translational modification of p53 in tumorigenesis. Nat Rev Cancer 4(10):793–805
84. Tweddle DA, Pearson AD, Haber M et al (2003) The p53 pathway and its inactivation in neuroblastoma. Cancer Lett 197(1–2):93–98
85. Coqueret O (2003) New roles for p21 and p27 cell-cycle inhibitors: a function for each cell compartment? Trends Cell Biol 13(2):65–70
86. Deng C, Zhang P, Harper JW, Elledge SJ, Leder P (1995) Mice lacking p21CIP1/WAF1 undergo normal development, but are defective in G1 checkpoint control. Cell 82(4): 675–684
87. Cheng M, Olivier P, Diehl JA et al (1999) The p21(Cip1) and p27(Kip1) CDK 'inhibitors' are essential activators of cyclin D-dependent kinases in murine fibroblasts. Embo J 18(6): 1571–1583
88. Sherr CJ, Roberts JM (1999) CDK inhibitors: positive and negative regulators of G1-phase progression. Genes Dev 13(12):1501–1512
89. Sanchez I, Dynlacht BD (2005) New insights into cyclins, CDKs, and cell cycle control. Semin Cell Dev Biol 16(3):311–321
90. Weinberg RA (1996) E2F and cell proliferation: a world turned upside down. Cell 85(4):457–459

91. Billon N, Carlisi D, Datto MB et al (1999) Cooperation of Sp1 and p300 in the induction of the CDK inhibitor p21WAF1/CIP1 during NGF-mediated neuronal differentiation. Oncogene 18(18):2872–2882
92. Yan GZ, Ziff EB (1997) Nerve growth factor induces transcription of the p21 WAF1/CIP1 and cyclin D1 genes in PC12 cells by activating the Sp1 transcription factor. J Neurosci 17(16):6122–6132
93. Koutsodontis G, Tentes I, Papakosta P, Moustakas A, Kardassis D (2001) Sp1 plays a critical role in the transcriptional activation of the human cyclin-dependent kinase inhibitor p21(WAF1/Cip1) gene by the p53 tumor suppressor protein. J Biol Chem 276(31): 29116–29125
94. Sumrejkanchanakij P, Tamamori-Adachi M, Matsunaga Y, Eto K, Ikeda MA (2003) Role of cyclin D1 cytoplasmic sequestration in the survival of postmitotic neurons. Oncogene 22(54):8723–8730
95. Sumrejkanchanakij P, Eto K, Ikeda MA (2006) Cytoplasmic sequestration of cyclin D1 associated with cell cycle withdrawal of neuroblastoma cells. Biochem Biophys Res Commun 340(1):302–308
96. Bartek J, Bartkova J, Lukas J (1996) The retinoblastoma protein pathway and the restriction point. Curr Opin Cell Biol 8(6):805–814
97. DelSal G, Loda M, Pagano M (1996) Cell cycle and cancer: critical events at the G1 restriction point. Crit Rev Oncog 7(1–2):127–142
98. Carey RG, Li B, DiCicco-Bloom E (2002) Pituitary adenylate cyclase activating polypeptide anti-mitogenic signaling in cerebral cortical progenitors is regulated by p57Kip2-dependent CDK2 activity. J Neurosci 22(5):1583–1591
99. Liu Y, Encinas M, Comella JX, Aldea M, Gallego C (2004) Basic helix-loop-helix proteins bind to TrkB and p21(Cip1) promoters linking differentiation and cell cycle arrest in neuroblastoma cells. Mol Cell Biol 24(7):2662–2672
100. Gallo R, Zazzeroni F, Alesse E et al (2002) REN: a novel, developmentally regulated gene that promotes neural cell differentiation. J Cell Biol 158(4):731–740
101. Quinn LM, Herr A, McGarry TJ, Richardson H (2001) The Drosophila Geminin homolog: roles for Geminin in limiting DNA replication, in anaphase and in neurogenesis. Genes Dev 15(20):2741–2754
102. Seo S, Kroll KL (2006) Geminin's double life: chromatin connections that regulate transcription at the transition from proliferation to differentiation. Cell Cycle 5(4):374–379
103. Seo S, Herr A, Lim JW, Richardson GA, Richardson H, Kroll KL (2005) Geminin regulates neuronal differentiation by antagonizing Brg1 activity. Genes Dev 19(14):1723–1734
104. Spencer ML, Theodosiou M, Noonan DJ (2004) NPDC-1, a novel regulator of neuronal proliferation, is degraded by the ubiquitin/proteasome system through a PEST degradation motif. J Biol Chem 279(35):37069–37078
105. Yano M, Okano HJ, Okano H (2005) Involvement of Hu and heterogeneous nuclear ribonucleoprotein K in neuronal differentiation through p21 mRNA post-transcriptional regulation. J Biol Chem 280(13):12690–12699
106. Chae JH, Stein GH, Lee JE (2004) NeuroD: the predicted and the surprising. Mol Cells 18(3):271–288
107. Schuurmans C, Armant O, Nieto M et al (2004) Sequential phases of cortical specification involve Neurogenin-dependent and -independent pathways. Embo J 23(14):2892–2902
108. Haubst N, Berger J, Radjendirane V et al (2004) Molecular dissection of Pax6 function: the specific roles of the paired domain and homeodomain in brain development. Development 131(24):6131–6140
109. Baek SH, Kioussi C, Briata P et al (2003) Regulated subset of G1 growth-control genes in response to derepression by the Wnt pathway. Proc Natl Acad Sci U S A 100(6):3245–3250
110. Kioussi C, Briata P, Baek SH et al (2002) Identification of a Wnt/Dvl/beta-Catenin --> Pitx2 pathway mediating cell-type-specific proliferation during development. Cell 111(5): 673–685

111. Oliver TG, Grasfeder LL, Carroll AL et al (2003) Transcriptional profiling of the Sonic hedgehog response: a critical role for N-myc in proliferation of neuronal precursors. Proc Natl Acad Sci U S A 100(12):7331–7336
112. Suzui M, Masuda M, Lim JT, Albanese C, Pestell RG, Weinstein IB (2002) Growth inhibition of human hepatoma cells by acyclic retinoid is associated with induction of p21(CIP1) and inhibition of expression of cyclin D1. Cancer Res 62(14):3997–4006
113. Geelen JA, Langman J (1977) Closure of the neural tube in the cephalic region of the mouse embryo. Anat Rec 189(4):625–640
114. Homma S, Yaginuma H, Oppenheim RW (1994) Programmed cell death during the earliest stages of spinal cord development in the chick embryo: a possible means of early phenotypic selection. J Comp Neurol 345(3):377–395
115. Barres BA, Hart IK, Coles HS et al (1992) Cell death and control of cell survival in the oligodendrocyte lineage. Cell 70(1):31–46
116. Oppenheim RW (1991) Cell death during development of the nervous system. Annu Rev Neurosci 14:453–501
117. Meikrantz W, Gisselbrecht S, Tam SW, Schlegel R (1994) Activation of cyclin A-dependent protein kinases during apoptosis. Proc Natl Acad Sci U S A 91(9):3754–3758
118. Kuerbitz SJ, Plunkett BS, Walsh WV, Kastan MB (1992) Wild-type p53 is a cell cycle checkpoint determinant following irradiation. Proc Natl Acad Sci U S A 89(16): 7491–7495
119. Xiong Y, Hannon GJ, Zhang H, Casso D, Kobayashi R, Beach D (1993) p21 is a universal inhibitor of cyclin kinases. Nature 366(6456):701–704
120. Freeman RS, Estus S, Johnson EM Jr (1994) Analysis of cell cycle-related gene expression in postmitotic neurons: selective induction of Cyclin D1 during programmed cell death. Neuron 12(2):343–355
121. Pardee AB (1989) G1 events and regulation of cell proliferation. Science 246(4930): 603–608
122. Haas-Kogan DA, Kogan SC, Levi D et al (1995) Inhibition of apoptosis by the retinoblastoma gene product. Embo J 14(3):461–472
123. Reed JC (1994) Bcl-2 and the regulation of programmed cell death. J Cell Biol 124(1–2):1–6
124. Ferrer I, Soriano E, del Rio JA, Alcantara S, Auladell C (1992) Cell death and removal in the cerebral cortex during development. Prog Neurobiol 39(1):1–43
125. Reznikov K, van der Kooy D (1995) Variability and partial synchrony of the cell cycle in the germinal zone of the early embryonic cerebral cortex. J Comp Neurol 360(3):536–554
126. Spreafico R, Frassoni C, Arcelli P, Selvaggio M, De Biasi S (1995) In situ labeling of apoptotic cell death in the cerebral cortex and thalamus of rats during development. J Comp Neurol 363(2):281–295
127. Thomaidou D, Mione MC, Cavanagh JF, Parnavelas JG (1997) Apoptosis and its relation to the cell cycle in the developing cerebral cortex. J Neurosci 17(3):1075–1085
128. Bursch W, Paffe S, Putz B, Barthel G, Schulte-Hermann R (1990) Determination of the length of the histological stages of apoptosis in normal liver and in altered hepatic foci of rats. Carcinogenesis 11(5):847–853
129. Greene LA, Liu DX, Troy CM, Biswas SC (2007) Cell cycle molecules define a pathway required for neuron death in development and disease. Biochim Biophys Acta 1772(4): 392–401
130. Bang AG, Papalopulu N, Goulding MD, Kintner C (1999) Expression of Pax-3 in the lateral neural plate is dependent on a Wnt-mediated signal from posterior nonaxial mesoderm. Dev Biol 212(2):366–380
131. Doniach T (1995) Basic FGF as an inducer of anteroposterior neural pattern. Cell 83(7):1067–1070
132. Muhr J, Graziano E, Wilson S, Jessell TM, Edlund T (1999) Convergent inductive signals specify midbrain, hindbrain, and spinal cord identity in gastrula stage chick embryos. Neuron 23(4):689–702

133. Storey KG, Goriely A, Sargent CM et al (1998) Early posterior neural tissue is induced by FGF in the chick embryo. Development 125(3):473–484
134. Bally-Cuif L, Hammerschmidt M (2003) Induction and patterning of neuronal development, and its connection to cell cycle control. Curr Opin Neurobiol 13(1):16–25
135. Megason SG, McMahon AP (2002) A mitogen gradient of dorsal midline Wnts organizes growth in the CNS. Development 129(9):2087–2098
136. Marquardt T, Pfaff SL (2001) Cracking the transcriptional code for cell specification in the neural tube. Cell 106(6):651–654
137. Jessell TM (2000) Neuronal specification in the spinal cord: inductive signals and transcriptional codes. Nat Rev Genet 1(1):20–29
138. Helms AW, Johnson JE (2003) Specification of dorsal spinal cord interneurons. Curr Opin Neurobiol 13(1):42–49
139. Mathis L, Kulesa PM, Fraser SE (2001) FGF receptor signalling is required to maintain neural progenitors during Hensen's node progression. Nat Cell Biol 3(6):559–566
140. Diez del Corral R, Storey KG (2004) Opposing FGF and retinoid pathways: a signalling switch that controls differentiation and patterning onset in the extending vertebrate body axis. Bioessays 26(8):857–869
141. Diez del Corral R, Breitkreuz DN, Storey KG (2002) Onset of neuronal differentiation is regulated by paraxial mesoderm and requires attenuation of FGF signalling. Development 129(7):1681–1691
142. Diez del Corral R, Olivera-Martinez I, Goriely A, Gale E, Maden M, Storey K (2003) Opposing FGF and retinoid pathways control ventral neural pattern, neuronal differentiation, and segmentation during body axis extension. Neuron 40(1):65–79
143. Lupo G, Harris WA, Lewis KE (2006) Mechanisms of ventral patterning in the vertebrate nervous system. Nat Rev Neurosci 7(2):103–114
144. Litingtung Y, Chiang C (2000) Control of Shh activity and signaling in the neural tube. Dev Dyn 219(2):143–154
145. Briscoe J, Ericson J (1999) The specification of neuronal identity by graded Sonic Hedgehog signalling. Semin Cell Dev Biol 10(3):353–362
146. Briscoe J, Ericson J (2001) Specification of neuronal fates in the ventral neural tube. Curr Opin Neurobiol 11(1):43–49
147. Wilson L, Maden M (2005) The mechanisms of dorsoventral patterning in the vertebrate neural tube. Dev Biol 282(1):1–13
148. Ruiz i Altaba A, Nguyen V, Palma V (2003) The emergent design of the neural tube: prepattern, SHH morphogen and GLI code. Curr Opin Genet Dev 13(5):513–521
149. Wijgerde M, McMahon JA, Rule M, McMahon AP (2002) A direct requirement for Hedgehog signaling for normal specification of all ventral progenitor domains in the presumptive mammalian spinal cord. Genes Dev 16(22):2849–2864
150. Wilson L, Gale E, Chambers D, Maden M (2004) Retinoic acid and the control of dorsoventral patterning in the avian spinal cord. Dev Biol 269(2):433–446
151. Pierani A, Brenner-Morton S, Chiang C, Jessell TM (1999) A sonic hedgehog-independent, retinoid-activated pathway of neurogenesis in the ventral spinal cord. Cell 97(7):903–915
152. Ingham PW (1998) Transducing Hedgehog: the story so far. Embo J 17(13): 3505–3511
153. Marigo V, Tabin CJ (1996) Regulation of patched by sonic hedgehog in the developing neural tube. Proc Natl Acad Sci U S A 93(18):9346–9351
154. Goodrich LV, Johnson RL, Milenkovic L, McMahon JA, Scott MP (1996) Conservation of the hedgehog/patched signaling pathway from flies to mice: induction of a mouse patched gene by Hedgehog. Genes Dev 10(3):301–312
155. Hynes M, Ye W, Wang K et al (2000) The seven-transmembrane receptor smoothened cell-autonomously induces multiple ventral cell types. Nat Neurosci 3(1):41–46
156. Ruiz i Altaba A (1999) Gli proteins and Hedgehog signaling: development and cancer. Trends Genet 15(10):418–425

157. Stamataki D, Ulloa F, Tsoni SV, Mynett A, Briscoe J (2005) A gradient of Gli activity mediates graded Sonic Hedgehog signaling in the neural tube. Genes Dev 19(5):626–641
158. Briscoe J, Chen Y, Jessell TM, Struhl G (2001) A hedgehog-insensitive form of patched provides evidence for direct long-range morphogen activity of sonic hedgehog in the neural tube. Mol Cell 7(6):1279–1291
159. Briscoe J, Pierani A, Jessell TM, Ericson J (2000) A homeodomain protein code specifies progenitor cell identity and neuronal fate in the ventral neural tube. Cell 101(4):435–445
160. Briscoe J, Sussel L, Serup P et al (1999) Homeobox gene Nkx2.2 and specification of neuronal identity by graded Sonic hedgehog signalling. Nature 398(6728):622–627
161. Muhr J, Andersson E, Persson M, Jessell TM, Ericson J (2001) Groucho-mediated transcriptional repression establishes progenitor cell pattern and neuronal fate in the ventral neural tube. Cell 104(6):861–873
162. Tanabe Y, William C, Jessell TM (1998) Specification of motor neuron identity by the MNR2 homeodomain protein. Cell 95(1):67–80
163. Pituello F, Medevielle F, Foulquier F, Duprat AM (1998) Activation of Pax6 depends on somitogenesis in the chick embryo cervical spinal cord. Development x 126(3):587–596
164. Swindell EC, Thaller C, Sockanathan S, Petkovich M, Jessell TM, Eichele G (1999) Complementary domains of retinoic acid production and degradation in the early chick embryo. Dev Biol 216(1):282–296
165. Maden M, Sonneveld E, van der Saag PT, Gale E (1998) The distribution of endogenous retinoic acid in the chick embryo: implications for developmental mechanisms. Development 125(21):4133–4144
166. Niederreither K, McCaffery P, Drager UC, Chambon P, Dolle P (1997) Restricted expression and retinoic acid-induced downregulation of the retinaldehyde dehydrogenase type 2 (RALDH-2) gene during mouse development. Mech Dev 62(1):67–78
167. Blentic A, Gale E, Maden M (2003) Retinoic acid signalling centres in the avian embryo identified by sites of expression of synthesising and catabolising enzymes. Dev Dyn 227(1):114–127
168. Berggren K, McCaffery P, Drager U, Forehand CJ (1999) Differential distribution of retinoic acid synthesis in the chicken embryo as determined by immunolocalization of the retinoic acid synthetic enzyme, RALDH-2. Dev Biol 210(2):288–304
169. Wilson L, Gale E, Maden M (2003) The role of retinoic acid in the morphogenesis of the neural tube. J Anat 203(4):357–368
170. Novitch BG, Wichterle H, Jessell TM, Sockanathan S (2003) A requirement for retinoic acid-mediated transcriptional activation in ventral neural patterning and motor neuron specification. Neuron 40(1):81–95
171. Niederreither K, Vermot J, Fraulob V, Chambon P, Dolle P (2002) Retinaldehyde dehydrogenase 2 (RALDH2)- independent patterns of retinoic acid synthesis in the mouse embryo. Proc Natl Acad Sci U S A 99(25):16111–16116
172. Mic FA, Haselbeck RJ, Cuenca AE, Duester G (2002) Novel retinoic acid generating activities in the neural tube and heart identified by conditional rescue of Raldh2 null mutant mice. Development 129(9):2271–2282
173. Zhao D, McCaffery P, Ivins KJ et al (1996) Molecular identification of a major retinoic-acid-synthesizing enzyme, a retinaldehyde-specific dehydrogenase. Eur J Biochem 240(1):15–22
174. Solomin L, Johansson CB, Zetterstrom RH et al (1998) Retinoid-X receptor signalling in the developing spinal cord. Nature 395(6700):398–402
175. Sockanathan S, Jessell TM (1998) Motor neuron-derived retinoid signaling specifies the subtype identity of spinal motor neurons. Cell 94(4):503–514
176. Liem KF Jr, Tremml G, Jessell TM (1997) A role for the roof plate and its resident TGFbeta-related proteins in neuronal patterning in the dorsal spinal cord. Cell 91(1):127–138
177. Liem KF Jr, Tremml G, Roelink H, Jessell TM (1995) Dorsal differentiation of neural plate cells induced by BMP-mediated signals from epidermal ectoderm. Cell 82(6): 969–979

178. Lee KJ, Jessell TM (1999) The specification of dorsal cell fates in the vertebrate central nervous system. Annu Rev Neurosci 22:261–294
179. Dickinson ME, Selleck MA, McMahon AP, Bronner-Fraser M (1995) Dorsalization of the neural tube by the non-neural ectoderm. Development 121(7):2099–2106
180. Garcia-Castro MI, Marcelle C, Bronner-Fraser M (2002) Ectodermal Wnt function as a neural crest inducer. Science 297(5582):848–851
181. Lee KJ, Dietrich P, Jessell TM (2000) Genetic ablation reveals that the roof plate is essential for dorsal interneuron specification. Nature 403(6771):734–740
182. Basler K, Edlund T, Jessell TM, Yamada T (1993) Control of cell pattern in the neural tube: regulation of cell differentiation by dorsalin-1, a novel TGF beta family member. Cell 73(4):687–702
183. Timmer JR, Wang C, Niswander L (2002) BMP signaling patterns the dorsal and intermediate neural tube via regulation of homeobox and helix-loop-helix transcription factors. Development 129(10):2459–2472
184. Nguyen VH, Trout J, Connors SA, Andermann P, Weinberg E, Mullins MC (2000) Dorsal and intermediate neuronal cell types of the spinal cord are established by a BMP signaling pathway. Development 127(6):1209–1220
185. Barth KA, Kishimoto Y, Rohr KB, Seydler C, Schulte-Merker S, Wilson SW (1999) Bmp activity establishes a gradient of positional information throughout the entire neural plate. Development 126(22):4977–4987
186. Wine-Lee L, Ahn KJ, Richardson RD, Mishina Y, Lyons KM, Crenshaw EB 3rd (2004) Signaling through BMP type 1 receptors is required for development of interneuron cell types in the dorsal spinal cord. Development 131(21):5393–5403
187. Muroyama Y, Fujihara M, Ikeya M, Kondoh H, Takada S (2002) Wnt signaling plays an essential role in neuronal specification of the dorsal spinal cord. Genes Dev 16(5): 548–553
188. Mannervik M, Nibu Y, Zhang H, Levine M (1999) Transcriptional coregulators in development. Science 284(5414):606–609
189. Lee SK, Pfaff SL (2001) Transcriptional networks regulating neuronal identity in the developing spinal cord. Nat Neurosci 4(Suppl):1183–1191
190. Barolo S, Posakony JW (2002) Three habits of highly effective signaling pathways: principles of transcriptional control by developmental cell signaling. Genes Dev 16(10): 1167–1181
191. Thaler JP, Koo SJ, Kania A et al (2004) A postmitotic role for Isl-class LIM homeodomain proteins in the assignment of visceral spinal motor neuron identity. Neuron 41(3):337–350
192. Shirasaki R, Pfaff SL (2002) Transcriptional codes and the control of neuronal identity. Annu Rev Neurosci 25:251–281
193. Sander M, Paydar S, Ericson J et al (2000) Ventral neural patterning by Nkx homeobox genes: Nkx6.1 controls somatic motor neuron and ventral interneuron fates. Genes Dev 14(17):2134–2139
194. Novitch BG, Chen AI, Jessell TM (2001) Coordinate regulation of motor neuron subtype identity and pan-neuronal properties by the bHLH repressor Olig2. Neuron 31(5):773–789
195. Lu QR, Sun T, Zhu Z et al (2002) Common developmental requirement for Olig function indicates a motor neuron/oligodendrocyte connection. Cell 109(1):75–86
196. William CM, Tanabe Y, Jessell TM (2003) Regulation of motor neuron subtype identity by repressor activity of Mnx class homeodomain proteins. Development 130(8):1523–1536
197. Rowitch DH, Lu QR, Kessaris N, Richardson WD (2002) An 'oligarchy' rules neural development. Trends Neurosci 25(8):417–422
198. Mizuguchi R, Sugimori M, Takebayashi H et al (2001) Combinatorial roles of olig2 and neurogenin2 in the coordinated induction of pan-neuronal and subtype-specific properties of motoneurons. Neuron 31(5):757–771
199. Ericson J, Rashbass P, Schedl A et al (1997) Pax6 controls progenitor cell identity and neuronal fate in response to graded Shh signaling. Cell 90(1):169–180

200. Mansouri A, Gruss P (1998) Pax3 and Pax7 are expressed in commissural neurons and restrict ventral neuronal identity in the spinal cord. Mech Dev 78(1–2):171–178
201. Warf BC, Fok-Seang J, Miller RH (1991) Evidence for the ventral origin of oligodendrocyte precursors in the rat spinal cord. J Neurosci 11(8):2477–2488
202. Timsit S, Martinez S, Allinquant B, Peyron F, Puelles L, Zalc B (1995) Oligodendrocytes originate in a restricted zone of the embryonic ventral neural tube defined by DM-20 mRNA expression. J Neurosci 15(2):1012–1024
203. Pringle NP, Richardson WD (1993) A singularity of PDGF alpha-receptor expression in the dorsoventral axis of the neural tube may define the origin of the oligodendrocyte lineage. Development 117(2):525–533
204. Ono K, Bansal R, Payne J, Rutishauser U, Miller RH (1995) Early development and dispersal of oligodendrocyte precursors in the embryonic chick spinal cord. Development 121(6):1743–1754
205. Miller RH (2002) Regulation of oligodendrocyte development in the vertebrate CNS. Prog Neurobiol 67(6):451–467
206. Pringle NP, Yu WP, Guthrie S et al (1996) Determination of neuroepithelial cell fate: induction of the oligodendrocyte lineage by ventral midline cells and sonic hedgehog. Dev Biol 177(1):30–42
207. Poncet C, Soula C, Trousse F et al (1996) Induction of oligodendrocyte progenitors in the trunk neural tube by ventralizing signals: effects of notochord and floor plate grafts, and of sonic hedgehog. Mech Dev 60(1):13–32
208. Soula C, Danesin C, Kan P, Grob M, Poncet C, Cochard P (2001) Distinct sites of origin of oligodendrocytes and somatic motoneurons in the chick spinal cord: oligodendrocytes arise from Nkx2.2-expressing progenitors by a Shh-dependent mechanism. Development 128(8):1369–1379
209. Alberta JA, Park SK, Mora J et al (2001) Sonic hedgehog is required during an early phase of oligodendrocyte development in mammalian brain. Mol Cell Neurosci 18(4):434–441
210. Orentas DM, Hayes JE, Dyer KL, Miller RH (1999) Sonic hedgehog signaling is required during the appearance of spinal cord oligodendrocyte precursors. Development 126(11):2419–2429
211. Park HC, Mehta A, Richardson JS, Appel B (2002) olig2 is required for zebrafish primary motor neuron and oligodendrocyte development. Dev Biol 248(2):356–368
212. Zhou Q, Wang S, Anderson DJ (2000) Identification of a novel family of oligodendrocyte lineage-specific basic helix-loop-helix transcription factors. Neuron 25(2):331–343
213. Lu QR, Yuk D, Alberta JA et al (2000) Sonic hedgehog--regulated oligodendrocyte lineage genes encoding bHLH proteins in the mammalian central nervous system. Neuron 25(2):317–329
214. Zhou Q, Anderson DJ (2002) The bHLH transcription factors OLIG2 and OLIG1 couple neuronal and glial subtype specification. Cell 109(1):61–73
215. Takebayashi H, Nabeshima Y, Yoshida S, Chisaka O, Ikenaka K (2002) The basic helix-loop-helix factor olig2 is essential for the development of motoneuron and oligodendrocyte lineages. Curr Biol 12(13):1157–1163
216. Gage FH (2002) Neurogenesis in the adult brain. J Neurosci 22(3):612–613
217. Luskin MB, Parnavelas JG, Barfield JA (1993) Neurons, astrocytes, and oligodendrocytes of the rat cerebral cortex originate from separate progenitor cells: an ultrastructural analysis of clonally related cells. J Neurosci 13(4):1730–1750
218. Doetsch F, Petreanu L, Caille I, Garcia-Verdugo JM, Alvarez-Buylla A (2002) EGF converts transit-amplifying neurogenic precursors in the adult brain into multipotent stem cells. Neuron 36(6):1021–1034
219. Alvarez-Buylla A, Garcia-Verdugo JM (2002) Neurogenesis in adult subventricular zone. J Neurosci 22(3):629–634
220. Doetsch F, Caille I, Lim DA, Garcia-Verdugo JM, Alvarez-Buylla A (1999) Subventricular zone astrocytes are neural stem cells in the adult mammalian brain. Cell 97(6):703–716

221. Doetsch F, Garcia-Verdugo JM, Alvarez-Buylla A (1999) Regeneration of a germinal layer in the adult mammalian brain. Proc Natl Acad Sci U S A 96(20):11619–11624
222. Ninkovic J, Gotz M (2007) Signaling in adult neurogenesis: from stem cell niche to neuronal networks. Curr Opin Neurobiol 17:338–344
223. Chojnacki A, Shimazaki T, Gregg C, Weinmaster G, Weiss S (2003) Glycoprotein 130 signaling regulates Notch1 expression and activation in the self-renewal of mammalian forebrain neural stem cells. J Neurosci 23(5):1730–1741
224. Zheng W, Nowakowski RS, Vaccarino FM (2004) Fibroblast growth factor 2 is required for maintaining the neural stem cell pool in the mouse brain subventricular zone. Dev Neurosci 26(2–4):181–196
225. Androutsellis-Theotokis A, Leker RR, Soldner F et al (2006) Notch signalling regulates stem cell numbers in vitro and in vivo. Nature 442(7104):823–826
226. Hoglinger GU, Rissling I, Metz A et al (2004) Enhancing recognition of early Parkinsonism in the community. Mov Disord 19(5):505–512
227. Hoglinger GU, Rizk P, Muriel MP et al (2004) Dopamine depletion impairs precursor cell proliferation in Parkinson disease. Nat Neurosci 7(7):726–735
228. Palma V, Lim DA, Dahmane N et al (2005) Sonic hedgehog controls stem cell behavior in the postnatal and adult brain. Development 132(2):335–344
229. Kuo CT, Mirzadeh Z, Soriano-Navarro M et al (2006) Postnatal deletion of Numb/Numblike reveals repair and remodeling capacity in the subventricular neurogenic niche. Cell 127(6):1253–1264
230. Liu X, Wang Q, Haydar TF, Bordey A (2005) Nonsynaptic GABA signaling in postnatal subventricular zone controls proliferation of GFAP-expressing progenitors. Nat Neurosci 8(9):1179–1187
231. Menn B, Garcia-Verdugo JM, Yaschine C, Gonzalez-Perez O, Rowitch D, Alvarez-Buylla A (2006) Origin of oligodendrocytes in the subventricular zone of the adult brain. J Neurosci 26(30):7907–7918
232. Thored P, Arvidsson A, Cacci E et al (2006) Persistent production of neurons from adult brain stem cells during recovery after stroke. Stem Cells 24(3):739–747
233. Ohab JJ, Fleming S, Blesch A, Carmichael ST (2006) A neurovascular niche for neurogenesis after stroke. J Neurosci 26(50):13007–13016
234. Nait-Oumesmar B, Decker L, Lachapelle F, Avellana-Adalid V, Bachelin C, Van Evercooren AB (1999) Progenitor cells of the adult mouse subventricular zone proliferate, migrate and differentiate into oligodendrocytes after demyelination. Eur J Neurosci 11(12):4357–4366
235. Hoglinger GU, Breunig JJ, Depboylu C et al (2007) The pRb/E2F cell-cycle pathway mediates cell death in Parkinson's disease. Proc Natl Acad Sci U S A 104(9):3585–3590
236. Curtis MA, Penney EB, Pearson AG et al (2003) Increased cell proliferation and neurogenesis in the adult human Huntington's disease brain. Proc Natl Acad Sci U S A 100(15):9023–9027
237. Nait-Oumesmar B, Picard-Riera N, Kerninon C et al (2007) Activation of the subventricular zone in multiple sclerosis: evidence for early glial progenitors. Proc Natl Acad Sci U S A 104(11):4694–4699
238. Durbec P, Cremer H (2001) Revisiting the function of PSA-NCAM in the nervous system. Mol Neurobiol 24(1–3):53–64
239. Magavi SS, Leavitt BR, Macklis JD (2000) Induction of neurogenesis in the neocortex of adult mice. Nature 405(6789):951–955
240. Gould E (2007) How widespread is adult neurogenesis in mammals? Nat Rev Neurosci 8(6):481–4888

Part IV
Cell Cycle and Differentiation in the Cardiovascular System

Chapter 9
Cell Cycle and Differentiation in the Cardiovascular System

W. Robb MacLellan

Abstract The control of cardiac myocyte growth is a highly regulated, developmentally dependent process. Cardiac myocytes rapidly proliferate during fetal life but exit the cell cycle soon after birth in mammals. Although the extent to which adult cardiac myocytes are capable of cell cycle reentry is controversial and whether species-specific differences exist, it appears that the vast majority of adult cardiac myocytes are terminally differentiated. The recent identification of endogenous cardiac stem cells capable of giving rise to cardiac myocyte-like cells may account for some of the previous discrepancies in published studies as to the proliferative potential of cardiac myocytes and challenge the dogma that the heart is a terminally differentiated organ. This chapter will review the recent advances that have been made in identifying candidate factors and signaling pathways that are involved in mediating cardiac cell cycle and proliferation.

Keywords Cardiac myocyte · Cell cycle regulators · E2F transcription factors · Myc-Mad signalling · FoxO transcription factors · Cardiac regeneration · Cardiac stem cells

1 Introduction

Cardiac myocytes rapidly proliferate during fetal life but in the perinatal period, proliferation slows and myocytes undergo an additional round of DNA synthesis and nuclear mitosis without cytokinesis (acytokinetic mitosis) that leaves the majority of adult cardiac myocytes binucleated [1, 2]. Adult cardiac myocytes do not reenter the cell cycle when exposed to growth signals and further increases in cardiac mass

W.R. MacLellan (✉)
The Cardiovascular Research Laboratory, David Geffen School of Medicine at UCLA, 675 C.E. Young Dr., MRL 3-645, Los Angeles, CA 90095-1760, USA
e-mail: rmaclellan@mednet.ucla.edu

A. Giordano, U. Galderisi (eds.), *Cell Cycle Regulation and Differentiation in Cardiovascular and Neural Systems*, DOI 10.1007/978-1-60327-153-0_9,

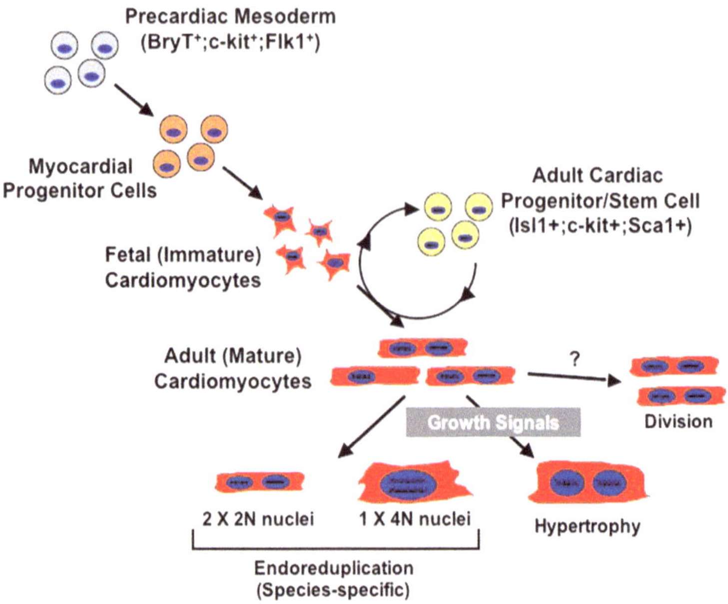

Fig. 9.1 Schematic model of cardiac myocyte growth and differentiation. Precardiac mesoderm gives rise to fetal cardiac myocytes that rapidly proliferate in utero but exit the cell cycle perinatally becoming binucleated. The potential outcomes of cell cycle reactivation in adult cardiac myocytes are shown. The relative importance of resident cardiac progenitor cell to myocardial regeneration in the adult heart is unknown but discussed in detail in the text

are achieved through an increase in cell size or hypertrophy. A model of growth and differentiation is shown in Fig. 9.1. Although much progress has been made in the last decade identifying and characterizing the factors that regulate cell cycle progression, the studies have usually been performed in vitro, using clonal cells despite the knowledge that the properties of these cells differ dramatically from primary cells [3, 4]. There have been several recent examples where examining the effects of "critical" or "ubiquitous" cell cycle regulators has revealed dramatic new insights into protein function [5, 6] or cell type-specific effects [7–9]. Even less is known regarding the role of these factors in more differentiated cell types. Since the onset of tissue-specific gene expression in developing myocardium occurs days before permanent cell cycle arrest, cardiac myocytes may provide a unique system to study cell cycle regulation during terminal differentiation.

Early studies that expressed oncogenic viral proteins in cardiac myocytes and then subsequently specific cell cycle regulators demonstrated the feasibility of mechanistic studies of cell cycle regulation in cardiac myocytes [10–12] but the interpretation of these results is limited by virtue of the fact that they relied on overexpressing supraphysiological protein levels and it is not clear whether the models used truly replicate normal cardiac biology. Other investigators have created in vivo models to study the effects of the factors felt to be critical for cell cycle regulation in adult post-mitotic ventricular muscle [13, 14]. Despite this progress in identifying and characterizing the protein networks that regulate proliferation and cell cycle exit in the heart, little is known regarding the factors that specifically impose the irreversible growth arrest that characterizes terminal differentiation. This chapter reviews the current knowledge of mechanisms that regulate cardiac cell cycle control, their relation to terminal differentiation, and prospects for myocardial regeneration.

2 Normal Cardiac Cell Cycle Progression

2.1 Cardiac Myocyte Proliferation During Development

Although many studies have examined rates of DNA synthesis in the adult heart, few have systematically analyzed cardiac proliferation rates in cardiac development. The model most amenable to these types of investigations is the mouse, where cardiac myocyte proliferation and DNA synthesis occur in two developmentally dependent phases [15]. Cardiac myocytes rapidly proliferate during fetal life with up to a third of the ventricular myocytes cycling at one time but this DNA synthesis progressively declines to near zero by the first postnatal day. This reduction in cell cycling is accompanied by a coordinated downregulation of positive-acting cell cycle regulators such as cyclin-dependent kinase (Cdk) 2 and 4, Cdc2, Cyclins A and E and upregulation of cell cycle inhibitors including the retinoblastoma gene product (Rb) and the Cdk inhibitors p21 and p27 [16–18]. During this fetal phase, karyokinesis and cytokinesis are matched, resulting in increased numbers of mononucleated cardiac myocytes. The second phase of DNA synthesis occurs early in the neonatal period, peaking 4–6 days after birth in mice. In this second phase, nuclear division occurs in the absence of myocyte cytokinesis, resulting in the hallmark binucleation of adult ventricular myocytes. This process of nuclear division in the absence of cellular division is a specific form of endoreduplication known as acytokinetic mitosis. This is not related to a defect in cardiomyocyte ability to express and assemble proteins required for an actomyosin contractile ring, but seems to be related to a defect in abscission [19]. In mice and rats, 85–90% of adult cardiomyocytes are binucleated [15, 20]. The degree of binucleation varies between species with human hearts having a binucleation rate ranging from 25 to 57% [21, 22] while in pigs the number of binucleated cardiomyocytes is approximately 32% [23]. The physiological consequences of binucleation are unknown and the bases for the species-specific differences are unclear but presumably it provides some metabolic or synthetic advantage.

2.2 Expression of Cell Cycle Regulators During Cardiac Development

The proliferative cell cycle, entails the highly regulated transduction of mitogenic signals to cyclically expressed proteins known as cyclins and, hence, to their catalytically active targets, the Cdks. In mammals, D-type cyclins in G1 associate with Cdk4/6, cyclins E and A sequentially partner with Cdk2 in S-phase, and cyclin B interacts with Cdc2 for entry into M-phase. There are two families of Cdk inhibitors, one is specific for Cdk4/6 (the INK4 family, comprising p15, p16, p18, and p19) and the other, the Cip/Kip family, inhibits all Cdks. Although activation of the cell cycle through induction of cyclin D has been traditionally thought to be related to Cdk4/6-dependent phosphorylation of the retinoblastoma family of "pocket" proteins, it is now known that it is in large part related to the titration of Cip/Kip proteins away from Cdk2 to the cyclin D-Cdk complexes. Cdk2, whose enzymatic activation thus depends on Cdk4/6, also phosphorylates pocket proteins, in sequential or processive fashion leading to release of E2F. Thus, together these regulators form a complex and interdependent network tightly regulating the cell cycle, in which the kinase activity of Cdk2 is ordinarily essential for DNA replication [24]. The factors that have been identified as playing a pivotal role in regulating this process at different developmental stages in cardiac myocytes are shown in Fig. 9.2 and reviewed below [25].

Although cyclins involved in G1, S, G2, and M-phase like D1, D2, D3, A, B1, and E and the cyclin-dependent kinases Cdc2, Cdk2, Cdk4, and Cdk6 are highly expressed in the developing heart, the specific role, if any, in cardiac cell cycle progression remains largely unknown [16, 26–28]. The most convincing data for

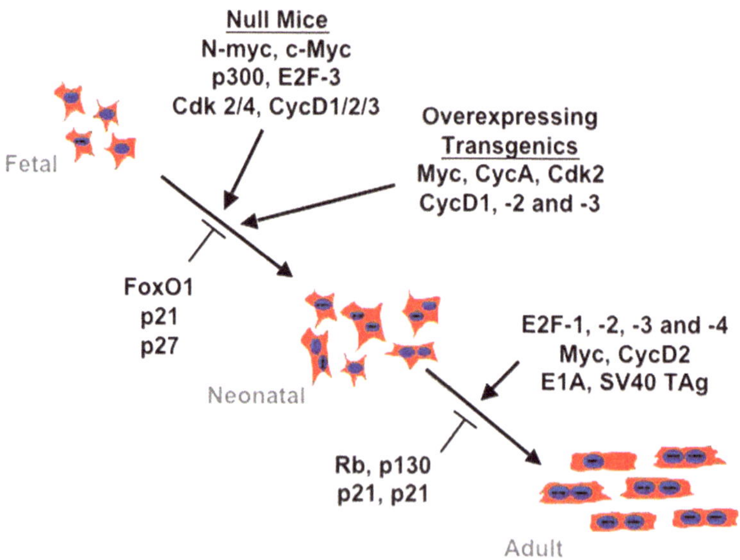

Fig. 9.2 Summary of factors implicated in regulating the cardiac myocyte cell cycle. A schematic model identifying the factors which have been implicated in regulating cardiac myocyte and the relevant developmental time point is shown. Whether the data was acquired from germline deletions (null mice) or cardiac-specific overexpressing transgenes is delineated

a specific role in cardiac development exists for members of the cyclin D family. Mice lacking all three mammalian D-cyclins (cyclin $D1^{-/-}$; $D2^{-/-}$; $D3^{-/-}$) die in utero secondary to heart abnormalities and defective hematopoiesis [29]. These CycD-null embryos displayed severely thinned ventricular walls with severely comprised compact zone and a high percentage of ventricular septal defects. Since cell cycle kinetics were normal in many tissues, it suggests a relatively cardiac-specific role for this family of proteins in cardiac cell cycle progression. Conversely, overexpressing cyclin D1, D2, or D3 in myocardium of transgenic mice increased cardiomyocyte DNA synthesis and proliferation [30]. Cardiac-specific transgenic mice with increased Cdk4 [13, 30] or Cdk2 [14] activity resulted in an increase in cardiac myocyte number and ongoing DNA synthesis in adult hearts, albeit at a very low rate, whereas deleting both Cdk2 and Cdk4 simultaneously results in embryonic lethality from heart defects [31]. Hearts from these double mutant mice displayed reduced global size, enlargement of atria, and thin ventricular walls associated with reduced cardiac myocyte proliferation. In contrast, the combined loss of Cdk4 and Cdk6 does not affect cell cycle progression; thus Cdk2 must be able to compensate, at least in part, for the lack of cyclin D-dependent kinases [32]. Although these results argue for a special role for CycD/Cdk4 complexes in normal cardiac development, these results should be interpreted with caution until more mechanistic studies are performed. Given the critical nature of the heart to development, knockouts of proteins critical to normal cell functioning may have their first manifestation during development as cardiac pathology.

The withdrawal of postnatal cardiomyocytes from the cell cycle is preceded by a coordinated change in the expression pattern in a host of cell cycle regulatory molecules. Expression of cyclins D1, D2, D3, A, B1, and E and their associated kinases is downregulated in cardiomyocytes after birth. Cyclins A, B, D1, E, and Cdc2 are undetectable in adult cardiomyocytes [16, 26–28]. This decrease in cyclin/Cdk expression is accompanied by a reciprocal upregulation of their inhibitors, CdkIs, ensuring cell cycle exit. Expression of Cip/Kip family members, p21 and p27, begins to increase in cardiac myocytes in the perinatal period and plateau in adult myocardium [17, 33]. Expression of the third member of the Cip/Kip family, p57, has been reported in the heart but it seems to be species specific [33, 34]. Evidence for an in vivo role for CdkIs is limited but mice lacking p27 showed a delay in cardiac myocyte cell cycle exit [35]. Deleting both p21 and p27 led to a synergistic increase in cycling cardiac myocytes and increased responsiveness to mitogenic serum in cardiac myocytes cultured from the double-knockout mice [36].

The primary target of G1 Cdks is Rb and this protein along with p107 and p130 comprises a family of proteins often called pocket proteins [37]. All three family members are expressed in heart but each family member has a specific temporal pattern. Rb is undetectable in the fetal mouse heart but is upregulated in late gestation and becomes the predominant family members expressed in adult terminally differentiated cardiac myocytes [38, 39]. p107 is expressed in a pattern opposite to Rb, highest in the embryonic, proliferative myocytes and lowest in the adult heart. p130 expression is intermediate. It peaks in the neonatal period and is subsequently downregulated and expressed at low levels in adult myocardium. These proteins are best known for their roles in inhibiting cell cycle progression through the regulation of E2F-responsive genes [40]. In their hypophosphorylated form, Rb proteins bind to E2F complexes, recruiting transcriptional repressors such as HDACs or the Jumonji, a repressor that plays a critical role in embryonic heart development [41].

The important role that pocket proteins play in regulating cardiac cell cycle appears to be model-dependent. Rb-null embryos die at day 14.5 p.c. with widespread cell death and aberrant cell cycling in a variety of tissues [42, 43] but normal hearts [44]. Likewise, cardiac-specific Rb-deficient mice develop normally without overt phenotypic abnormalities and normal hearts both morphologically and biochemically [38]. However, $Rb^{-/-}$ embryonic stem (ES) cells displayed a delay in the expression of cardiac-specific transcription factors and subsequently cardiac differentiation. LEK1, a murine homolog of the cardiomyogenic factor 1, interacts with Rb in ES cells inhibiting its activity and promoting cardiac myocytes to proliferation [45]. ES cells deficient in LEK1 recapitulated the delay in cardiac differentiation seen in $Rb^{-/-}$ ES cells [46]. The reason for the discrepancy between the in vivo and ES cell work is not simply a matter of timing of Rb deletion as suggested for skeletal muscle. Mox2-Cre deleted Rb mice, where the protein is deleted prior to cardiac commitment, also have normal hearts [47].

Germline deletions of p107 and p130 were initially reported as viable and phenotypically normally [48] but a subsequent report by different investigators where p130 deletion was created on a different strain background resulted in embryonic lethality. These p130-deficient mice on a Balbc/J background developed thin-walled, hypoplastic ventricles [49]. While these experiments seem to suggest that p130 has a unique, strain-dependent role in cardiac development, the original p130-deficient mice [48] do not display a similar phenotype in a Balbc/J background [50]. Nonetheless, marked abnormalities of cardiac myocyte cell cycling and differentiation were uncovered when both Rb and p130 were deleted in cardiac muscle. Mice deficient in Rb and p130 specifically in the heart demonstrated markedly enlarged hearts with persistent myocyte cycling even in adult ventricles [38]. These data suggest that Rb and p130 have overlapping functional roles in vivo to suppress cell cycle activators and maintain quiescence in postnatal cardiac muscle. Whether Rb family members mediate the inability of adult myocytes to reenter the cell cycle in response to growth stimuli is unknown but data from studies in skeletal muscle would suggest terminal differentiation is mediated by a pocket protein independent pathway [51].

2.3 Transcriptional Control of Cardiac Cell Cycle E2F Family of Transcription Factors

The E2F family is comprised of eight members E2F-1 through E2F-8 [52]. Based on structural and functional characteristics they are often subdivided into activator E2Fs (E2F-1, E2F-2, and E2F-3) and repressor E2Fs (E2F-4, E2F-5, E2F-6, E2F-7, and E2F-8). Relatively little is known regarding the specific properties or expression pattern of individual family members in the heart. E2F-1 and E2F-3 are downregulated as the cardiac myocytes exit the cell cycle in the perinatal period [38, 53] while E2F-5 is upregulated [38]. Some [38], but not all [54], studies have suggested E2F-4 is also upregulated in the adult myocardium similar to skeletal muscle [55]. Deleting E2F-1, E2F-4, or E2F-5 did not cause any discernable cardiac defect [5, 7, 56, 57]; however, it is likely that significant functional redundancy exists in this family [6, 7]. In contrast, the majority of E2F-3-deficient embryos died in utero. The few that survived adulthood developed progressive signs of congestive heart failure [58]. The mechanism underlying the cardiac defect was not determined but it does suggest, albeit indirectly, E2F-3 may have a distinct role regulating cardiac function in vivo. When E2F-1 expressed de novo in adult myocardium it induced cardiac myocyte DNA synthesis and apoptosis [59]. Recent studies confirmed that forced expression of E2F-1 and E2F-3 in neonatal cardiomyocytes provokes apoptosis concomitant with cell cycle progression [60]. However, forced E2F-2 and E2F-4 expression induced S-phase entry without apoptosis. Although both E2F-2 and E2F-4 could induce cell cycle reentry only E2F-2 resulted in mitotic cell division of cardiomyocytes. Expression of E2F-5 had no effect on cell cycle progression. The basis for the divergent effects of E2F members in cardiac development and proliferation and the relevant gene targets are unresolved.

2.4 The Myc-Mad Signaling Network

Myc is the prototypical member of a family of sequence-specific DNA-binding proteins that are postulated to act as "third messengers" for ligand-dependent signals and are implicated in the regulation of growth in a variety of tissues [61]. The Myc family, which includes c-Myc, N-Myc, and L-Myc, are transcription factors of the basic helix–loop–helix-leucine zipper (bHLHZ) family of proteins that activate transcription as part of a heteromeric complex with a protein termed Max. Myc normally forms heterodimers with Max, a ubiquitous bHLHZ protein, which is an obligate partner in mediating its functions [62]. However, Myc must compete for Max-binding with Mad family members (Mad1, Mxi1, Mad3, and Mad4), which suppress Myc function. Transcriptional activation is exclusively mediated by Myc:Max complexes, whereas Max:Mad or Max:Mxi complexes mediate transcriptional repression through identical binding sites [63]. Family members share a common DNA binding motif (E-Boxes) and have been demonstrated to be capable of functionally complementing each other, suggesting that the Myc gene family must have evolved to facilitate differential patterns of expression [64]. Myc has been implicated in regulating growth, differentiation, apoptosis, and metabolism in a wide variety of organisms and cell types [65] and is one of the few factors implicated in controlling both cell size and number [66, 67]. This is of importance in the heart since Myc is expressed in embryonic ventricular myocytes [68].

Myc is expressed in embryonic ventricular myocytes and Myc-deficient mice die prematurely at E10.5 with cardiac defects [69]. Myc-null embryos globally smaller and retarded in development compared with their littermates suggesting a general role for Myc in cellular proliferation. They displayed heart enlargement and pericardial effusions but a detailed analysis of the cardiac phenotype was never performed and whether the observed defects are due to a primary effect of Myc-deficiency on myocytes size or number versus an indirect effect secondary to Myc's effects on other organ systems is unknown. Germline deletion of Myc family member, N-myc, results in embryonic lethality with a specific cardiac phenotype. There was a reduction in cardiac myocyte number, most evident in the subepicardial compact layer [70].

Conversely, forced expression of Myc in the developing heart led to myocyte hyperplasia associated with a twofold increase in myocyte number at birth [71]. Although proliferation did not continue in the adult heart in this model, it was seen in inducible, cardiac-restricted transgenic mice where Myc reexpression could be activated in adult ventricular myocytes [72]. In this model Myc activation resulted in DNA synthesis, leading to increased nuclei number and DNA content [72]. No obvious change was seen in myocyte number, implying that Myc may be sufficient under these conditions to induce both S-phase reentry and karyokinesis, but not cytokinesis. In contrast to studies using E1A or E2F-1, no apoptosis was observed. Myc regulates a number of candidate genes implicated in cell growth including a several cell cycle genes [73, 74]. Myc activation in the heart is accompanied by the upregulation of CycD2 and cyclin-dependent kinase Cdk2 and Cdk4 activities which are important for cell cycle progression [72]. Consistent with results from nonmyocytes, Myc-dependent cell cycle reentry was not seen in CycD2-deficient hearts and was dependent on Cdk2 activity [75]. This is consistent with the observation that forced expression of CycD2 in the heart led to cell cycle activation in cardiac myocytes [30, 76].

2.5 Forkhead Box O (FoxO) Transcription Factors

The FoxO family of transcription factors, which includes FoxO1, FoxO3a, FoxO4, and FoxO6, are the mammalian orthologs of the *Caenorhabditis elegans* DAF-16. This family of proteins modulates the expression of genes involved in many cellular processes including cell cycle and differentiation. Thus it is not surprising that a complex regulatory network has arisen to control their function. They undergo inhibitory phosphorylation by protein kinases such as Akt, SGK, IKK, and Cdk2 in response to external and internal stimuli but can also be activated by kinases such as JNK and MST1 under stress conditions. They are also targets of the ubiquitin-protease system and polyubiquitylation of FoxO1 or FoxO3a leads to their degradation by the proteosome but monoubiquitylation of FoxO4 facilitates its nuclear localization and augments its transcriptional activity. Thus, FoxO activity is tightly controlled by complex regulatory pathways under normal physiological conditions so it is not surprising that they play an important role in regulating cardiac growth.

FoxOs have been implicated in regulating a number of cellular processes including cardiac growth [77, 78]. FoxO1 and FoxO3 are upregulated in heart in a temporal pattern that matches the exit of cardiac myocytes from the cell cycle and they are inactivated in the adult heart with hypertrophic signals [78]. Forced expression of FoxO1 in the developing heart inhibits cardiac myocyte proliferation and leads to premature upregulation of CdkIs p21, p27, and p57 [79]. Conversely, transgenic overexpression of a dominant negative FoxO1 increased myocyte proliferation and decreased expression of these same CdkIs. FoxO1 binding to the p21 promoter correlated directly with p21 transcription and inversely with myocyte proliferation. FoxOs can also indirectly effect cell cycle progression as they antagonize Myc activity. FoxO3a upregulates Mxi1-SRα, which inhibits Myc activity, and downregulates Myc target genes [80]. Activation of FoxO3a in response to inhibition of Akt also resulted in activation of Mxi1-SRα expression consistent with the finding that Myc-induced proliferation and transformation require Akt-mediated phosphorylation of FoxO proteins [81]. Since binding of CBP/p300 to FoxO factors is essential for FoxO-mediated transcription, this might account for some of the effect that p300 disrupting E1A mutants had on cell cycle progression in cultured cardiac myocytes [10].

2.6 p300

p300 is a transcriptional adaptor that was originally cloned as an E1A-binding factor [82] and has intrinsic histone acetyltransferase (HAT) activity which contributes to its ability to activate transcription [83]. p300 is crucial for normal development as mice deficient in this protein die in utero displaying cardiac, neurological, and yolk sac abnormalities [84]. The hearts of p300-null mutants were hypoplastic with poor trabeculation and reduced rates of DNA synthesis. This corroborates the previous in vitro data on the importance for p300 in cardiac differentiation [10, 12] and suggests that p300 is critical for normal cardiac development.

The mechanism underlying p300's effects on cardiac growth is unclear, but it is known to function as a coactivator for several cardiac transcription factors including NK-4/tinman in Drosophila [85], and GATA [86], MEF2D [87, 88], FoxO [89], COUP-TF [90], and C/EBPbeta [91]. It is through its interaction with C/EBPbeta that p300 is recruited to the promoters of E2F-regulated growth-related genes, which is required for high-level induction [91]. p300 interacts with cell cycle machinery at multiple levels. p300 also directly interacts with anaphase-promoting complex, which coordinates the temporal progression of eukaryotic cells through mitosis and the subsequent G1-phase of the cell cycle [92]. This interaction stimulates intrinsic p300 acetyltransferase activity and potentiates p300-dependent transcription. E1A mutants that selectively bind and inactivate p300 are sufficient for induction of S-phase in cardiac myocytes, whereas variants defective for both p300 and Rb are inactive [10–12]. DNA synthesis induced by p300-interacting E1A mutants can be blocked by p21 or dominant-negative Cdk2, suggesting it may be operated through regulating E2F-dependent gene expression in cardiac myocytes as well [93]. Although E1A readily provoked G1 exit in myocytes that do not respond to mitogenic serum, the cells accumulated in G2/M, without increasing cell number. p300 is critical for cell cycle regulation in skeletal muscle as well, since disruption of p300 function by neutralizing antibodies or dominant negatives blocks both differentiation and cell cycle arrest in skeletal myocytes [94, 95]. It is unknown whether p300-dependent induction of CdkIs plays a causative role in cardiac muscle cell cycle exit, but a requirement for p300 in the developmental upregulation of p21 possibly through FoxO factors has been reported in other cell types [96].

3 Pathological Cardiac Cell Cycle Progression

3.1 Cardiac Cell Cycle Reactivation in the Adult Heart

The capacity of heart to regenerate itself and the degree to which the adult mammalian cardiomyocytes can reenter the cell cycle, divide, or renew themselves from resident stem cells or precursor cells remain controversial [97, 98]. It has been reported that adult cardiac myocytes from some species, including humans, retain the capacity to reenter the cell cycle, at least under pathophysiological conditions [99]. However, determination of DNA synthesis rates in rodents where there is less chance for methodological artifacts indicates that the number of cardiomyocytes entering cell cycle in the normal adult heart is very low [100, 101]. Only 1 in 200,000 ventricular cardiac myocytes showed evidence of DNA synthesis in uninjured adult mice hearts [102]. This data is consistent with genetic fate mapping studies in vivo to track the fate of adult mouse cardiomyocytes that concluded there was essentially no turnover or replacement of adult mammalian cardiomyocytes during normal aging [103]. Although DNA synthesis does not seem to increase significantly in the mouse heart (0.004%) after injury [30], there is accumulating evidence that it likely does to a limited extent in the adult human heart. However, while restricted cell cycle reentry may occur in the injured human versus mouse hearts (1–4% [104] human versus 0.0014% mouse [105] cardiac myocytes), the ultimate fate of these myocytes and whether species-specific differences really exist with respect to proliferative capacity remain unresolved.

Proliferation and DNA synthesis are not synonymous and even if ventricular myocytes enter cell cycle exit, this need not culminate in cytokinesis. The possible outcomes of cell cycle reentry in the adult heart are shown in Fig. 9.1. While cardiac myocyte division might occur, DNA synthesis can also result in endoreduplication (increased DNA content per nuclei) or endomitosis (nuclear division without cytokinesis) [106]. This obviously occurs during normal cardiac development since most adult cardiac myocytes are binucleated. It also seems to occur in human hypertrophied and failing hearts where increased DNA content per nuclei and nuclei per myocyte in cardiomyopathic human hearts is well described [107–110]. Analogous changes have not been detected in several murine models of hypertrophy and may reflect the differences in potential for cell cycle reentry between these species [101]. Consistent with this concept, if cell cycle reentry is induced in adult mouse cardiac myocytes an increase in ploidy and nuclei per myocyte can be seen [72]. Even in humans, cardiomyocyte entry into the cell cycle after myocardial infarction appears to be transient and that as opposed to cytokinesis and proliferation, it leads to endoreduplication [111]. Presently, there is little convincing evidence that a contractile ring necessary for division is formed in adult cardiac myocytes from any species.

An alternate explanation for the discrepancies in published studies as to the proliferative potential of cardiac myocytes is the recent identification of stem cells capable of giving rise to cardiac myocyte-like cells that express cardiac-specific markers but retain the capacity to undergo several rounds of cell division before permanently withdrawing from the cell cycle [112, 113]. Recent genetic fate mapping has suggested that stem cells or precursor cells contribute to the replacement of adult mammalian cardiomyocytes primarily after injury [103]. Thus, it is a possibility that DNA synthesis in cells that possess cardiac-specific markers might reflect the recruitment of endogenous stem cells to a cardiac fate. Finally, cell cycle reentry could culminate in apoptosis since this was the result associated with many attempts to force myocytes to proliferate [59]. For all of these reasons, the detection of DNA synthesis in terminally differentiated myocardium cannot be taken, by itself, as evidence for the resumption of proliferative growth. Unfortunately, direct, definitive methodologies to accurately assess myocyte number do not exist and thus determination of accurate myocyte proliferation rates will require the development of new technologies.

3.2 Molecular Basis for Terminal Differentiation

Many differentiated tissues undergo cell cycle arrest as part of their differentiation pathway but not all cells undergo permanent arrest (terminal differentiation) and notable examples exist of highly specialized cell types having the capacity for regeneration [114]. Despite others and our work, the mechanisms underlying the permanent growth arrest in adult cardiac muscle are poorly understood. Rb has been implicated in mediating the irreversibility of cell cycle arrest associated with terminal differentiation in various lineages including skeletal muscle [115], adipocytes [116], and macrophages [117]. Early studies in vitro using fibroblasts induced to transdifferentiate into skeletal myocytes by overexpression of MyoD demonstrated that $Rb^{-/-}$ but not p107- or p130-null skeletal myocytes have a defect in cell cycle exit and maintenance of quiescence [118, 119]. Thus Rb appeared uniquely required for normal myogenic cell cycle control and full differentiation. These results were partially confirmed in vivo by deleting a floxed Rb allele either in proliferating myoblasts or after differentiation [120]. Deleting Rb prior to myogenic differentiation with Myf5-Cre resulted in severe defect in cell cycle, differentiation, and apoptosis but if Rb was deleted after differentiation, the cells formed normal multinucleated myotubes that did not enter S-phase in response to serum stimulation. It was subsequently shown that serum could not induce cell cycle reentry in differentiated myotubes even if all three pocket proteins had been removed [51]. Thus, it was concluded that Rb plays a crucial role in the switch from proliferation to differentiation in skeletal myocytes but was dispensable for the maintenance of the terminally differentiated state. However, studies demonstrating that Rb is required for maintaining quiescence in differentiated cells suggest Rb's role in maintaining terminal differentiation may be tissue specific [121].

Cardiac muscle terminal differentiation has also been suggested to display a similar dependence on pocket proteins. Using viral proteins SV40 large T antigen and E1A, which can block pocket protein function, investigators have demonstrated that they promote G1 exit in cardiac myocytes [10, 122]. This effect could be reproduced using mutants that primarily inhibit Rb-family members [10, 123, 124]. Likewise, transgenic mice that overexpressed G1 Cdks, Cdk4 [13], or Cdk2 [14], displayed an increase in cardiac myocyte number and ongoing DNA synthesis in adult hearts. However, inactivating Cdks were overexpressed at a developmental time point when cardiac myocytes still have proliferative potential and thus do not differentiate between a requirement for pocket protein for cardiac myocyte cell cycle exit or maintenance of the terminally differentiated state. As well, the work with G1 Cdk transgenic suggests that one or more pocket proteins are necessary in cardiac muscle but do not discriminate among them. To address these questions, cardiac-restricted Rb-deficient mice were created. They develop normally and do not display cardiac cell cycle defects even after physiological and pharmacological growth signals [38]. In contrast, deleting both Rb and p130-null led to defects in cardiac cell cycle exit and differentiation suggesting that p130 functionally substitutes for the loss of Rb [38]. Whether Rb and p130 are also necessary for maintaining quiescence in cardiac myocytes is unknown since they have not been deleted after terminal differentiation analogous to the studies in skeletal muscle [51]. Nonetheless, since activation and inactivation of pocket protein family members during the cell cycle is a readily reversible process there must be additional mechanisms at work.

Terminal differentiation is not the only situation under which adult cells undergo an irreversible cell cycle arrest. Senescent cells are also unable to reenter the cell cycle or express genes required for proliferation when stimulated with growth signals [125, 126]. Rb is a regulator of senescence [127, 128] and similar to studies on terminally differentiated cells overexpression of E1A or SV_{40} large T Ag proteins that block pocket protein function overrides the block to G1 exit in senescent cells, whereas mutant viral proteins defective in pocket protein binding do not [129]. Although inactivation of Rb by itself does not affect cell cycle arrest in senescence, cells lacking Rb along with p107 and p130 proteins fail to senesce in culture [130, 131]. The ability of Rb to effect irreversible growth arrest in senescent cells has been attributed to its ability to control heterochromatin structure leading to stable silencing of E2F target genes [132]. Senescence involves a number of phenotypic changes including the appearance of distinct heterochromatic structure in the nucleus. Rb regulates the formation of heterochromatin in the nuclei of senescent cells by recruiting acetylated histone H3 and heterochromatin protein 1 (HP1) to E2F responsive promoters, which then stably silence the expression of

E2F responsive genes and produce a permanent insensitivity to mitogenic signals [132]. Interestingly, both differentiated skeletal and cardiac myocyte nuclei display heterochromatin formation [133].

4 Practical Implications and Therapeutic Perspectives

4.1 Potential for Cardiac Regeneration

Many mammalian tissues respond to injury by activating committed progenitor cells or stem cells or through proliferation of differentiated cells such as liver or endothelial cells [134]. In contrast, adult mammalian cardiomyocytes have always been thought to have a very limited potential for self-renewal although specific exceptions exist. A robust cardiac regenerative response was reported in MRL mice, a that strain has dramatically enhanced capacity to heal surgical wounds [135], but this has since been disputed by several other groups [136, 137]. In contrast, there is general agreement that lower vertebrates are capable of regenerating myocardium after injury [138–141]. Amphibians, such as newt, were the first adult vertebrates identified that are capable of regenerating their organs. This ability to regenerate large sections of the body is widespread in Metazoan phylogeny [142]. This process of tissue regeneration was felt to consist of three steps, dedifferentiation of differentiation cells, and proliferation of the dedifferentiated cells followed by subsequent redifferentiation of the multipotent cells into the differentiated cell types that were lost. Mechanistic studies in newt are difficult and this work was first performed prior to the identification of endogenous stem cells, so recently investigators have begun to reexamine the origin of these proliferating new cardiac myocytes. Whether they truly arose from fully differentiated cardiomyocytes that dedifferentiated or were the result of recruitment of progenitor cells is unclear. Similar to newt, zebrafish are able to fully regenerate their heart after injury. Originally it was reported that cardiac regeneration in zebrafish followed the classic pattern described in newt and was dependent on the robust proliferation of cardiomyocytes localized at the leading epicardial edge of the new myocardium [143]. However, using an elegant genetic model in zebrafish these same authors reported that regeneration proceeds through two coordinated stages involving cardiac progenitor cells [144]. First a blastema is formed, comprised of progenitor cells that express precardiac markers, undergo differentiation, and proliferate. Second, epicardial tissue surrounding both cardiac chambers induces developmental markers and rapidly expands, creating a new epithelial cover for the exposed myocardium. A subpopulation of these epicardial cells undergoes epithelial-to-mesenchymal transition and provides new vasculature to regenerating muscle. This study questions the long-held belief that there is a fundamental difference in the proliferative capacity of amphibian versus mammalian heart and raises the question whether the difference lies in the regenerative potential of the resident endogenous cardiac progenitor cells. We will discuss the attempts at regenerating cardiac muscle using both of these strategies below.

4.2 Strategies to Reactivate the Cell Cycle in Adult Cardiac Myocytes

A number of approaches to promote cardiac myocyte proliferation by manipulating cell cycle regulators have been examined. Since proliferating cardiac myocytes express high levels and activity of cell cycle promoting factors and low levels of the Cdk inhibitors p21 and p27 [16, 145], attempts to induce cardiomyocyte proliferation have focused on reexpressing cell cycle activators or removing inhibitory factors to promote cell cycle progression. The gene products of DNA tumor viruses are known to stimulate entry of quiescent or differentiated cells into S-phase and the first attempts at activating the cardiac myocyte cell cycle overexpressed viral oncoproteins such as adenovirus E1A and SV40 large T antigen (SV40) to override cell cycle checkpoints. Targeted expression of SV40 T-Ag was sufficient to induce sustained cycling of cardiomyocytes in both embryonic and adult heart [122, 146]. Transgenic mice expressing SV40 large T antigen specifically in the ventricles developed ventricular hyperplasia [124] or cardiomyopathy [147] depending on the developmental timing of expression. Similarly, overexpressing T antigen targeted to the atria developed atrial tumors [122]. De novo expression of E1A or its downstream effector, E2F-1, although activating DNA synthesis resulted in widespread apoptosis and a block at G2M that limits its usefulness as a regeneration strategy [10]. Although a number of studies have shown that forced expression of cell cycle promoting factors can stimulate DNA synthesis in cardiomyocytes and even karyokinesis, cytokinesis remains an elusive goal in the adult heart.

Attempts at directly activating cell cycle in vivo using various cell cycle related proteins have had mixed results. Cardiac-specific Cdk2 overexpressing mice showed significantly myocyte hyperplasia along with increased levels of Cdk4 and cyclins A, D3, and E but limited ongoing DNA synthesis in the adult hearts [14]. Overexpression of CycD1, D2, or D3 in transgenic mice was sufficient to stimulate DNA synthesis in adult myocardium under baseline conditions although at very low levels. When these mice were subjected to myocardial infarction infarct size was initially similar; however, it decreased by about 30% in CycD2 overexpressing mice at later time points suggesting regeneration had occurred [30]. Similar results were seen in cyclin A2 overexpressing transgenic mice where there was elevated cardiac myocyte DNA synthesis and mitotic index [148]. These mice exhibited decreased ventricular dilation and improved cardiac function after myocardial infarction. New cardiomyocyte formation was noted in the infarcted zones as well as cell cycle reentry of periinfarct myocardium with an increase in DNA synthesis and mitotic indices [149]. In contrast to these studies where a constitutively active transgene was used that is expressed throughout development, de novo activation of Myc in adult myocardium resulted in robust cell cycle reentry and nuclear mitosis but whether cytokinesis occurred is questionable [71]. Thus a major limitation of these types of studies in general is the inability when analyzing the effects of constitutively active transgenes, to differentiate between dedifferentiation and proliferation of existing myocytes, enhanced proliferation of an endogenous cardiac progenitor cell or that a subset of cardiac myocytes in the transgenic hearts never underwent terminal differentiation because of the persistent expression of a positive cell cycle activator. Highlighting this issue, cardiac-specific expression of nuclear-targeted Akt in transgenics have also been reported to increase cardiac myocyte cell cycling; however, these investigators concluded it was due in large part to expansion of the presumptive cardiac progenitor cell population [150]. Finally, in vitro studies where cardiac myocyte G1 exit was induced after differentiation resulted in a block in G2M with cytokinesis. Thus, even if these molecules can overcome the block to cell cycle reentry in adult cardiac myocytes similar to the results with inducible Myc transgenics, there appears to be an equally potent impediment to myocyte cytokinesis.

A novel approach to overcoming the block to cardiac myocyte cell cycle reentry and proliferation was reported using a combination of p38 MAP kinase inhibitors and fibroblast growth factor (FGF). Four different p38 isoforms have been identified; the predominant isoform expressed in the adult heart is p38α, while p38β and p38γ are expressed at low levels, and p38δ is not expressed in heart [151, 152]. Genetic activation of p38 in vivo reduced fetal cardiomyocyte proliferation, whereas targeted disruption of p38α along with growth factor stimulation of cultured adult myocytes promoted cardiomyocyte cell cycle reentry [153]. This actual magnitude of the effect was a modest increase in mitotic cardiac nuclei index of 0.14%. Nonetheless, treatment with FGF1 in combination with a p38 MAP kinase inhibitor after myocardial injury led to improved cardiac function and an increase in mitotic index in vivo suggesting a role for this pathway in maintaining terminal differentiation in addition to regulating cardiac cell cycle [154]. Complicating the interpretation of p38α's role in adult myocardium is a contradictory report that cardiac-specific transgenic mice expressing a dominant-negative mutant p38α generates a hypertrophic rather than proliferative response in adult hearts [155]. Thus, further characterization of p38α mutant mice and its molecular interaction with growth factor signaling will be necessary to clarify p38's role in cardiomyocyte proliferation and terminal differentiation.

4.3 Recruitment and Expansion of Endogenous Cardiac Stem Cells

The dogma that the heart is a post-mitotic non-regenerating organ has recently been challenged by several groups. At least four resident cardiac stem cells or progenitor cell population with the capacity to differentiate into cardiac myocytes have been identified [112, 113, 156, 157]. Whether these are truly four distinct populations or whether there is overlap between will require the development of better markers or genetic tracking systems. One endogenous cardiac stem cell population was characterized by expression of c-kit, the receptor for stem cell factor. They have been identified in cardiac niches present in adult mice, rat, and human myocardium [112, 158]. c-kit^{+} cells can differentiate into cells that are phenotypically indistinguishable from cardiomyocytes as well as smooth muscle cells and endothelial cells, indicative of their possible pluripotency. When injected into the border zone of hearts with new infarcts, cardiac c-kit$^+$ cells led to bands of regenerating myocardium, contributed to endothelium and vascular smooth muscle, and improved the function of the heart [112]. A second, independent heart-derived cardiac progenitor cell expressing stem cell antigen 1 (Sca-1$^+$) on its surface has also been reported [113]. These cells have been reported to be distinct from isl1$^+$ and c-kit$^+$ cells in the heart and transplanted Sca-1$^+$ donor cells can also engraft and differentiate into cardiac myocytes after myocardial infarction. Whether they can also differentiate into smooth muscle or endothelial cells is unknown at this time. Further approximately one half of the engraft Sca-1$^+$ donor cells expressing cardiac-specific markers had fused with host cardiomyocytes and only 50% appeared to differentiate de novo into cardiac myocytes without fusion. Another cardiac-derived subpopulation with progenitor potential that likely overlaps with the Sca-1$^+$ cells is a rare population of cells termed side population cells [156, 159]. They were isolated from mouse hearts based on their ability to exclude Hoechst dye, which was shown to be dependent on the expression of the ABCG2 transport protein, a member of the family of ATP-binding cassette transporters [156]. These cells are present throughout cardiac development, also express Sca-1$^+$, but are rare and their ability to differentiate into cardiomyocytes and contribute to functional repair of the damaged myocardium has not been fully evaluated. Finally, investigators demonstrated that a subpopulation of cells in the anterior pharynx expressing the homeobox gene *islet-1* (*isl1*) contribute to formation of the outflow tract, the atria, and the right ventricle [160]. These *isl1*$^+$ cells do not express Sca-1, CD31, or c-kit, though they express Nkx 2.5 and GATA4 but they can differentiate into cardiomyocytes both in vivo and in vitro [157]. *isl1*$^+$ could also be used to identify a multipotent cardiovascular progenitor in embryonic stem cell cultures along with two other markers (Nkx2.5 and flk1), which is capable of differentiating into all three cardiovascular cell types [161]. *isl1*$^+$ cells have been identified in the hearts of newborn rodents and humans where they remain undifferentiated but expression of *isl1* is lost when these cells differentiate into cardiomyocyte [157]. However, *isl1*$^+$ cells are extremely rare in the adult heart and their capacity to repair damaged myocardial tissue and form functional myocytes remains to be determined.

Although this data argues that the postnatal heart has one or more populations of resident stem or progenitor cells that might be utilized to regenerate the heart after injury, significant regeneration is not observed following myocardial infarction despite evidence that one or more of these endogenous stem or progenitor cells are present in adult rodent or human hearts [113, 157, 158, 162]. Thus these cells, by themselves, are not capable of mounting a robust response to repopulate damaged myocardium as is seen in the newt and zebrafish. Instead, myocardial injury leads to loss of myocardium, fibrosis, and scar formation. Therefore, the endogenous cardiac stem cells are either not responsive to local growth signals or unable to migrate and differentiate in response to infarction. Investigators have utilized HGF to mobilize endogenous cardiac stem cells and IGF-1 to promote their survival and proliferation in vivo reporting that this strategy regenerated myocardium and improved cardiac function after infarct [163]. Translation of these results to human studies will require improved delivery systems and better understanding of the signals involved in the proliferation and migration of human cardiac stem cells.

5 Summary and Conclusions

This review has attempted to summarize the advances that have been made in our understanding of cardiac cell cycle regulation. The ability to genetically dissect cardiac growth control pathways in vivo, although sometimes contradictory to previous in vitro work, has provided new insights and dramatically altered our view of this process. These advances are not only theoretical, since investigators and clinicians have begun to use this knowledge to design strategies to regenerate myocardium. However, majors hurdles remain before manipulating the cardiac cell cycle will be feasible as a therapeutic intervention. Adult cardiac myocytes undergo cytokinesis poorly and maneuvers that induce cell cycle reactivation are often accompanied by unwanted cell death. The field of cardiac regeneration has exploded since the identification of stem and progenitor cells capable of differentiating into cardiac myocytes and while much research presently focuses on the identification of these stem cells, ultimately the information gained on the regulation of the cardiac cell cycle will prove invaluable as methodologies are developed for the expansion of these cell populations.

References

1. Li F, Wang X, Bunger PC, Gerdes AM (1997) Formation of binucleated cardiac myocytes in rat heart: I. Role of actin-myosin contractile ring. J Mol Cell Cardiol 29(6):1541–1551
2. Li F, Wang X, Gerdes AM (1997) Formation of binucleated cardiac myocytes in rat heart: II. Cytoskeletal organization. J Mol Cell Cardiol 29(6):1553–1565
3. Bodnar AG, Ouellette M, Frolkis M et al (1998) Extension of life-span by introduction of telomerase into normal human cells. Science 279(5349):349–352
4. Morales CP, Holt SE, Ouellette M et al (1999) Absence of cancer-associated changes in human fibroblasts immortalized with telomerase. Nat Genet 21(1):115–118
5. Yamasaki L, Jacks T, Bronson R, Goillot E, Harlow E, Dyson NJ (1996) Tumor induction and tissue atrophy in mice lacking E2F-1. Cell 85(4):537–548
6. Field SJ, Tsai FY, Kuo F et al (1996) E2F-1 functions in mice to promote apoptosis and suppress proliferation. Cell 85(4):549–561
7. Lindeman GJ, Dagnino L, Gaubatz S et al (1998) A specific, nonproliferative role for E2F-5 in choroid plexus function revealed by gene targeting. Genes Dev 12(8):1092–1098
8. Tsutsui T, Hesabi B, Moons DS et al (1999) Targeted disruption of CDK4 delays cell cycle entry with enhanced p27(Kip1) activity. Mol Cell Biol 19(10):7011–7019
9. Rane SG, Dubus P, Mettus RV et al (1999) Loss of Cdk4 expression causes insulin-deficient diabetes and Cdk4 activation results in beta-islet cell hyperplasia. Nat Genet 22(1):44–52
10. Kirshenbaum LA, Schneider MD (1995) Adenovirus E1A represses cardiac gene transcription and reactivates DNA synthesis in ventricular myocytes, via alternative pocket protein- and p300-binding domains. J Biol Chem 270:7791–7794

11. Bishopric NH, Zeng GQ, Sato B, Webster KA (1997) Adenovirus E1A inhibits cardiac myocyte-specific gene expression through its amino terminus. J Biol Chem 272(33): 20584–20594
12. Liu Y, Kitsis RN (1996) Induction of DNA synthesis and apoptosis in cardiac myocytes by E1A oncoprotein. J Cell Biol 133(2):325–334
13. Soonpaa MH, Koh GY, Pajak L et al (1997) Cyclin D1 overexpression promotes cardiomyocyte DNA synthesis and multinucleation in transgenic mice. J Clin Invest 99(11): 2644–2654
14. Liao HS, Kang PM, Nagashima H et al (2001) Cardiac-specific overexpression of cyclin-dependent kinase 2 increases smaller mononuclear cardiomyocytes. Circ Res 88(4):443–450
15. Soonpaa MH, Kim KK, Pajak L, Franklin M, Field LJ (1996) Cardiomyocyte DNA synthesis and binucleation during murine development. Am J Physiol 271(5 Pt 2):H2183–H2189
16. Brooks G, Poolman RA, McGill CJ, Li JM (1997) Expression and activities of cyclins and cyclin-dependent kinases in developing rat ventricular myocytes. J Mol Cell Cardiol 29(8):2261–2271
17. Li JM, Brooks G (1997) Downregulation of cyclin-dependent kinase inhibitors p21 and p27 in pressure-overload hypertrophy. Am J Physiol 273(3 Pt 2):H1358–H1367
18. Wang J, Nadal-Ginard B (1995) Regulation of cyclins and p34CDC2 expression during terminal differentiation of C2C12 myocytes. Biochem Biophys Res Commun 206(1):82–88
19. Engel FB, Schebesta M, Keating MT (2006) Anillin localization defect in cardiomyocyte binucleation. J Mol Cell Cardiol 41(4):601–612
20. Clubb FJJ, Bishop SP (1984) Formation of binucleated myocardial cells in the neonatal rat. An index for growth hypertrophy. Lab Invest 50(5):571–577
21. Schmid G, Pfitzer P (1985) Mitoses and binucleated cells in perinatal human hearts. Virchows Arch B Cell Pathol Incl Mol Pathol 48(1):59–67
22. Olivetti G, Cigola E, Maestri R et al (1996) Aging, cardiac hypertrophy and ischemic cardiomyopathy do not affect the proportion of mononucleated and multinucleated myocytes in the human heart. J Mol Cell Cardiol 28(7):1463–1477
23. Grabner W, Pfitzer P (1974) Number of nuclei in isolated myocardial cells of pigs. Virchows Arch B Cell Pathol 15(4):279–294
24. Bouchard C, Thieke K, Maier A et al (1999) Direct induction of cyclin D2 by Myc contributes to cell cycle progression and sequestration of p27. EMBO J 18(19):5321–5333
25. Perez-Roger I, Kim SH, Griffiths B, Sewing A, Land H (1999) Cyclins D1 and D2 mediate myc-induced proliferation via sequestration of p27(Kip1) and p21(Cip1) EMBO J 18(19):5310–5320
26. Yoshizumi M, Lee WS, Hsieh CM et al (1995) Disappearance of cyclin A correlates with permanent withdrawal of cardiomyocytes from the cell cycle in human and rat hearts. J Clin Invest 95(5):2275–2280
27. Flink IL, Oana S, Maitra N, Bahl JJ, Morkin E (1998) Changes in E2F complexes containing retinoblastoma protein family members and increased cyclin-dependent kinase inhibitor activities during terminal differentiation of cardiomyocytes. J Mol Cell Cardiol 30(3):563–578
28. Kang MJ, Koh GY (1997) Differential and dramatic changes of cyclin-dependent kinase activities in cardiomyocytes during the neonatal period. J Mol Cell Cardiol 29(7):1767–1777
29. Kozar K, Ciemerych MA, Rebel VI et al (2004) Mouse development and cell proliferation in the absence of D-cyclins. Cell 118(4):477–491
30. Pasumarthi KB, Nakajima H, Nakajima HO, Soonpaa MH, Field LJ (2005) Targeted expression of cyclin D2 results in cardiomyocyte DNA synthesis and infarct regression in transgenic mice. Circ Res 96(1):110–118
31. Berthet C, Klarmann KD, Hilton MB et al (2006) Combined loss of Cdk2 and Cdk4 results in embryonic lethality and Rb hypophosphorylation. Dev Cell 10(5):563–573
32. Malumbres M, Sotillo R, Santamaria D et al (2004) Mammalian cells cycle without the D-type cyclin-dependent kinases Cdk4 and Cdk6. Cell 118(4):493–504

33. Burton PB, Yacoub MH, Barton PJ (1999) Cyclin-dependent kinase inhibitor expression in human heart failure. A comparison with fetal development. Eur Heart J 20(8):604–611
34. Matsuoka S, Edwards MC, Bai C, Parker S, Zhang P, Baldini A (1995) p51^{KIP2}, a structurally distinct member of the p21^{CIP1} Cdk inhibitor family, is a candidate tumor suppressor gene. Genes Dev 9:650–662
35. Poolman RA, Li JM, Durand B, Brooks G (1999) Altered expression of cell cycle proteins and prolonged duration of cardiac myocyte hyperplasia in p27KIP1 knockout mice. Circ Res 85(2):117–127
36. Zhan S, Schneider MD (2000) p21(CIP1) and p27(Kip1) mediate perinatal cell cyle exit in mouse myocardium. Circulation 102[18], II–139
37. Classon M, Salama S, Gorka C, Mulloy R, Braun P, Harlow E (2000) Combinatorial roles for pRB, p107, and p130 in E2F-mediated cell cycle control. Proc Natl Acad Sci U S A 97(20):10820–10825
38. MacLellan WR, Garcia A, Oh H et al (2005) Overlapping roles of pocket proteins in the myocardium are unmasked by germ line deletion of p130 plus heart-specific deletion of Rb. Mol Cell Biol 25(6):2486–2497
39. Jiang Z, Zacksenhaus E, Gallie BL, Phillips RA (1997) The retinoblastoma gene family is differentially expressed during embryogenesis. Oncogene 14(15):1789–1797
40. Cobrinik D (2005) Pocket proteins and cell cycle control. Oncogene 24(17):2796–2809
41. Jung J, Kim TG, Lyons GE, Kim HR, Lee Y (2005) Jumonji regulates cardiomyocyte proliferation via interaction with retinoblastoma protein. J Biol Chem 280(35):30916–30923
42. Lee EYHP, Chang C-Y, Hu N et al (1992) Mice deficient for Rb are nonviable and show defects in neurogenesis and haematopoiesis. Nature 359:288–294
43. Clarke AR, Maandag ER, van Roon M et al (1992) Requirement for a functional Rb-1 gene in murine development. Nature 359(6393):328–330
44. Jacks T, Fazeli A, Schmitt EM, Bronson RT, Goodell MA, Weinberg RA. Effects of an *Rb* mutation in the mouse. Nature 359:295–300
45. Ashe M, Pabon-Pena L, Dees E, Price KL, Bader D (2004) LEK1 is a potential inhibitor of pocket protein-mediated cellular processes. J Biol Chem 279(1):664–676
46. Papadimou E, Menard C, Grey C, Puceat M (2005) Interplay between the retinoblastoma protein and LEK1 specifies stem cells toward the cardiac lineage. EMBO J 24(9):1750–1761
47. Wu L, de Bruin A, Saavedra HI et al (2003) Extra-embryonic function of Rb is essential for embryonic development and viability. Nature 421(6926):942–947
48. Cobrinik D, Lee MH, Hannon G et al (1996) Shared role of the pRB-related p130 and p107 proteins in limb development. Genes Dev 10(13):1633–1644
49. LeCouter JE, Kablar B, Whyte PF, Ying C, Rudnicki MA (1998) Strain-dependent embryonic lethality in mice lacking the retinoblastoma- related p130 gene. Development 125(23):4669–4679
50. Mao S, MacLellan WR (2007) Unpublished observation
51. Camarda G, Siepi F, Pajalunga D et al (2004) A pRb-independent mechanism preserves the postmitotic state in terminally differentiated skeletal muscle cells. J Cell Biol 167(3): 417–423
52. Cam H, Dynlacht BD (2003) Emerging roles for E2F: beyond the G1/S transition and DNA replication. Cancer Cell 3(4):311–316
53. Vara D, Bicknell KA, Coxon CH, Brooks G (2003) Inhibition of E2F Abrogates the development of cardiac myocyte hypertrophy. J Biol Chem 278(24):21388–21394
54. Vara D, Bicknell K, La Thangue NB, Brooks G (2001) Differential regulation of E2F family members during normal cardiac development and myocyte hypertrophy. Circulation 104, II–904
55. Corbeil HB, Whyte P, Branton PE (1995) Characterization of transcription factor E2F complexes during muscle and neuronal differentiation. Oncogene 11(5):909–920
56. Humbert PO, Rogers C, Ganiatsas S et al (2000) E2F4 is essential for normal erythrocyte maturation and neonatal viability. Mol Cell 6(2):281–291

57. Rempel RE, Saenz-Robles MT, Storms R et al (2000) Loss of E2F4 activity leads to abnormal development of multiple cellular lineages. Mol Cell 6(2):293–306
58. Cloud JE, Rogers C, Reza TL et al (2002) Mutant mouse models reveal the relative roles of E2F1 and E2F3 in vivo. Mol Cell Biol 22(8):2663–2672
59. Agah R, Kirshenbaum LA, Abdellatif M et al (1997) Adenoviral delivery of E2F-1 directs cell cycle reentry and p53- independent apoptosis in postmitotic adult myocardium in vivo. J Clin Invest 100(11):2722–2728
60. Ebelt H, Hufnagel N, Neuhaus P et al (2005) Divergent siblings: E2F2 and E2F4 but not E2F1 and E2F3 induce DNA synthesis in cardiomyocytes without activation of apoptosis. Circ Res 96(5):509–517
61. Evan GI, Littlewood TD (1993) The role of c-myc in cell growth. Curr Opin Genet Dev 3(1):44–49
62. Gallant P (2006) Myc/Max/Mad in invertebrates: the evolution of the Max network. Curr Top Microbiol Immunol 302:235–253
63. Rottmann S, Luscher B (2006) The Mad side of the Max network: antagonizing the function of Myc and more. Curr Top Microbiol Immunol 302:63–122
64. Malynn BA, de Alboran IM, O'Hagan RC et al (2000) N-myc can functionally replace c-myc in murine development, cellular growth, and differentiation. Genes Dev 14(11):1390–1399
65. Pelengaris S, Khan M, Evan G (2002) c-MYC: more than just a matter of life and death. Nat Rev Cancer 2(10):764–776
66. Stocker H, Hafen E (2000) Genetic control of cell size. Curr Opin Genet Dev 10(5):529–535
67. Trumpp A, Refaeli Y, Oskarsson T et al (2001) c-Myc regulates mammalian body size by controlling cell number but not cell size. Nature 414(6865):768–773
68. Starksen NF, Simpson PC, Bishopric N et al (1986) Cardiac myocyte hypertrophy is associated with c-myc protooncogene expression. Proc Natl Acad Sci U S A 83(21):8348–8350
69. Davis AC, Wims M, Spotts GD, Hann SR, Bradley A (1993) A null c-myc mutation causes lethality before 10.5 days of gestation in homozygotes and reduced fertility in heterozygous female mice. Genes Dev 7(4):671–682
70. Moens CB, Stanton BR, Parada LF, Rossant J (1993) Defects in heart and lung development in compound heterozygotes for two different targeted mutations at the N-myc locus. Development 119(2):485–499
71. Jackson T, Allard MF, Sreenan CM, Doss LK, Bishop SP, Swain JL (1990) The c-myc protooncogene regulates cardiac development in transgenic mice. Mol Cell Biol 10(7):3709–3716
72. Xiao G, Mao S, Baumgarten G et al (2001) Inducible activation of c-Myc in adult myocardium In vivo provokes cardiac myocyte hypertrophy and reactivation of DNA synthesis. Circ Res 89(12):1122–1129
73. Dang CV (1999) c-Myc target genes involved in cell growth, apoptosis, and metabolism. Mol Cell Biol 19(1):1–11
74. Gomez-Roman N, Grandori C, Eisenman RN, White RJ (2003) Direct activation of RNA polymerase III transcription by c-Myc. Nature 421(6920):290–294
75. Zhong W, Mao S, Tobis S et al (2006) Hypertrophic growth in cardiac myocytes is mediated by Myc through a Cyclin D2-dependent pathway. EMBO J 25(16):3869–3879
76. Busk PK, Hinrichsen R, Bartkova J et al (2005) Cyclin D2 induces proliferation of cardiac myocytes and represses hypertrophy. Exp Cell Res 304(1):149–161
77. Zhang W, Patil S, Chauhan B et al (2006) FoxO1 regulates multiple metabolic pathways in the liver: effects on gluconeogenic, glycolytic, and lipogenic gene expression. J Biol Chem 281(15):10105–10117
78. Ni YG, Berenji K, Wang N et al (2006) Foxo transcription factors blunt cardiac hypertrophy by inhibiting calcineurin signaling. Circulation 114(11):1159–1168
79. Evans-Anderson HJ, Alfieri CM, Yutzey KE (2008) Regulation of cardiomyocyte proliferation and myocardial growth during development by FOXO transcription factors. Circ Res 102(6):686–694

80. Delpuech O, Griffiths B, East P et al (2007) Induction of Mxi1-SR alpha by FOXO3a contributes to repression of Myc-dependent gene expression. Mol Cell Biol 27(13):4917–4930
81. Bouchard C, Marquardt J, Bras A, Medema RH, Eilers M (2004) Myc-induced proliferation and transformation require Akt-mediated phosphorylation of FoxO proteins. EMBO J 23(14):2830–2840
82. Arany Z, Newsome D, Oldread E, Livingston DM, Eckner R (1995) A family of transcriptional adaptor proteins targeted by the E1A oncoprotein. Nature 374(6517):81–84
83. Ogryzko VV, Schiltz RL, Russanova V, Howard BH, Nakatani Y (1996) The transcriptional coactivators p300 and CBP are histone acetyltransferases. Cell 87(5):953–959
84. Yao T-P, Oh S, Fuchs M et al (1998) Gene dosage-dependent embryonic development and proliferation defects in mice lacking the transcriptional integrator p300. Cell 93: 361–372
85. Choi CY, Lee YM, Kim YH et al (1999) The homeodomain transcription factor NK-4 acts as either a transcriptional activator or repressor and interacts with the p300 coactivator and the Groucho corepressor. J Biol Chem 274(44):31543–31552
86. Kakita T, Hasegawa K, Morimoto T, Kaburagi S, Wada H, Sasayama S (1999) p300 protein as a coactivator of GATA-5 in the transcription of cardiac- restricted atrial natriuretic factor gene. J Biol Chem 274(48):34096–34102
87. Slepak TI, Webster KA, Zang J et al (2000) Control of cardiac-specific transcription by p300 through myocyte enhancer factor 2-D. J Biol Chem 276(10):7575–85
88. Poizat C, Sartorelli V, Chung G, Kloner RA, Kedes L (2000) Proteasome-mediated degradation of the coactivator p300 impairs cardiac transcription Mol Cell Biol 20(23):8643–8654
89. van der Heide LP, Smidt MP (2005) Regulation of FoxO activity by CBP/p300-mediated acetylation. Trends Biochem Sci 30(2):81–86
90. Pipaon C, Tsai SY, Tsai MJ (1999) COUP-TF upregulates NGFI-A gene expression through an Sp1 binding site. Mol Cell Biol 19(4):2734–2745
91. Wang H, Larris B, Peiris TH et al (2007) C/EBPbeta activates E2F-regulated genes in vivo via recruitment of the coactivator CREB-binding protein/P300. J Biol Chem 282(34): 24679–24688
92. Turnell AS, Stewart GS, Grand RJ et al (2005) The APC/C and CBP/p300 cooperate to regulate transcription and cell-cycle progression. Nature 438(7068):690–695
93. Akli S, Zhan S, Abdellatif M, Schneider MD (1999) E1A can provoke G1 exit that is refractory to p21 and independent of activating cdk2. Circ Res 85(4):319–328
94. Sartorelli V, Huang J, Hamamori Y, Kedes L (1997) Molecular mechanisms of myogenic coactivation by p300: direct interaction with the activation domain of MyoD and with the MADS box of MEF2C. Mol Cell Biol 17(2):1010–1026
95. Puri PL, Avantaggiati ML, Balsano C et al (1997) p300 is required for MyoD-dependent cell cycle arrest and muscle- specific gene transcription. EMBO J 16(2):369–383
96. Kawasaki H, Eckner R, Yao TP et al (1998) Distinct roles of the co-activators p300 and CBP in retinoic-acid- induced F9-cell differentiation. Nature 393(6682):284–289
97. Rubart M, Field LJ (2006) Cardiac regeneration: repopulating the heart. Annu Rev Physiol 68:29–49
98. Anversa P, Kajstura J, Leri A, Bolli R (2006) Life and death of cardiac stem cells: a paradigm shift in cardiac biology. Circulation 113(11):1451–1463
99. Anversa P, Kajstura J (1998) Ventricular myocytes are not terminally differentiated in the adult mammalian heart. Circ Res 83(1):1–14
100. Rumyantsev PP, Borisov A (1987) DNA synthesis in myocytes from different myocardial compartments of young rats in norm, after experimental infarction and in vitro. Biomed Biochim Acta 46(8–9):S610–S615
101. Soonpaa MH, Field LJ (1998) Survey of studies examining mammalian cardiomyocyte DNA synthesis. Circ Res 83(1):15–26
102. Soonpaa MH, Field LJ (1997) Assessment of cardiomyocyte DNA synthesis in normal and injured adult mouse hearts. Am J Physiol 272(1 Pt 2):H220–H226

103. Hsieh PC, Segers VF, Davis ME et al (2007) Evidence from a genetic fate-mapping study that stem cells refresh adult mammalian cardiomyocytes after injury. Nat Med 13(8): 970–974
104. Beltrami AP, Urbanek K, Kajstura J et al (2001) Evidence that human cardiac myocytes divide after myocardial infarction. N Engl J Med 344(23):1750–1757
105. Soonpaa MH, Field LJ (1994) Assessment of cardiomyocyte DNA synthesis during hypertrophy in adult mice. Am J Physiol 266:H1439–H1445
106. Jackson CW (1990) Megakaryocyte endomitosis: a review. Int J Cell Cloning 8(4):224–226
107. Yan SM, Finato N, Di Loreto C, Beltrami CA (1999) Nuclear size of myocardial cells in end-stage cardiomyopathies. Anal Quant Cytol Histol 21(2):174–180
108. Goodman LC, Epling S, Kelly S, Lee S, Fishbein MC (1990) DNA flow cytometry of myocardial cell nuclei in paraffin-embedded, human autopsy, cardiac tissue. Am J Cardiovasc Pathol 3(1):55–59
109. Herget GW, Neuburger M, Plagwitz R, Adler CP (1997) DNA content, ploidy level and number of nuclei in the human heart after myocardial infarction. Cardiovasc Res 36(1): 45–51
110. Beltrami CA, Di Loreto C, Finato N, Yan SM (1997) DNA content in end-stage heart failure. Adv Clin Path 1(1):59–73
111. Meckert PC, Rivello HG, Vigliano C, Gonzalez P, Favaloro R, Laguens R (2005) Endomitosis and polyploidization of myocardial cells in the periphery of human acute myocardial infarction. Cardiovasc Res 67(1):116–123
112. Beltrami AP, Barlucchi L, Torella D et al (2003) Adult cardiac stem cells are multipotent and support myocardial regeneration. Cell 114(6):763–776
113. Oh H, Bradfute SB, Gallardo TD et al (2003) Cardiac progenitor cells from adult myocardium: homing, differentiation, and fusion after infarction. Proc Natl Acad Sci U S A 100(21):12313–12318
114. Fausto N (2000) Liver regeneration. J Hepatol 32(1 Suppl):19–31
115. Gu W, Schneider JW, Conderstil G, Kaushai S, Mahdavi V, Nadal-Ginard B (1993) Interaction of myogenic factors and the retinoblastoma protein mediates muscle cell commitment and differentiation. Cell 72:309–324
116. Chen PL, Riley DJ, Chen Y, Lee WH (1996) Retinoblastoma protein positively regulates terminal adipocyte differentiation through direct interaction with C/EBPs. Genes Dev 10(21):2794–2804
117. Chen PL, Riley DJ, Chen-Kiang S, Lee WH (1996) Retinoblastoma protein directly interacts with and activates the transcription factor NF-IL6. Proc Natl Acad Sci U S A 93(1): 465–469
118. Novitch BG, Mulligan GJ, Jacks T, Lassar AB (1996) Skeletal muscle cells lacking the retinoblastoma protein display defects in muscle gene expression and accumulate in S and G2 phases of the cell cycle. J Cell Biol 135(2):441–456
119. Novitch BG, Spicer DB, Kim PS, Cheung WL, Lassar AB (1999) pRb is required for MEF2-dependent gene expression as well as cell- cycle arrest during skeletal muscle differentiation. Curr Biol 9(9):449–459
120. Huh MS, Parker MH, Scime A, Parks R, Rudnicki MA (2004) Rb is required for progression through myogenic differentiation but not maintenance of terminal differentiation. J Cell Biol 166(6):865–876
121. Mantela J, Jiang Z, Ylikoski J, Fritzsch B, Zacksenhaus E, Pirvola U (2005) The retinoblastoma gene pathway regulates the postmitotic state of hair cells of the mouse inner ear. Development 132(10):2377–2388
122. Field LJ (1988) Atrial natriuretic factor-SV40 T antigen transgenes produce tumors and cardiac arrhythmias in mice. Science 239(4843):1029–1033
123. Jahn L, Sadoshima J, Greene A, Parker C, Morgan KG, Izumo S (1996) Conditional differentiation of heart- and smooth muscle-derived cells transformed by a temperature-sensitive mutant of SV40 T antigen. J Cell Sci 109(Pt 2):397–407

124. Katz EB, Steinhelper ME, Delcarpio JB, Daud AI, Claycomb WC, Field LJ (1992) Cardiomyocyte proliferation in mice expressing alpha-cardiac myosin heavy chain-SV40 T-antigen transgenes. Am J Physiol 262(6 Pt 2):H1867–H1876
125. Dimri GP, Hara E, Campisi J (1994) Regulation of two E2F-related genes in presenescent and senescent human fibroblasts. J Biol Chem 269(23):16180–16186
126. Dimri GP, Testori A, Acosta M, Campisi J (1996) Replicative senescence, aging and growth-regulatory transcription factors. Biol Signals 5(3):154–162
127. Mallette FA, Goumard S, Gaumont-Leclerc MF, Moiseeva O, Ferbeyre G (2004) Human fibroblasts require the Rb family of tumor suppressors, but not p53, for PML-induced senescence. Oncogene 23(1):91–99
128. Ferbeyre G, de Stanchina E, Querido E, Baptiste N, Prives C, Lowe SW (2000) PML is induced by oncogenic ras and promotes premature senescence. Genes Dev 14(16): 2015–2027
129. Shay JW, Pereira-Smith OM, Wright WE (1991) A role for both RB and p53 in the regulation of human cellular senescence. Exp Cell Res 196(1):33–39
130. Dannenberg JH, van Rossum A, Schuijff L, te RH (2000) Ablation of the retinoblastoma gene family deregulates G(1) control causing immortalization and increased cell turnover under growth-restricting conditions. Genes Dev 14(23):3051–3064
131. Sage J, Mulligan GJ, Attardi LD et al (2000) Targeted disruption of the three Rb-related genes leads to loss of G(1) control and immortalization. Genes Dev 14(23):3037–3050
132. Narita M, Nunez S, Heard E et al (2003) Rb-mediated heterochromatin formation and silencing of E2F target genes during cellular senescence. Cell 113(6):703–716
133. Terranova R, Sauer S, Merkenschlager M, Fisher AG (2005) The reorganisation of constitutive heterochromatin in differentiating muscle requires HDAC activity. Exp Cell Res 310(2):344–356
134. Brockes JP, Kumar A (2005) Appendage regeneration in adult vertebrates and implications for regenerative medicine. Science 310(5756):1919–1923
135. Leferovich JM, Bedelbaeva K, Samulewicz S et al (2001) Heart regeneration in adult MRL mice. Proc Natl Acad Sci U S A 98(17):9830–9835
136. Abdullah I, Lepore JJ, Epstein JA, Parmacek MS, Gruber PJ (2005) MRL mice fail to heal the heart in response to ischemia-reperfusion injury. Wound Repair Regen 13(2): 205–208
137. Oh YS, Thomson LE, Fishbein MC, Berman DS, Sharifi B, Chen PS (2004) Scar formation after ischemic myocardial injury in MRL mice. Cardiovasc Pathol 13(4):203–206
138. Nag AC, Healy CJ, Cheng M (1979) DNA synthesis and mitosis in adult amphibian cardiac muscle cells in vitro. Science 205(4412):1281–1282
139. Oberpriller JO, Oberpriller JC, Matz DG, Soonpaa MH (1995) Stimulation of proliferative events in the adult amphibian cardiac myocyte. Ann N Y Acad Sci 752:30–46
140. Matz DG, Oberpriller JO, Oberpriller JC (1998) Comparison of mitosis in binucleated and mononucleated newt cardiac myocytes. Anat Rec 251(2):245–255
141. Bettencourt-Dias M, Mittnacht S, Brockes JP (2003) Heterogeneous proliferative potential in regenerative adult newt cardiomyocytes. J Cell Sci 116(Pt 19):4001–4009
142. Dinsmore CE (1992) The foundations of contemporary regeneration research: historical perspectives. Monogr Dev Biol 23:1–27
143. Poss KD, Wilson LG, Keating MT (2002) Heart regeneration in zebrafish. Science 298(5601):2188–2190
144. Lepilina A, Coon AN, Kikuchi K et al (2006) A dynamic epicardial injury response supports progenitor cell activity during zebrafish heart regeneration. Cell 127(3):607–619
145. Poolman RA, Gilchrist R, Brooks G (1998) Cell cycle profiles and expressions of p21CIP1 AND P27KIP1 during myocyte development. Int J Cardiol 67(2):133–142
146. Sen A, Dunnmon P, Henderson SA, Gerard RD, Chien KR (1988) Terminally differentiated neonatal rat myocardial cells proliferate and maintain specific differentiated functions following expression of SV40 large T antigen. J Biol Chem 263(35):19132–19136

147. DeLeon J, Federoff HJ, Dickson DW, Vikstrom KL, Fishman GI (1994) Cardiac and skeletal myopathy in beta myosin heavy-chain simian virus 40 tsA58 transgenic mice. Proc Natl Acad Sci U S A 91(2):519–523
148. Chaudhry HW, Dashoush NH, Tang H et al (2004) Cyclin A2 mediates cardiomyocyte mitosis in the postmitotic myocardium. J Biol Chem 279(34):35858–35866
149. Cheng RK, Asai T, Tang H et al (2007) Cyclin A2 induces cardiac regeneration after myocardial infarction and prevents heart failure. Circ Res 100(12):1741–1748
150. Gude N, Muraski J, Rubio M et al (2006) Akt promotes increased cardiomyocyte cycling and expansion of the cardiac progenitor cell population. Circ Res 99(4):381–388
151. Liang Q, Molkentin JD (2003) Redefining the roles of p38 and JNK signaling in cardiac hypertrophy: dichotomy between cultured myocytes and animal models. J Mol Cell Cardiol 35(12):1385–1394
152. Wang Y, Huang S, Sah VP et al (1998) Cardiac muscle cell hypertrophy and apoptosis induced by distinct members of the p38 mitogen-activated protein kinase family. J Biol Chem 273(4):2161–2168
153. Engel FB, Schebesta M, Duong MT et al (2005) p38 MAP kinase inhibition enables proliferation of adult mammalian cardiomyocytes. Genes Dev 19(10):1175–1187
154. Engel FB, Hsieh PC, Lee RT, Keating MT (2006) FGF1/p38 MAP kinase inhibitor therapy induces cardiomyocyte mitosis, reduces scarring, and rescues function after myocardial infarction. Proc Natl Acad Sci U S A 103(42):15546–15551
155. Braz JC, Bueno OF, Liang Q et al (2003) Targeted inhibition of p38 MAPK promotes hypertrophic cardiomyopathy through upregulation of calcineurin-NFAT signaling. J Clin Invest 111(10):1475–1486
156. Martin CM, Meeson AP, Robertson SM et al (2004) Persistent expression of the ATP-binding cassette transporter, Abcg2, identifies cardiac SP cells in the developing and adult heart. Dev Biol 265(1):262–275
157. Laugwitz KL, Moretti A, Lam J et al (2005) Postnatal isl1+ cardioblasts enter fully differentiated cardiomyocyte lineages. Nature 433(7026):647–653
158. Urbanek K, Quaini F, Tasca G et al (2003) Intense myocyte formation from cardiac stem cells in human cardiac hypertrophy. Proc Natl Acad Sci U S A 100(18):10440–10445
159. Pfister O, Mouquet F, Jain M et al (2005) CD31- but not CD31+ cardiac side population cells exhibit functional cardiomyogenic differentiation. Circ Res 97(1):52–61
160. Cai CL, Liang X, Shi Y et al (2003) Isl1 identifies a cardiac progenitor population that proliferates prior to differentiation and contributes a majority of cells to the heart. Dev Cell 5(6):877–889
161. Moretti A, Caron L, Nakano A et al (2006) Multipotent embryonic isl1+ progenitor cells lead to cardiac, smooth muscle, and endothelial cell diversification. Cell 127(6):1151–1165
162. Urbanek K, Torella D, Sheikh F et al (2005) Myocardial regeneration by activation of multipotent cardiac stem cells in ischemic heart failure. Proc Natl Acad Sci U S A 102(24):8692–8697
163. Urbanek K, Rota M, Cascapera S et al (2005) Cardiac stem cells possess growth factor-receptor systems that after activation regenerate the infarcted myocardium, improving ventricular function and long-term survival. Circ Res 97(7):663–673

Chapter 10
Cell Cycle and Differentiation in Vessels

Amalia Forte, Marilena Cipollaro, and Antonino Cascino

Abstract The vasculature is the first organ to arise during development. Blood vessels assure metabolic homeostasis by supplying oxygen and nutrients and removing waste products. Vasculature also mediates signals for correct organogenesis. Not surprisingly therefore, vessels are critical for organ growth in the embryo and for repair of wounded tissue in the adult. In this chapter we will focus in particular on molecular factors involved in differentiation and proliferation of arterial smooth muscle cells, endothelial cells, and activated myofibroblasts. Moreover, we will provide a brief overview of the current experimental evidence about the contribution of stem cells to arterial structure in pathophysiological conditions, with particular attention to injury-induced arterial restenosis.

Keywords Artery · Smooth muscle cells · Endothelial cells · Myofibroblasts · Stem cells · Restenosis · Differentiation · Cell cycle · TGF-β1 · VEGF

1 Blood Vessel Development

Vasculature is the organ that develops first during embryogenesis, since all other organs depend on a vascular supply for delivery of oxygen and nutrients and may also receive signals for directing organogenesis from the invading vasculature [1].

Blood vessels are divided into distinct arterial and venous structural systems. Arteries are composed by a layer of endothelial cells (ECs) (tunica intima) and multiple layers of smooth muscle cells (SMCs), with elastic fibers in the wall (tunica media) and sheathed by fibrous connective tissue (tunica adventitia). Arteries

---

M. Cipollaro (✉)
Department of Experimental Medicine, Section of Biotechnology and Molecular Biology, Second University of Naples, Via L. De Crecchio, 7, Naples, Italy
e-mail: marilena.cipollaro@unina2.it

In memory of Prof. Antonino Cascino who passed on April 2006

A. Giordano, U. Galderisi (eds.), *Cell Cycle Regulation and Differentiation in Cardiovascular and Neural Systems*, DOI 10.1007/978-1-60327-153-0_10,

are followed by arterioles, which are smaller arterial branches leading to capillaries. Veins contain thinner and less elastic layers than arteries and are followed by venules, which are minute vessels draining blood from capillaries into veins. Between these two systems, capillaries are tiny vessels devoted to exchange of material (oxygen, nutrients, waste) between blood and tissues [2].

Vascular structure is influenced both by genetic factors and by epigenetic factors, such as hemodynamics and oxygen tension.

The process of vessel development, growth, and remodeling in the embryo provides important insights into mechanisms that regulate vessel function and SMC growth in the adult.

The process of blood vessel development in the embryo is termed angiogenesis. During embryonal angiogenesis, the angioblasts, i.e., EC precursors derived from nascent mesodermal cells, form a primary capillary plexus that is then remodeled into an organized hierarchy of arteries, veins, and capillaries [3]. In more detail, once that the primitive EC tubes are formed, the endothelium secretes factors that lead to the recruitment of primordial SMCs to cover the new tube, to form the tunica media, and to inhibit proliferation and migration of ECs [4].

In the adult, three different terms can be used to described different processes: arteriogenesis, angiogenesis, and vasculogenesis. Adult angiogenesis is referred to collateral vessel sprouting and growth from pre-existing vessels resulting in new capillary networks [3]. Adult angiogenesis can have different origins and has been shown to occur in either physiological conditions (e.g., wound healing, reproductive cycle) or in pathological situations (inflammation, ischemia, tumor vascularization). In many cases, angiogenesis is initiated as a reaction to hypoxia. In angiogenesis, the formation of new blood vessels from existing ones by sprouting is enforced by intussusception of blood-borne cells.

Angiogenesis is enforced by vasculogenesis, a process that comprises the recruitment of blood-borne bone marrow-derived cells and the subsequent intussusception of these cells in the newly forming vessels [5].

Adult arteriogenesis describes the growth of functional collateral arteries from pre-existing arterio-arteriolar anastomoses [6]. Initial triggers of adult arteriogenesis are physical forces, such as altered shear stress.

We still do not fully understand how angioblasts form the organized structure of arteries, veins, and capillaries. Arterial\venous segregation appears to be a critical trigger for capillary formation. Moreover, studies focusing on EC identity also suggest that interactions between ECs and SMCs promote the structural maturation of arterial blood vessels (see below for details), but the detailed mechanisms have not been fully addressed.

In this chapter we will focus in particular on molecular factors involved in differentiation and proliferation of arterial SMCs, ECs, and activated myofibroblasts (MFs). Moreover, we will provide a brief overview of the current experimental evidence about the contribution of stem cells to arterial structure in pathophysiological conditions.

Studies indicate a molecular distinction of arterial and venous cell differentiation. For example, it has been demonstrated that different members of the ephrin\Eph system are implicated in the differentiation of arterial or venous ECs [7, 8]. We will not focus on venous cell differentiation due to space constraints. Specific information about vein development can be found in other excellent reviews [2].

The creation of knock-out and of transgenic mice has been fundamental for the comprehension of molecular pathways involved in cell differentiation and phenotype modulation during vascular morphogenesis and in response to changes of local environmental cues. Relevant indications came also from the analysis of genetic defects at the basis of human arteriovenous malformations, often caused by loss of arterial or venous vascular cell identity.

2 Endothelial Cells: Cell Cycle and Differentiation

Vascular endothelium is considered as an organ, in which ECs play an intense synthetic, regulatory, and secretory activity.

As a matter of fact, endothelium disruption leads to an intense reparative reaction and to a disregulation of the activity and status of other vascular cell populations. In particular, ECs play a key role in a number of pathological conditions, such as restenosis, atherosclerosis, hypertension. Previously mentioned angiogenesis also involves EC proliferation, migration, and differentiation.

During the last years, several arterial EC markers have been identified. Among them, we would like to mention Notch 4 and 5 [9, 10], Connexin37 [11] and CXCR4 [12].

In particular, it has been demonstrated that Notch and vascular endothelial growth factor (VEGF) play a critical role in arteriogenesis. The VEGF-A gene is subjected to alternative splicing and consequently can originate multiple isoforms, including, in the mouse, VEGF120 (soluble), VEGF188 (on cell surface or bound to extracellular matrix), and VEGF164 (with intermediate properties) [13]. Different studies demonstrated that VEGF isoforms have distinct effects on primary vascular patterning [14]. There are three receptors for VEGF identified to date: Flk1\VEGFR2 (the primary receptor), Flt1\VEGFR1 (thought to be a decoy receptor), and neuropilin-1 (Nrp1) (a coreceptor for VEGF164 isoform) [15 17].

VEGF is a secreted factor crucial for EC differentiation, proliferation, migration, and survival. It activates Notch signaling and also triggers a positive feedback mechanism for arteriogenesis by inducing Nrp1 expression. In more detail, it has been demonstrated that high levels of VEGF from peripheral nerves induce expression of arterial markers, including Nrp1, which in turn increases the sensitivity of $Nrp1^{+}Flk1^{+}$ ECs to VEGF164 to obtain further arteriogenesis [18]. In addition, endothelial-specific knock-out of Nrp1 induces a defect of arterial EC differentiation, together with an impairment of SMC coverage, suggesting that arterial differentiated ECs are required for maturation of arterial SMCs during arteriogenesis. Studies conducted in zebrafish demonstrate that VEGF may activate Notch signaling and consequently modify EC fate and vascular structure. These observations are in agreement with in vitro studies conducted on human arterial ECs, showing that exogenous VEGF induces Notch1 expression [19]. Notch in turn induces arterial endothelial specification by suppressing venous cell fate.

Among factors involved both in differentiation and proliferation of SMCs and ECs, transforming growth factor (TGF)-β1 plays an important role. Interestingly, it has been supposed that defects seen in recruitment and differentiation in SMCs are likely indirectly caused by defects in EC function. Indeed, endothelial-specific knock-out of TGF-β1 receptors I and II phenocopied the complete knock-out mouse embryos, indicating that the defects in recruitment and differentiation of SMCs in these mice are an indirect consequence of the primary defect in ECs [20]. Consequently, we will refer TGF-β1 roles in angiogenesis mainly to ECs.

TGF-β1 family is composed of over 35 members identified in the human genome [21], all having profound effects on developmental processes. These proteins exert their effect by binding to specific serine/threonine kinase type I (also termed activin-receptor-like kinase, ALK, 1 to 7) and type II receptor complexes.

Smads are signal molecules that play a pivotal role in transducing the TGF-β signal from the membrane-bound receptors to the nucleus.

TGF-β1 induces a heteromeric complex of type I and II receptors. On ligand binding, a type II receptor interacts with a type I receptor and phosphorylates its GS domain. Then, the activated type I receptor activates R-Smad, that in turn forms heteromeric complexes with Smad, that translocate into the nucleus. Within the nucleus, the heteromeric Smad complexes, in collaboration with transcription factors and cofactors, participate in the regulation of target gene expression [20].

The importance of TGF-β1 signaling in angiogenesis and vascular remodeling has been highlighted by many studies, mainly in knock-out mice [22]. In previous experiments conducted in vitro, it has been demonstrated that TGF-β1 can exert bifunctional effects on ECs, as it can both stimulate or inhibit their proliferation. In particular, low doses of TGF-β1 stimulate endothelial proliferation, while high doses inhibit it. Similar contrasting effects of TGF-β1 have been observed for EC migration and extracellular matrix formation [23] and degradation by matrix metalloproteinase (MMP) 2 and 9. TGF-β1 can also act in a paracrine manner by stimulating the chemotaxis of monocytes and the release of proangiogenic cytokines that in turn can activate ECs. Further experiments aimed at clarifying the role of TGF-β1 in angiogenesis and based on selective receptor disruption or activation revealed that the ratio between ALK5-Smad2/3 pathway and ALK1-Smad1/5 pathway activation by TGF-β1 in the endothelium eventually determine whether it is stimulated or remains quiescent, since these two receptors induce different target genes, involved in angiogenesis or in EC maturation [24–26]. ALK1 and ALK5 are also able to interact at various levels [24], providing a new basis for further researches to clarify TGF-β1 pathways and cross talks with other angiogenic and antiangiogenic agents.

It has been also proposed that TGF-β1 may be involved in transdifferentiation of ECs into SMCs, as ECs can express early SMC markers in presence of TGF-β1 or after overexpression of ALK5 [27]. Nevertheless, additional factors are supposed to be necessary to induce late SMC differentiation markers [28].

It has been demonstrated that TGF-β1 signaling can play a role not only in developmental angiogenesis but also in the adult, since it acts as a potent activator of tumor progression and metastasis through the stimulation of angiogenesis [29].

Finally, a regulatory genomic DNA segment that drives ALK1 expression in arteries, indicating a possible implication of TGF-β1 not only in angiogenesis but also in vascular remodeling in vivo [30] has been recently identified. In this concern, it has been demonstrated that TGF-β1 levels are rapidly increased within 6–24 h in experimental balloon injury models [31]. Moreover, it has been demonstrated that neointima formation, extracellular matrix deposition, and SMC proliferation are increased by overexpressing TGF-β1 and decreased by inhibition in balloon injury models [32, 33]. All these studies indicate a role for TGF-β1 signaling in lesion formation, but it is unclear if it has a beneficial or detrimental effect at all. In fact, it has also been demonstrated that SMCs from advanced human plaque contain mutations in the TGF-β1 type II receptor that decrease the sensitivity of these cells to TGF-β1 [34]. Grainger DJ et al. [35] demonstrated that patients with unstable angina have decreased plasma levels of TGF-β1. Finally, it has been demonstrated that in Apo E–/– mice, the neutralization of TGF-β1 through antibodies accelerates the development of atherosclerosis, and the lesions displayed increased inflammatory cells and decreased collagen content [36]. These latter findings suggest that TGF-β1 may contribute to matrix production within lesions but also acts to reduce inflammation. Overall results described above seem to indicate that TGF-β1 may have a protective effect in pathological settings.

3 Smooth Muscle Cells: Cell Cycle and Differentiation

SMCs are highly specialized cells whose principal function is contraction and regulation of blood vessel tone and consequently of blood pressure and blood flow distribution. A large part of studies focusing on SMC cycle and differentiation has been lead by Prof. Gary K. Owens. SMCs within adult animals retain remarkable plasticity and can undergo profound and reversible changes in phenotype in response to variations of local environmental cues. However, SMCs can also undergo more subtle changes in phenotype, including alterations in calcium handling [37]. The extensive plasticity of fully mature SMCs likely evolved in higher organisms because it conferred a survival advantage, since high plasticity makes SMCs able to participate to vascular repair. Consequently, mutations that compromised the ability of SMCs to participate to vascular repair were likely detrimental to the organism and did not persist. However, an unfortunate consequence of SMC high degree of plasticity is that cells often response to external signals with abnormal reactions, leading to development and\or progression of vascular disease, e.g., restenosis and atherosclerosis [37]. In particular, it has been demonstrated that SMCs have an important role in atherosclerosis and, in particular, in the evolution of the atherosclerotic plaque [38], in plaque rupture, and in thrombus deposition.

SMCs can exhibit a wide range of different phenotypes during development as well as in adult organisms. In more detail, during early stages of vasculogenesis, SMCs undergo rapid cell proliferation, are highly migratory, and show very high rates of synthesis of extracellular matrix components, such as cadherins, collagen, elastin, proteoglycans, and integrins, that constitute a major portion of blood vessel mass. Moreover, during the development, SMCs form a number of gap junctions with ECs, aimed at vascular maturation and vessel remodeling [39]. In contrast, in adult blood vessels, SMCs fully carry out their contractile function, supported by the expression of an appropriate repertoire of receptors, ion channels, calcium regulatory proteins, signal transduction molecules, and contractile proteins, and exhibit a very low proliferation rate, are largely nonmigratory, and have very low rates of synthesis of extracellular matrix components. Under vascular injury, "contractile" SMCs undergo transient modifications and switch to a "synthetic" phenotype, participating to vascular repair, and then returning to the previous contractile status. A broad range of different phenotypes are comprised between the highly synthetic and the contractile phenotypes, all characterized by specific sets of SMC-specific markers. SMC phenotype switching is dependent on the complex interaction of many local environmental cues (e.g., mechanical injury, stress, hemodynamic force, inflammatory stimuli, oxygen tension, cell\cell contacts, diffusible factors, neurotransmitters). The constant integration of local signals determines the pattern of gene expression appropriate for the circumstance.

It has been recently reported by Brisset AC et al. [40] that SMC populations isolated from normal porcine coronary artery media exhibit distinct phenotypes: spindle-shaped and rhomboid. These two phenotypes have been demonstrated to be involved in different processes, as rhomboid SMCs have been demonstrated to be present in high proportion in stent-induced intimal thickening formation. It has also been demonstrated that S100A4 can be considered a specific marker of rhomboid SMCs, with a precise but not yet well-defined functional role, as its silencing decreases rhomboid SMC proliferation.

As efficaciously demonstrated by GK Owens and collaborators, differentiation and proliferation are not necessarily mutually exclusive processes and many factors other than the SMC proliferation status influences their differentiation state [41]. Many studies support this point. For example, in atherosclerotic lesions SMCs show a very low proliferation rate, similar to that of fully differentiated SMCs, but they are highly phenotypically modulated, as indicated by marked reduction of expression of SMC marker genes [42]. Moreover, during embryonic and post-natal development, SMCs show a very high proliferation rate, but at the same time they undergo a very rapid induction of expression of a number of SMC differentiation marker genes [43]. Interesting evidences supporting the contemporary occurrence of differentiation and proliferation of SMCs come from in vitro experiments on rat aortic SMCs: treatment of post-confluent cells with PDGF-BB is associated with a rapid downregulation of many SMC marker genes, including SM α-actin and SM-MHC, even if PDGF-BB stimulated only a transient mitogenic effect, returning to control values within 36 h, despite daily treatment of cells with this factor [44]. The same authors observed also that cultured SMCs can be sustained in a highly de-differentiated phenotype indefinitely by treatment with PDGF-BB. Upon its removal, SMC marker genes are rapidly reinduced. Interestingly, GK Owens and collaborators showed that the concentration of PDGF-BB required for SMC phenotypic modulation was 10-fold lower than that required to elicit a growth response: this means that it is possible to downregulate SM α-actin expression in cultured SMCs without inducing cell cycle entry. In contrast, bFGF and fetal bovine serum (FBS) had little effect on SMC differentiation marker gene expression in post-confluent cultured SMCs despite stimulating a proliferative reaction. Moreover, thrombin-induced proliferation was associated with increased expression of SMC marker genes [44, 45]. Overall results indicate that PDGF-BB is an effective negative regulator of SMC differentiation, and that its effects on differentiation are not secondary to growth stimulation (see below for further details). Similar results in vivo are awaited to confirm these interesting data.

Many studies also demonstrated a heterogeneity of both phenotype and function of SMC population within the same vessel and among different kind of vessels [46, 47], presumably with different properties in phenotype switching and reactions to local environmental cues, such as differentiation and growth factors. Both developmental (cells of different embryonic origin and/or SMC stem cells) and environmental (e.g., hemodynamic force variations, metabolic changes) factors influence SMC heterogeneity.

The precise factors and mechanisms that regulate SMC differentiation in vivo are poorly understood. Many data are currently available about SMC differentiation factors in vitro, but these results should be considered with caution since it has been demonstrated that cultured SMCs often fail to adequately recapitulate regulatory pathways that are critical in vivo [43].

It has been demonstrated that the expression of SMC marker genes (e.g., SM22α, desmin, SM α-actin, SM-MHC) is dependent on one or more CArG elements (i.e., the $CC(AT)_6GG$ motif) within their promoter and\or intronic sequences [48, 49]. Interestingly, it has also been observed that mutations of these conserved elements had different effects on distinct subsets of SMCs [49]. This indicates that SM subtypes use different selective CArG-dependent transcription regulatory schemes, consistent with the high degree of plasticity of these cells.

CArG elements bind the serum responsive factor (SRF) as a dimer, a ubiquitously expressed transcription factor able to regulate both growth-responsive and cell-specific genes in SMCs and in other cell types by recruiting several cofactors [50].

Among molecules involved in stimulation of SMC differentiation, myocardin, an extremely potent SRF coactivator recently discovered by Wang D et al. [51] should first be mentioned. It is selectively expressed in cardiac and vascular differentiated SMCs in vivo, and complementary knock-out and adenoviral-mediated overexpression studies [52, 53] demonstrated that myocardin is able to selectively induce the expression of all CArG-dependent differentiated SMC marker genes, including SM α-actin, SM-MHC, SM22α, and calponin. It has been demonstrated that the leucin zipper motif of myocardin may bridge adjacent CArG elements and unmask myocardin's activation domain [52]. In more detail, it has been demonstrated that upon chromatin structure relaxation through histone hyperacetylation promoted by stimulation with retinoic acid, SRF becomes able to bind CArG elements in SMC-specific gene promoters and recruits myocardin and other possible coactivators, inducing the expression of multiple differentiated SMC genes [54]. Recent studies demonstrated that different SMC subtypes exhibit differential dependence on myocardin and/or myocardin-related transcriptional factors (MRTFs) [55].

Conversely, PDGF-BB has been demonstrated to be associated with a rapid downregulation of SMC marker genes [56] and thus it can be considered a potent negative regulator of SMC differentiation. Other suppressors of SMC differentiation, distinct from PDGF-BB, include the forkhead transcription factor FOXO4, upregulated after vascular injury and able to suppress SMC differentiation marker gene expression by interacting with and inhibiting the activity of myocardin and of SRF [57]. A similar SMC differentiation-suppressive function has been attributed to atypical homeodomain protein (HOP), able to inhibit the ability of myocardin to activate SRF-dependent transcription of SM22α, and able to interact directly with SRF, thus modulating its DNA binding. It has also been demonstrated that HOP can increase histone deacetylase 2 (HDAC2) binding to SM22α promoter, thus inducing deacetylation of histone H4 and transcriptional repression [58]. Finally, among repressors of SMC differentiation, we would lie to mention the transcription factor Yin Yang-1 (YY1), induced in SMCs in response to vascular injury [59], and FHL2/DRAL/Slim3 [60], both able to bind and inhibit SRF. Further studies are necessary to analyze the contribution of all these factors in vivo.

It is well known that vascular injury induces a decrease or a complete transient loss of expression of differentiated SMC marker genes, such as SM-α actin and SM-MHC and thus a phenotype switching. Experiments conducted by Owens GK et al. revealed that repression of SMC marker genes is mediated at transcriptional level. In more detail, this group used SM-α actin, SM22, and SM-MHC LacZ transgenic mice for vascular injury experiments. While above-mentioned transgenes were highly expressed in the media of uninjured mouse carotid arteries, after vascular injury there was an almost complete loss of the three transgenes, thus demonstrating for the first time that SMC phenotype switching in vivo was mediated at least in part by transcriptional repression [61]. The same authors also observed that 14 days after injury, subpopulations of cells began to exhibit redifferentiation, as indicated by increased expression of SM-α actin and SM-MHC, further documenting the dynamic nature of SMC phenotypic modulation in vivo. For what concerns the molecular elements involved in transcriptional regulation of expression of SMC marker genes, it has been identified a *cis* G/C-rich regulatory element in the SM-MHC promoter able to bind the transcription factors Sp1 and Sp3 and that functioned as a repressor in vitro. Sp1 and Sp3 are both increased after vascular injury [62]. It has also been found that G/C-rich elements are located between previously mentioned CArG elements of many SMC marker gene promoters. On this basis, the model proposed by researchers implies that phenotypic modulation of SMCs may be regulated by injury-induced increase of expression of Sp1 transcription factors that bind to the C/G repressor element, disrupt cooperative interaction between CArG elements, and consequently inhibit SMC marker gene expression. Interestingly, it has been found that Sp1 is also involved in the activation of genes characteristic of modulated SMCs, such as PDGF-BB [63]. This observation suggests that some pathways or factors can simultaneously turn off and turn on group of genes during SMC phenotype switching.

From the studies described above it emerges that the phenotype switching of SMCs from a differentiated state is actively regulated and is not simply a function of loss of normal positive differentiation cues. It is also clear that multiple repressive pathways contribute to SMC phenotype switching, and that they are not mutually exclusive but are combined.

Similar in depth studies focused on post-translational mechanisms of regulation of phenotype switching of SMCs are currently lacking but would be very important to add new information to the field of SMC phenotype modulation in reaction to microenvironmental changes.

Some of the above-mentioned studies, related in particular to myocardin mechanisms of action and to suppressors of SMC differentiation, implied a fundamental role for histone modification, and in particular for histone acetylation, in regulation of SMC differentiation. The histone post-translational modifications control the chromatin-binding properties of SRF and cofactors and consequently the subsets of genes expressed in SMCs and the related phenotype in response to microenvironmental cues. This represents an example of how cellular differentiation is dynamically controlled at chromatin level within the context of variable conditions [64].

As previously stated in the introductive paragraph, the use of knock-out mice has been of great importance for the comprehension of the role played by a number of molecular factors and cofactors in regulation of vascular cell differentiation and phenotype switching. Nevertheless, the development of conditional and/or chimeric knock-out mice is necessary for the comprehension of the role played in vivo by a number of transcriptional factors and transcriptional repressors, since conventional knock-out of these genes is associated with early embryonic or prenatal lethality [65, 66].

4 The Myofibroblasts: Origins and Functions

MFs, also defined as "mesenchyme-like interstitial cells," are specialized cells originating in pathophysiological conditions and contributing to tissue repair during wound healing. They are responsible for granulation tissue contraction and soft tissue retractions typical of fibrocontractive diseases. Their main activities are the production and modification of the extracellular matrix (ECM), secretion of angiogenic and pro-inflammatory factors, and generation of tensile force.

MFs are not specific of the vascular wall. Nevertheless, MFs have been demonstrated to play a key role in vascular diseases, and as such will be described in this chapter.

Discovery of this peculiar cell type is related to pioneering research led by Prof. G. Gabbiani [67]. MFs are known to participate to arterial remodeling in a variety of vascular pathophysiological processes, including restenosis induced by angioplasty or surgical injury (e.g., arterial graft). In vascular tissue, MFs are known to derive from adventitial fibroblasts and from transdifferentiation of SMCs resident in the tunica media.

Interestingly, vascular MFs can also derive from circulating fibrocytes, i.e., mesenchymal progenitors that exhibit mixed morphological and molecular characteristics of hematopoietic stem cells, monocytes, and fibroblasts. Fibrocytes likely represent the obligate intermediate stage of differentiation into mature mesenchymal cells of a bone marrow-derived precursor of the monocyte lineage under permissive conditions. On in vitro stimulation with pro-fibrotic cytokines and growth factors, human fibrocytes produce large quantities of extracellular matrix components (e.g., vimentin and collagens I and III). Fibrocytes contribute to the remodeling response by secreting MMPs and further differentiating into cells identical to the contractile MFs that emerge at the tissue sites during repair processes and in fibrotic lesions. Studies in various animal models of wound healing or fibrotic diseases have confirmed the ability of fibrocytes to differentiate into mature mesenchymal cells in vivo and have suggested a causal link between fibrocyte accumulation and ongoing tissue fibrogenesis or vascular remodeling in response to tissue damage or hypoxia. The available data indicate that human fibrocytes serve as a source of mature mesenchymal cells during reparative processes and in fibrotic disorders or stromal reactions predominantly associated with a persistent inflammatory infiltrate or with the selective recruitment of monocytes induced by ischemic changes [68].

Other studies highlighted that MFs can derive from transdifferentiation of macrophages during inflammatory process. Furthermore, MFs can derive from transdifferentiation of ECs. Of interest, the MFs have been proposed also to transdifferentiate into an endothelial-like cell as well as into synthetic phenotype SMCs during intimal thickening [69].

Activated MFs and SMCs share a number of common markers including SM α-actin (α-SMA), SM22, vinculin, h-calponin. This is not surprising, since these two cell types have common functional properties, such as force development and contraction, as well as the production of extracellular matrix. Conversely, SM-MHC and the transcription factor myocardin have been found only in SMCs and not in MFs [70] and thus can be considered as SMCs specific markers. Nevertheless, these observations should be confirmed in vivo. Of interest, recent studies revealed that SMCs and MFs use distinct transcriptional mechanisms for α-SMA expression [71]. In more detail, results of small interfering RNA-induced knockdown experiments in mice showed that RTEF-1 regulated α-SMA transcription in MFs, but not in differentiated SMCs. Moreover, quantitative chromatin immunoprecipitation assays revealed that RTEF-1 bound to the MCAT element-containing region within the α-SMA promoter in MFs, whereas transcriptional enhancer factor (TEF)-1 was bound to the same region in differentiated SMCs. These results provide novel evidence that, although both SMCs and MFs express α-SMA, they use distinct transcriptional control mechanisms for regulating its expression.

Nevertheless, it should be also underlined that there are still some unresolved questions regarding the properties that distinguish SMCs from MFs and the lineage and relationships between these two cell types.

Force generated by MFs to produce tissue remodeling involves mechanisms different from those classically attributed to SMC contraction [72, 73]. Three major ultrastructural features discriminate MFs from quiescent fibroblasts in tissues: (1) bundles of contractile microfilaments, (2) extensive cell-to-matrix attachment sites, and (3) intercellular adherens and gap junctions [74].

4.1 Myofibroblast Differentiation Pathways

Expression of α-SMA renders fibroblasts highly contractile and hallmarks MF differentiation. Less frequently, acquisition of contractile properties by MFs is reflected also in other SMC-specific proteins, as myosin heavy chains, desmin, and caldesmon [75].

The mechanisms of force generation by the MFs are clearly different than those occurring in classical striated and smooth muscle contraction, depending on the formation and contraction of stress fibers. It has been identified α-SMA as a mechanosensitive protein that is recruited to stress fibers under high tension. Generation of this threshold tension requires the anchoring of stress fibers at sites of 8–30-μm-long "supermature" focal adhesions (suFAs), which exert a stress approximately fourfold higher (approximately 12 nN/μm^2) on micropatterned deformable substrates than 2–6-μm-long classical FAs. Inhibition of suFA formation by growing MFs on substrates with a compliance of $<$ or $=$ 11 kPa and on rigid micropatterns of 6-μm-long classical FA islets confines α-SMA to the cytosol [76]. These authors propose that the different molecular composition and higher phosphorylation of FAs on supermature islets, compared with FAs on classical islets, accounts for higher stress resistance.

TGF-β1 is a prototypical tumor-suppressor cytokine with cytostatic and pro-apoptotic effects on most target cells. TGF-β1 and downstream Smad family of signal transducers play an important role in tissue fibrosis and matrix remodeling. In particular, TGF-β1 is a well-known inducer of MF differentiation, as it has been demonstrated that it is able to induce expression of α-SMA in differentiating MFs, due to the presence of a TGF-β1 response element in α-SMA promoter [77].

Activated TGF-β1 binds to a heteromeric receptor complex consisting of one TGF-β1 type I and one type II receptor. Both of these receptors possess tyrosine kinase activity.

A distinctive feature of TGF-β1 is the ability to control cell adhesion and migration by modulating the adhesion molecule repertoire [78], as well as the synthesis of ECM components such as fibronectin (Fn) and collagen [79]. Furthermore, Serini G et al. [80] well demonstrated that ECM can influence TGF-β1 expression and effects on target fibroblasts and consequently on MF differentiation. All these data suggest a feedback loop in vivo between ECM and TGF-β1.

Fn is a dimeric glycoprotein widely distributed in plasma and ECM. Fn polymorphism is due to alternative splicing of the type III segments ED-A, ED-B, and IIICS. In vitro, TGF-β1 preferentially promotes the accumulation of the ED-A Fn isoform [81]. It has been demonstrated that ED-A probably induces a conformational change of Fn, which in turn increases the accessibility of the amino acid sequence RGD to interaction with the integrin α5β1 [82], generating intracellular signals in a conformational sensitive manner. The ED-A Fn can thus exert a cooperative function with TGF-β1 to activate α-SMA expression in differentiating MFs.

For what concerns the intracellular pathways triggered by TGF-β1, this cytokine is known to induce MF differentiation in association with the delayed activation of focal adhesion kinase (FAK) and protein kinase B (PKB/AKT).

Horowitz et al. [83] demonstrated that FAK and AKT are independently regulated by early activation of SMAD3 and p38 MAPK, respectively. Pharmacological or genetic approaches that disrupt SMAD3 signaling block TGF-β1-induced activation of FAK, but not AKT; in contrast, disruption of early p38 MAPK signaling abrogates AKT activation, but does not alter FAK activation.

TGF-β1 is able to activate AKT in cells expressing mutant FAK or in cells treated with an RGD-containing peptide that interferes with integrin signaling, inhibits FAK activation, and induces anoikis (apoptosis induced by loss of adhesion signaling). TGF-β1 protects MFs from anoikis, in part, by activation of the PI3K–AKT pathway. Thus, TGF-β1 coordinately and independently activates the FAK and AKT protein kinase pathways to confer an anoikis-resistant phenotype to MFs. Activation of these pro-survival/anti-anoikis pathways in MFs likely contributes to essential roles of TGF-β1 in tissue fibrosis and it is adhesion dependent and focal adhesion kinase (FAK) mediated.

Another study, focused on cultured rat aorta fibroblasts, revealed that TGF-β1 is able to induce MF differentiation not only through PKB/AKT but also through a protein kinase C alpha (PKC alpha) [84].

A recent paper by Greenberg et al. [85] reconstructed the signaling pathway involved in the FAK-dependent regulation of MF differentiation. In more detail, it has been showed that TGF-β1 induces MF differentiation and increases expression of integrins and Fn. Fn binding to integrins activates FAK, that in turn leads to increased cell surface expression of fibroblast growth factor receptor (FGFR). Binding of FGF to its receptors triggers intracellular signal transmission, which negatively regulates the levels of α-SMA. This means that normally FAK activation by TGF-β1 may contribute to a negative feedback mechanism to prevent excessive MF differentiation, ameliorating fibrosis, and tissue remodeling.

In the context of regulation of MF differentiation, it is also known that prostaglandin E(2) (PGE(2)) inhibits MF differentiation via E prostanoid receptor 2 (EP2) signaling and cAMP elevation, but whether PGE(2) does so by interfering with TGF-β1 signaling was unknown. Recently, it has been demonstrated that PGE(2) does not interfere with TGF-β1-induced Smad phosphorylation or its translocation to the nucleus. Rather, PGE(2) has dramatic effects on cell shape and cytoskeletal architecture and disrupts the formation of appropriate focal adhesions. PGE(2) treatment diminishes TGF-β1-induced phosphorylation of paxillin, STAT-3, and FAK and, in turn, limits activation of the protein kinase B (PKB/Akt) pathway. These alterations do not, however, result in increased apoptosis. Interestingly, the effects of PGE(2) stimulation alone do not always mirror the effects of PGE(2) in the presence of TGF-β1, indicating that the context for EP2 signaling is different in the presence of TGF-β1. Taken together, these results demonstrate that PGE(2) has the potential to limit TGF-β1-induced MF differentiation via adhesion-dependent, but Smad-independent, pathways [86].

4.2 What Stimulates Myofibroblast Differentiation?

For what concerns the external stimuli and agents that are able to trigger the differentiation of adventitial fibroblasts to MFs, recent studies demonstrate that they can be quite heterogeneous. Vascular injury, causing a loss of interaction between vascular cells and ECM, can stimulate the formation of MFs aimed at a rapid wound healing. TGF-β1, above described as the inducer of MF differentiation, is synthesized as latent precursor complexed with latent TGF-β1 latent proteins (LTBP) in the matrix. TGF-β1 is activated when LTBP is removed extracellularly via proteolytic cleavage. The TGF-β1 activators found so far are proteins intimately associated with the wound healing response, such as MMP-2 and MMP-9, thrombospondin-1, integrin $\alpha_V\beta 6$ [87].

It has also been demonstrated that angiotensin II is able to induce MF differentiation through a pathway that involves NADPH oxidase generation of reactive oxygen species (ROS) and activation of p38MAPK and JNK pathways [88]. Endothelin-1, another vasoactive peptide, is also able to induce differentiation to MFs of cultured lung fibroblasts through the ETA receptor [89].

4.3 Myofibroblast Ultimate Fate

The fate of recruited/activated MFs in injured vascular tissue may ultimately determine whether normal healing occurs or progression to end stage fibrosis ensues.

For what concerns the question whether MF differentiation is terminal or reversible, several studies demonstrated a downregulation of α-SMA in cultured differentiated MFs in response to different factors, like TGF-β1 antagonists, growth factors, cell density, without evident increased apoptosis. In vivo, however, reversal of MFs has not been detected, while massive apoptosis normally occurs after wound healing [90]. If this not happens, then MFs increase in number and induce marked remodeling and fibrosis. Normal MF apoptosis can be countered by persistence of TGF-β1 expression and ECM deposition which promote the pro-survival/anti-apoptotic phenotype, that can lead to tissue chronic fibrosis [91]. The combinatorial activation of the adhesion-dependent focal adhesion kinase pathway and the soluble growth factor-mediated AKT pathway (see below for details) confers anoikis/apoptosis resistance to differentiated MFs [83]. Zhang et al. [91] have also demonstrated a selective susceptibility of MFs to nitric oxide-induced apoptosis in vitro. Thus, the combinatorial effects of reduced growth factor expression increased ECM turnover, and nitric oxide generation may set the stage for triggering MF apoptosis during the resolution of tissue repair and remodeling [74]. Alternatively to growth factor signals, two mechanisms are conceivable to account for MF disappearance at the end of physiological tissue repair: cell release from stress and increased formation of cell–cell contacts. In more detail, it has been hypothesized that in properly repaired tissue, the ECM regains its original mechanical properties and returns to stress-shield the resident fibroblasts. To support this hypothesis, it has been demonstrated that the prevention of tension release of granulation tissue by splinting the wound inhibits MF apoptosis [92]. Similarly, it has been demonstrated that apoptosis is induced by relaxing fibroblasts attached to collagen gels [93]. The mechanisms that link stress release to induction of apoptosis are probably similar to those that control fibroblast to MF transition, i.e., stress perception pathways involving cell–matrix junctions. Cell–cell contacts and cadherins, other than cell–matrix contacts, have been demonstrated to be also involved in transmission of survival signals in MFs [94]: in this context, it has been demonstrated that formation of homotypic cadherin junctions is a possible signal of the presence of MF accumulation in late granulation tissue. Consistently, corneal MFs in dense culture decrease the expression of α-SMA and de-differentiate into fibroblasts [95], probably due to contact-induced desensitization to TGF-β1 [96].

5 The Role of Stem Cells in Vascular Pathophysiology

Recent evidence has shown that vascular function not only depends on cells within the vessels but is also significantly modulated by circulating cells derived from the bone marrow [97]. Stem cells hold a great potential for the regeneration of damaged tissues in cardiovascular diseases. In particular, in the past, it was believed that the regeneration of injured endothelium and media in arteries was due to migration and proliferation of neighboring ECs and SMCs. Recent studies clearly indicated that different stem cell populations, derived from bone marrow and characterized by different markers and with different behaviors, contribute to vascular remodeling after injury [98]. Moreover, different studies indicate that the contribution of bone marrow-derived cells to (re)stenosis depends on the type of model of injury [99].

Finally, we would like to mention that it has been demonstrated that pluripotent embryonic stem cells are also able to differentiate into vascular ECs in primates, thus revealing primate-specific vascular developmental mechanisms [100]. In this context, Yamamoto et al. demonstrated that shear stress can induce mouse embryonic stem cell proliferation and the expression of EC specific markers, such as Flk-1, Flt-1, cadherin, but not SMC marker α-actin [101].

Both hematopoietic and nonhematopoietic bone marrow-derived stem cells have been demonstrated to participate in vascular repair after injury. Nevertheless, it should be mentioned that other tissue-specific niches, as well, distinct from bone marrow, are probably a source of vascular progenitor cells, whose circulation in blood is triggered by vascular damage. In particular, Tintut et al. [102] demonstrated that mesenchymal stem cells (MSCs) with self-renewal, lineage plasticity, and a unique differentiation repertoire are also contained in the artery wall. These vascular MSCs apparently lack adipogenic lineage in their differentiation repertoire, and the authors suggest that this vascular cell population could represent a stage of commitment one generation below the MSCs in the mesengenic lineage hierarchy.

Recruitment and incorporation of vascular cell precursors at the injury site require a coordinated sequence of events, including chemoattraction, cell adhesion, and transmigration, and finally differentiation to ECs or SMCs. All these steps involve a large variety of molecules, including integrins, chemokines, MMPs, and kinases.

This last paragraph will provide an overview of the current evidence of the role played by different vascular cell precursors in models of vascular injury. In particular, we will overview the data available for MSCs, endothelial progenitor cells (EPCs), smooth muscle precursors, mesoangioblasts (Mabs), hemangioblasts, and monocyte lineage cells (MLCs). Such division can be considered arbitrary and is based mainly on the descriptions and definitions provided by the papers we analyzed.

Further studies aimed at defining univocal markers and hierarchy of bone marrow-derived stem cells would be required, as well as a well-defined nomenclature to be adopted by all researchers.

5.1 Mesenchymal Stem Cells

MSCs have been first identified in adult bone marrow [103]. Subsequent studies demonstrated that MSCs are widely distributed in vivo since they have been isolated also from other tissues, including lung, adipose tissue, skeletal muscle, trabecular bone, synovium, and the human umbilical cord perivascular cells derived from the Wharton's jelly [104– 106]. In particular, in human long-term bone marrow culture, a subset of adherent cells resembles immature SMC in cytoskeletal features such as α-SMA and vimentin filaments [107]. A clone of mouse bone marrow smooth muscle-like cells has also been isolated [108].

MSCs can be distinguished from HSCs on the basis of their cell surface antigens and can be separated from HSCs by their propensity to adhere to cell culture plastics. In fact, MSCs do not express CD34, which conversely is a marker of hematopoietic cells. MSCs in bone marrow have been described as a very rare population; in particular, Wexler estimated the frequency of MSCs in bone marrow nucleated cell populations as 1 in 3.4 x 10^4 cells [109].

Studies demonstrated that MSCs extensively proliferate in vitro while preserving a normal karyotype and telomerase activity on several passages [110].

Bone marrow-derived MSCs can potentially differentiate along different mesenchymal lineages including those forming bone, cartilage, fat, ligament/tendon, muscle, neurons, astrocytes, and bone marrow stroma that supports hematopoiesis [111]. MSC potentiality has been first described by Prockop in a study showing that transplanted marrow cells engraft nonhematopoietic connective tissues such as spleen and liver [112, 113].

Many studies have focused so far on MSCs due to their intrinsic ability to differentiate into functional cell types able to repair the diseased or injured tissue in which they are localized. This trend to adopt the local identity may be correlated to local cytokines and matrix factors, as well as to adequate contact with host cells.

MSCs injected in blood flow immediately after injury in a model of rat common carotid arteriotomy [114] are able to home at the injury site, as demonstrated by the presence of labeled MSCs in the area around adventitial *vasa vasorum* detected 1 week after injury (Forte A et al., unpublished data).

Han [115] demonstrated that bone marrow-derived MSCs contribute to neointima formation only in a model of severe injury of iliac artery in chimeric mice and not in arteries submitted to minimal damage. In particular, these authors demonstrated that about 56% of α-actin-positive cells detectable in a large neointima induced by scratch injury were derived from bone marrow. These cells resemble fetal/immature vascular SMCs. These data are in agreement with findings published by Tanaka [99], suggesting that bone marrow cells substantially contribute to lesion formation when arteries are subjected to severe injuries and that these cells expressed α-SMA but not markers for highly differentiated SMCs.

5.2 Endothelial Progenitor Cells

EPCs are a specific subset of circulating bone marrow-derived cell population and are characterized by coexpression of Sca-1 and VEGF receptor 2 (VEGFR2 or Flk-1). It is now known that EPCs are a heterogeneous population, derived chiefly from HSCs, and consist of cells at different stages of maturation, ranging from early CD133+ VEGFR2+ to more mature CD34+ VEGFR2+ phenotypes. An exhaustive description of EPC markers is in Sata et al. [116]. Despite the differences, these cells have the defining characteristic of being able to differentiate into more mature forms and being able to line the internal elastic membrane of blood vessels, they thus play a key role in neovascularization. Asahara and collaborators [97] were the first group to demonstrate that CD34+ hematopoietic cells purified from adults were able to differentiate to an endothelial phenotype (expressing various specific markers, such as vWF) and were named EPCs. Also, nonhematopoietic cells are able to differentiate to EPCs, expressing CD34 and specific endothelial markers.

5.3 Smooth Muscle Progenitor Cells

In comparison to EPCs, only a few papers are fully focused on putative smooth muscle progenitor cells and on their markers. Simper et al. demonstrated for the first time in human that SMCs can derive from blood smooth progenitor cells cultured in endothelial growth medium supplemented with platelet-derived growth factor BB [117]. These cells were positive for α-actin, myosin heavy chain, calponin, CD34, Flt-1, and Flk-1 VEGF receptors, as well as for α5β1 integrin. It should be underlined that CD34 is a surface marker known to be absent from adult human SMCs. Moreover, smooth progenitor cells do not express Tie-2 receptor, a receptor tyrosine kinase, consistent with an angioblastic lineage distinct from ECs that has been described as Tie-2 receptor positive [118].

Deb described the integrin profile of this cell population, underlying that smooth muscle precursors are characterized by a high expression of β1 integrin but do not express other integrins, which are conversely typical of the EPCs [119]. The first evidence that circulating smooth muscle precursors were able to participate in neointima formation in coronaries in a model of heterotopic cardiac transplantation between wild-type mice and LacZ mice came from Saiura [120]. Similarly, Shimizu demonstrated in a model of graft arterial disease that intimal smooth muscle-like cells derived from bone marrow and were positive to α-actin, calponin, and SM1 [121].

Interestingly, Kobayashi demonstrated that multipotent MSCs differentiated toward SMC phenotype under the stimulus represented by shear and compressive stress induced by blood flow in a model in vitro [122].

5.4 Mesoangioblasts

Recently, a novel type of vessel-associated stem cell named mesoangioblast (Mab), that can differentiate into different mesoderm cell types, has been described [123].

Mabs are physically associated with the embryonic dorsal aorta in avian and mammalian species and express the key marker of angiopoietic progenitors, such as Sca-1, Kit, Flk-1, and CD34, as well as genes typical of mesoderm, including receptors and signaling molecules for classical mesoderm inducers, such as BMP, Wnt, and Notch [124]. Presumably, they derive from a primitive angioblast [125] and are able to efficiently differentiate into endothelium, smooth, and cardiac muscle in vitro [126] and when transplanted in vivo. In particular, it has been demonstrated that the differentiation of Mabs into smooth muscle is dependent on expression of msx2 and necdin, two transcription factors able to induce, in turn, a number of smooth muscle markers [127, 128]. In vivo experiments demonstrated that Mabs are as effective as bone marrow progenitor cells in reducing postinfarction left ventricular disfunction [129]. These authors also demonstrated that Mabs in this model differentiate into smooth muscle, whereas the production of endothelium is extremely rare. Finally, Mabs are able to produce several growth factors in the heart that stimulate the proliferation of SMCs, but not of ECs.

Mabs are also able to home inside damaged muscle through the general circulation and proliferate there to reconstruct the tissue [130]. High-mobility group box 1 protein (HMGB1) has been recently identified as a chemoattractant of Mabs at the injury site [124].

Nevertheless, a positive or negative role of Mabs in vascular injury-induced remodeling has not been clearly established yet.

5.5 Hemangioblasts

Hematopoietic cells and ECs develop from mesoderm via a transitional progenitor named hemangioblast. Gene-targeting studies using embryonic stem cells have identified Flk-1 and Scl as important regulatory molecules that specify both hematopoietic and vascular outcome [131]. Flk-1 is the VEGF receptor 1 and acts as a receptor tyrosine kinase. Scl is a basic helix-loop-helix transcription factor.

Hemangioblast is present not only during the embryonic development, but its activity persists into adult life. For example, human AC133+ cells from granulocyte–CSF-mobilized peripheral blood can differentiate both into hematopoietic and ECs in culture. Moreover, these AC133+ cells can form new blood vessels in vivo [132]. As a result of this recent finding, questions are currently arising about the roles and the factors influencing the hemangioblast activity in adults and which could be the power of this cell population in therapeutic strategies involving neovascularization. No studies specifically targeting the analysis of the role of adult hemangioblasts in restenosis and to their potential applications are currently available in literature.

5.6 Bone Marrow Monocyte Lineage Cells

Recent findings suggest that peripheral blood-derived mature CD34- CD14+ monocytes are able to transdifferentiate into ECs under angiogenic conditions [133] and play a role in neovascularization via leukocyte–leukocyte interaction via CD34+ cells [134]. Others demonstrated that bone marrow-derived monocyte mononuclear cells (MLCs) differentiate into neocapillaries in ischemic limb or myocardium [135, 136]. On this basis, Fujiyama demonstrated the effectiveness of bone-derived MLCs in stenosis prevention in a model of carotid angioplasty in immunodeficient nude rats [137] (see subsequent paragraph for details).

Another study by Ohtani and colleagues demonstrated that the blockade of VEGF by soluble Flt-1 in a model of intraluminal injury in rabbits, mice, and rats inhibits the recruitment of bone marrow-MLCs with a reduction of neointima formation after injury [138]. Consequently, the exact role of monocytes, as well as of VEGF, in stenosis progression remains unclear.

References

1. Lammert E, Cleaver O, Melton D (2003) Role of endothelial cells in early pancreas and liver development. Mech Dev 120:59–64
2. Hirashima M, Suda T (2006) Differentiation of arterial and venous endothelial cells and vascular morphogenesis. Endothelium 13:137–145

3. Risau W. Mechanisms of angiogenesis. Nature 386:671–674
4. Carmeliet P (2000) Mechanisms of angiogenesis and arteriogenesis. Nat Med 6:389–395
5. Carmeliet P (2003) Angiogenesis in health and disease. Nat Med 9:653–660
6. Heil M, Eitenmuller I, Schmitz-Rixen T, Schaper W (2006) Arteriogenesis versus angiogenesis: similarities and differences. J Cell Mol Med 10:45–55
7. Adams RH, Wilkinson GA, Weiss C, Diella F, Gale NW, Deutsch U, Risau W, Klein R (1999) Roles of ephrinB ligands and EphB receptors in cardiovascular development: demarcation of arterial/venous domains, vascular morphogenesis, and sprouting angiogenesis. Genes Dev 13:295–306
8. Wang HU, Chen ZF, Anderson DJ (1998) Molecular distinction and angiogenic interaction between embryonic arteries and veins revealed by ephrin-B2 and its receptor Eph-B4. Cell 93:741–753
9. Shirayoshi Y, Yuasa Y, Suzuki T, Sugaya K, Kawase E, Ikemura T, Nakatsuji N (1997) Protooncogene of int-3, a mouse Notch homologue, is expressed in endothelial cells during early embryogenesis. Genes Cells 2:213–224
10. Lawson ND, Scheer N, Pham VN, Kim CH, Chitnis AB, Campos-Ortega JA, Weinstein BM (2001) Notch signaling is required for arterial-venous differentiation during embryonic vascular development. Development 128:3675–3683
11. Shin D, Anderson DJ (2005) Isolation of arterial-specific genes by subtractive hybridization reveals molecular heterogeneity among arterial endothelial cells. Dev Dyn 233: 1589–1604
12. Ara T, Tokoyoda K, Okamoto R, Koni PA, Nagasawa T (2005) The role of CXCL12 in the organ-specific process of artery formation. Blood 105:3155–3161
13. Park JE, Keller GA, Ferrara N (1993) The vascular endothelial growth factor (VEGF) isoforms: differential deposition into the subepithelial extracellular matrix and bioactivity of extracellular matrix-bound VEGF. Mol Biol Cell 4:1317–1326
14. Ruhrberg C (2003) Growing and shaping the vascular tree: multiple roles for VEGF. Bioessays 25:1052–1060
15. Hiratsuka S, Minowa O, Kuno J, Noda T, Shibuya M (1998) Flt-1 lacking the tyrosine kinase domain is sufficient for normal development and angiogenesis in mice. Proc Natl Acad Sci U S A 95:9349–9354
16. Shalaby F, Ho J, Stanford WL, Fischer KD, Schuh AC, Schwartz L, Bernstein A, Rossant J (1997) A requirement for Flk1 in primitive and definitive hematopoiesis and vasculogenesis. Cell 89:981–990
17. Soker S, Takashima S, Miao HQ, Neufeld G, Klagsbrun M (1998) Neuropilin-1 is expressed by endothelial and tumor cells as an isoform-specific receptor for vascular endothelial growth factor. Cell 92:735–745
18. Mukouyama YS, Gerber HP, Ferrara N, Gu C, Anderson DJ (2005) Peripheral nerve-derived VEGF promotes arterial differentiation via neuropilin 1-mediated positive feedback. Development 132:941–952
19. Liu ZJ, Shirakawa T, Li Y, Soma A, Oka M, Dotto GP, Fairman RM, Velazquez OC, Herlyn M (2003) Regulation of Notch1 and Dll4 by vascular endothelial growth factor in arterial endothelial cells: implications for modulating arteriogenesis and angiogenesis. Mol Cell Biol 23:14–25
20. Bertolino P, Deckers M, Lebrin F, ten Dijke P (2005) Transforming growth factor-beta signal transduction in angiogenesis and vascular disorders. Chest 128:585S–590S
21. Chang H, Brown CW, Matzuk MM (2002) Genetic analysis of the mammalian transforming growth factor-beta superfamily. Endocr Rev 23:787–823
22. Goumans MJ, Lebrin F, Valdimarsdottir G (2003) Controlling the angiogenic switch: a balance between two distinct TGF-b receptor signaling pathways. Trends Cardiovasc Med 13:301–307
23. Pepper MS, Vassalli JD, Orci L, Montesano R (1993) Biphasic effect of transforming growth factor-beta 1 on in vitro angiogenesis. Exp Cell Res 204:356–363

24. Goumans MJ, Valdimarsdottir G, Itoh S, Rosendahl A, Sideras P, ten Dijke P (2002) Balancing the activation state of the endothelium via two distinct TGF-beta type I receptors. EMBO J 21:1743–1753
25. Goumans MJ, Valdimarsdottir G, Itoh S, Lebrin F, Larsson J, Mummery C, Karlsson S, ten Dijke P (2003) Activin receptor-like kinase (ALK)1 is an antagonistic mediator of lateral TGFbeta/ALK5 signaling. Mol Cell 12:817–828
26. Ota T, Fujii M, Sugizaki T, Ishii M, Miyazawa K, Aburatani H, Miyazono K (2002) Targets of transcriptional regulation by two distinct type I receptors for transforming growth factor-beta in human umbilical vein endothelial cells. J Cell Physiol 193:299–318
27. Frid MG, Kale VA, Stenmark KR (2002) Mature vascular endothelium can give rise to smooth muscle cells via endothelial-mesenchymal transdifferentiation: in vitro analysis. Circ Res 90:1189–1196
28. Hautmann MB, Adam PJ, Owens GK (1999) Similarities and differences in smooth muscle alpha-actin induction by TGF-beta in smooth muscle versus non-smooth muscle cells. Arterioscler Thromb Vasc Biol 19:2049–2058
29. Waite KA, Eng C (2003) From developmental disorder to heritable cancer: it's all in the BMP/TGF-beta family. Nat Rev Genet 4:763–773
30. Seki T, Hong KH, Yun J, Kim SJ, Oh SP (2004) Isolation of a regulatory region of activin receptor-like kinase 1 gene sufficient for arterial endothelium-specific expression. Circ Res 94:e72–77
31. Majesky MW, Lindner V, Twardzik DR, Schwartz SM, Reidy MA (1991) Production of transforming growth factor beta 1 during repair of arterial injury. J Clin Invest 88: 904–910
32. Schulick AH, Taylor AJ, Zuo W, Qiu CB, Dong G, Woodward RN, Agah R, Roberts AB, Virmani R, Dichek DA (1998) Overexpression of transforming growth factor beta1 in arterial endothelium causes hyperplasia, apoptosis, and cartilaginous metaplasia. Proc Natl Acad Sci U S A 95:6983–6988
33. Smith JD, Bryant SR, Couper LL, Vary CP, Gotwals PJ, Koteliansky VE, Lindner V (1999) Soluble transforming growth factor-beta type II receptor inhibits negative remodeling, fibroblast transdifferentiation, and intimal lesion formation but not endothelial growth. Circ Res 84:1212–1222
34. McCaffrey TA, Du B, Consigli S, Szabo P, Bray PJ, Hartner L, Weksler BB, Sanborn TA, Bergman G, Bush HL Jr (1997) Genomic instability in the type II TGF-beta1 receptor gene in atherosclerotic and restenotic vascular cells. J Clin Invest 100:2182–2188
35. Grainger DJ, Kemp PR, Metcalfe JC, Liu AC, Lawn RM, Williams NR, Grace AA, Schofield PM, Chauhan A (1995) The serum concentration of active transforming growth factor-beta is severely depressed in advanced atherosclerosis. Nat Med 1:74–79
36. Mallat Z, Gojova A, Marchiol-Fournigault C, Esposito B, Kamate C, Merval R, Fradelizi D, Tedgui A (2001) Inhibition of transforming growth factor-beta signaling accelerates atherosclerosis and induces an unstable plaque phenotype in mice. Circ Res 89:930–934
37. Owens GK, Kumar MS, Wamhoff BR (2004) Molecular regulation of vascular smooth muscle cell differentiation in development and disease. Physiol Rev 84:767–801
38. Ross R (1999) Atherosclerosis is an inflammatory disease. Am Heart J 138:S419–420
39. Hungerford JE, Little CD (1999) Developmental biology of the vascular smooth muscle cell: building a multilayered vessel wall. J Vasc Res 36:2–27
40. Brisset AC, Hao H, Camenzind E, Bacchetta M, Geinoz A, Sanchez JC, Chaponnier C, Gabbiani G, Bochaton-Piallat ML (2007) Intimal smooth muscle cells of porcine and human coronary artery express S100A4, a marker of the rhomboid phenotype in vitro. Circ Res 100:1055–1062
41. Owens GK (1995) Regulation of differentiation of vascular smooth muscle cells. Physiol Rev 75:487–517
42. Wilcox JN (1992) Analysis of local gene expression in human atherosclerotic plaques. J Vasc Surg 15:913–916

43. Owens GK, Vernon SM, Madsen CS (1996) Molecular regulation of smooth muscle cell differentiation. J Hypertens Suppl 14:S55–S64
44. Blank RS, Owens GK (1990) Platelet-derived growth factor regulates actin isoform expression and growth state in cultured rat aortic smooth muscle cells. J Cell Physiol 142: 635–642
45. Corjay MH, Thompson MM, Lynch KR, Owens GK (1989) Differential effect of platelet-derived growth factor- versus serum-induced growth on smooth muscle alpha-actin and nonmuscle beta-actin mRNA expression in cultured rat aortic smooth muscle cells. J Biol Chem 264:10501–10506
46. Frid MG, Moiseeva EP, Stenmark KR (1994) Multiple phenotypically distinct smooth muscle cell populations exist in the adult and developing bovine pulmonary arterial media in vivo. Circ Res 75:669–681
47. Hao H, Ropraz P, Verin V, Camenzind E, Geinoz A, Pepper MS, Gabbiani G, Bochaton-Piallat ML (2002) Heterogeneity of smooth muscle cell populations cultured from pig coronary artery. Arterioscler Thromb Vasc Biol 22:1093–1099
48. Mack CP, Owens GK (1999) Regulation of smooth muscle alpha-actin expression in vivo is dependent on CArG elements within the 5' and first intron promoter regions. Circ Res 84:852–861
49. Manabe I, Owens GK (2001) CArG elements control smooth muscle subtype-specific expression of smooth muscle myosin in vivo. J Clin Invest 107:823–834
50. Miano JM, Long X, Fujiwara K (2007) Serum response factor: master regulator of the actin cytoskeleton and contractile apparatus. Am J Physiol Cell Physiol 292:C70–C81
51. Wang D, Chang PS, Wang Z, Sutherland L, Richardson JA, Small E, Krieg PA, Olson EN (2001) Activation of cardiac gene expression by myocardin, a transcriptional cofactor for serum response factor. Cell 105:851–862
52. Wang Z, Wang DZ, Pipes GC, Olson EN (2003) Myocardin is a master regulator of smooth muscle gene expression. Proc Natl Acad Sci U S A 100:7129–7134
53. Li S, Wang DZ, Wang Z, Richardson JA, Olson EN (2003) The serum response factor coactivator myocardin is required for vascular smooth muscle development. Proc Natl Acad Sci U S A 100:9366–9370
54. Manabe I, Owens GK (2001) Recruitment of serum response factor and hyperacetylation of histones at smooth muscle-specific regulatory regions during differentiation of a novel P19-derived in vitro smooth muscle differentiation system. Circ Res 88: 1127–1134
55. Li J, Zhu X, Chen M, Cheng L, Zhou D, Lu MM, Du K, Epstein JA, Parmacek MS (2005) Myocardin-related transcription factor B is required in cardiac neural crest for smooth muscle differentiation and cardiovascular development. Proc Natl Acad Sci USA 102:8916–8921
56. Holycross BJ, Blank RS, Thompson MM, Peach MJ, Owens GK (1992) Platelet-derived growth factor-BB-induced suppression of smooth muscle cell differentiation. Circ Res 71:1525–1532
57. Liu ZP, Wang Z, Yanagisawa H, Olson EN (2005) Phenotypic modulation of smooth muscle cells through interaction of Foxo4 and myocardin. Dev Cell 9:261–270
58. Kook H, Lepore JJ, Gitler AD, Lu MM, Wing-Man Yung W, Mackay J, Zhou R, Ferrari V, Gruber P, Epstein JA (2003) Cardiac hypertrophy and histone deacetylase-dependent transcriptional repression mediated by the atypical homeodomain protein Hop. J Clin Invest 112:863–871
59. Santiago FS, Lowe HC, Bobryshev YV, Khachigian LM (2001) Induction of the transcriptional repressor Yin Yang-1 by vascular cell injury. Autocrine/paracrine role of endogenous fibroblast growth factor-2. J Biol Chem 276:41143–41149
60. Philippar U, Schratt G, Dieterich C, Muller JM, Galgoczy P, Engel FB, Keating MT, Gertler F, Schule R, Vingron M, Nordheim A (2004) The SRF target gene Fhl2 antagonizes RhoA/MAL-dependent activation of SRF. Mol Cell 16:867–880

61. Dandre F, Owens GK (2004) Platelet-derived growth factor-BB and Ets-1 transcription factor negatively regulate transcription of multiple smooth muscle cell differentiation marker genes. Am J Physiol Heart Circ Physiol 286:H2042–2051
62. Madsen CS, Regan CP, Owens GK (1997) Interaction of CArG elements and a GC-rich repressor element in transcriptional regulation of the smooth muscle myosin heavy chain gene in vascular smooth muscle cells. J Biol Chem 272:29842–29851
63. Rafty LA, Khachigian LM (1998) Zinc finger transcription factors mediate high constitutive platelet-derived growth factor-B expression in smooth muscle cells derived from aortae of newborn rats. J Biol Chem 273:5758–5764
64. McDonald OG, Owens GK (2007) Programming smooth muscle plasticity with chromatin dynamics. Circ Res 100:1428–1441
65. Marin M, Karis A, Visser P, Grosveld F, Philipsen S (1997) Transcription factor Sp1 is essential for early embryonic development but dispensable for cell growth and differentiation. Cell 89:619–628
66. Leveen P, Pekny M, Gebre-Medhin S, Swolin B, Larsson E, Betsholtz C (1994) Mice deficient for PDGF B show renal, cardiovascular, and hematological abnormalities. Genes Dev 8:1875–1887
67. Gabbiani G, Ryan GB, Majne G (1971) Presence of modified fibroblasts in granulation tissue and their possible role in wound contraction. Experientia 27:549–550
68. Bellini A, Mattoli S (2007) The role of the fibrocyte, a bone marrow-derived mesenchymal progenitor, in reactive and reparative fibroses. Lab Invest 87:858–870
69. Wilcox JN, Cipolla GD, Martin FH, Simonet L, Dunn B, Ross CE, Scott NA (1997) Contribution of adventitial myofibroblasts to vascular remodeling and lesion formation after experimental angioplasty in pig coronary arteries. Ann N Y Acad Sci 811:437–447
70. Yoshida T, Sinha S, Dandre F, Wamhoff BR, Hoofnagle MH, Kremer BE, Wang DZ, Olson EN, Owens GK (2003) Myocardin is a key regulator of CArG-dependent transcription of multiple smooth muscle marker genes. Circ Res 92:856–864
71. Gan Q, Yoshida T, Li J, Owens GK (2007) Smooth muscle cells and myofibroblasts use distinct transcriptional mechanisms for smooth muscle alpha-actin expression. Circ Res 101:883–892
72. Hinz B, Gabbiani G (2003) Mechanisms of force generation and transmission by myofibroblasts. Curr Opin Biotechnol 14:538–546
73. Bogatkevich GS, Tourkina E, Abrams CS, Harley RA, Silver RM, Ludwicka-Bradley A (2003) Contractile activity and smooth muscle alpha-actin organization in thrombin-induced human lung myofibroblasts. Am J Physiol Lung Cell Mol Physiol 285:L334–343
74. Hinz B, Phan SH, Thannickal VJ, Galli A, Bochaton-Piallat ML, Gabbiani G (2007) The Myofibroblast. One Function, Multiple Origins. Am J Pathol 170:1807–1816
75. Serini G, Gabbiani G (1999) Mechanisms of myofibroblast activity and phenotypic modulation. Exp Cell Res 250:273–283
76. Goffin JM, Pittet P, Csucs G, Lussi JW, Meister JJ, Hinz B (2006) Focal adhesion size controls tension-dependent recruitment of alpha-smooth muscle actin to stress fibers. J Cell Biol 172:259–268
77. Hautmann MB, Madsen CS, Owens GK (1997) A transforming growth factor beta (TGFbeta) control element drives TGFbeta-induced stimulation of smooth muscle alpha-actin gene expression in concert with two CArG elements. J Biol Chem 272:10948–10956
78. Zambruno G, Marchisio PC, Marconi A, Vaschieri C, Melchiori A, Giannetti A, De Luca M (1995) Transforming growth factor-beta 1 modulates beta 1 and beta 5 integrin receptors and induces the de novo expression of the alpha v beta 6 heterodimer in normal human keratinocytes: implications for wound healing. J Cell Biol 129:853–865
79. Roberts AB, Sporn MB, Assoian RK, Smith JM, Roche NS, Wakefield LM, Heine UI, Liotta LA, Falanga V, Kehrl JH et al (1986) Transforming growth factor type beta: rapid induction of fibrosis and angiogenesis in vivo and stimulation of collagen formation in vitro. Proc Natl Acad Sci USA 83:4167–4171

80. Serini G, Bochaton-Piallat ML, Ropraz P, Geinoz A, Borsi L, Zardi L, Gabbiani G (1998) The fibronectin domain ED-A is crucial for myofibroblastic phenotype induction by transforming growth factor-beta1. J Cell Biol 142:873–881
81. Kocher O, Kennedy SP, Madri JA (1990) Alternative splicing of endothelial cell fibronectin mRNA in the IIICS region. Functional significance. Am J Pathol 137:1509–1524
82. Manabe R, Ohe N, Maeda T, Fukuda T, Sekiguchi K (1997) Modulation of cell-adhesive activity of fibronectin by the alternatively spliced EDA segment. J Cell Biol 139: 295–307
83. Horowitz JC, Rogers DS, Sharma V, Vittal R, White ES, Cui Z, Thannickal VJ (2007) Combinatorial activation of FAK and AKT by transforming growth factor-beta1 confers an anoikis-resistant phenotype to myofibroblasts. Cell Signal 19:761–771
84. Gao PJ, Li Y, Sun AJ, Liu JJ, Ji KD, Zhang YZ, Sun WL, Marche P, Zhu DL (2003) Differentiation of vascular myofibroblasts induced by transforming growth factor-beta1 requires the involvement of protein kinase Calpha. J Mol Cell Cardiol 35:1105–1112
85. Greenberg RS, Bernstein AM, Benezra M, Gelman IH, Taliana L, Masur SK (2006) FAK-dependent regulation of myofibroblast differentiation. FASEB J 20:1006–1008
86. Thomas PE, Peters-Golden M, White ES, Thannickal VJ, Moore BB (2007) PGE(2) inhibition of TGF-beta1-induced myofibroblast differentiation is Smad-independent but involves cell shape and adhesion-dependent signaling. Am J Physiol Lung Cell Mol Physiol 293:L417–428
87. Leask A, Abraham DJ (2004) TGF-beta signaling and the fibrotic response. FASEB J 18:816–827
88. Shen WL, Gao PJ, Che ZQ, Ji KD, Yin M, Yan C, Berk BC, Zhu DL (2006) NAD(P)H oxidase-derived reactive oxygen species regulate angiotensin-II induced adventitial fibroblast phenotypic differentiation. Biochem Biophys Res Commun 339:337–343
89. Shi-Wen X, Chen Y, Denton CP, Eastwood M, Renzoni EA, Bou-Gharios G, Pearson JD, Dashwood M, du Bois RM, Black CM, Leask A, Abraham DJ (2004) Endothelin-1 promotes myofibroblast induction through the ETA receptor via a rac/phosphoinositide 3-kinase/Akt-dependent pathway and is essential for the enhanced contractile phenotype of fibrotic fibroblasts. Mol Biol Cell 15:2707–2719
90. Desmouliere A, Redard M, Darby I, Gabbiani G (1995) Apoptosis mediates the decrease in cellularity during the transition between granulation tissue and scar. Am J Pathol 146:56–66
91. Zhang HY, Phan SH (1999) Inhibition of myofibroblast apoptosis by transforming growth factor beta(1). Am J Respir Cell Mol Biol 21:658–665
92. Carlson MA, Longaker MT, Thompson JS (2003) Wound splinting regulates granulation tissue survival. J Surg Res 110:304–309
93. Grinnell F, Zhu M, Carlson MA, Abrams JM (1999) Release of mechanical tension triggers apoptosis of human fibroblasts in a model of regressing granulation tissue. Exp Cell Res 248:608–619
94. Yap AS, Kovacs EM (2003) Direct cadherin–activated cell signaling: a view from the plasma membrane. J Cell Biol 160:11–16
95. Petridou S, Maltseva O, Spanakis S, Masur SK (2000) TGF-beta receptor expression and smad2 localization are cell density dependent in fibroblasts. Invest Ophthalmol Vis Sci 41:89–95
96. Hinz B (2007) Formation and function of the myofibroblast during tissue repair. J Invest Dermatol 127:526–537
97. Asahara T, Murohara T, Sullivan A, Silver M, van der Zee R, Li T, Witzenbichler B, Schatteman G, Isner JM (1997) Isolation of putative progenitor endothelial cells for angiogenesis. Science 275:964–967
98. Forte A, Cipollaro M, Cascino A, Galderisi U (2007) Pathophysiology of stem cells in restenosis. Histol Histopathol 22:547–557
99. Tanaka K, Sata M, Hirata Y, Nagai R (2003) Diverse contribution of bone marrow cells to neointimal hyperplasia after mechanical vascular injuries. Circ Res 93:783–790

100. Yamashita JK (2004) Differentiation and diversification of vascular cells from embryonic stem cells. Int J Hematol 80:1–6
101. Yamamoto K, Sokabe T, Watabe T, Miyazono K, Yamashita JK, Obi S, Ohura N, Matsushita A, Kamiya A, Ando J (2005) Fluid shear stress induces differentiation of Flk-1-positive embryonic stem cells into vascular endothelial cells in vitro. Am J Physiol Heart Circ Physiol 288:H1915–1924
102. Tintut Y, Alfonso Z, Saini T, Radcliff K, Watson K, Bostrom K, Demer LL (2003) Multilineage potential of cells from the artery wall. Circulation 108:2505–2510
103. Friedenstein AJ, Gorskaja JF, Kulagina NN (1976) Fibroblast precursors in normal and irradiated mouse hematopoietic organs. Exp Hematol 4:267–274
104. Noth U, Schupp K, Heymer A, Kall S, Jakob F, Schutze N, Baumann B, Barthel T, Eulert J, Hendrich C (2005) Anterior cruciate ligament constructs fabricated from human mesenchymal stem cells in a collagen type I hydrogel. Cytotherapy 7:447–455
105. Tuan RS, Boland G, Tuli R (2003) Adult mesenchymal stem cells and cell-based tissue engineering. Arthritis Res Ther 5:32–45
106. Sarugaser R, Lickorish D, Baksh D, Hosseini MM, Davies JE (2005) Human umbilical cord perivascular (HUCPV) cells: a source of mesenchymal progenitors. Stem Cells 23:220–229
107. Charbord P, Lerat H, Newton I, Tamayo E, Gown AM, Singer JW, Herve P (1990) The cytoskeleton of stromal cells from human bone marrow cultures resembles that of cultured smooth muscle cells. Exp Hematol 18:276–282
108. Peled A, Zipori D, Abramsky O, Ovadia H, Shezen E (1991) Expression of alpha-smooth muscle actin in murine bone marrow stromal cells. Blood 78:304–309
109. Wexler SA, Donaldson C, Denning-Kendall P, Rice C, Bradley B, Hows JM (2003) Adult bone marrow is a rich source of human mesenchymal 'stem' cells but umbilical cord and mobilized adult blood are not. Br J Haematol 121:368–374
110. Pittenger MF, Mackay AM, Beck SC, Jaiswal RK, Douglas R, Mosca JD, Moorman MA, Simonetti DW, Craig S, Marshak DR (1999) Multilineage potential of adult human mesenchymal stem cells. Science 284:143–147
111. Gao J, Dennis JE, Muzic RF, Lundberg M, Caplan AI (2001) The dynamic in vivo distribution of bone marrow-derived mesenchymal stem cells after infusion. Cells Tissues Organs 169:12–20
112. Prockop DJ (1997) Marrow stromal cells as stem cells for nonhematopoietic tissues. Science 276:71–74
113. Prockop DJ (1998) Marrow stromal cells as stem cells for continual renewal of non-hematopoietic tissues and as potential vectors for gene therapy. J Cell Biochem Suppl 30–31:284–285
114. Forte A, Di Micco G, Galderisi U, Guarino FM, Cipollaro M, De Feo M, Gregorio R, Bianco MR, Vollono C, Esposito F, Berrino L, Angelini F, Renzulli A, Cotrufo M, Rossi F, Cascino A (2001) Molecular analysis of arterial stenosis in rat carotids. J Cell Physiol 186:307–313
115. Han CI, Campbell GR, Campbell JH (2001) Circulating bone marrow cells can contribute to neointimal formation. J Vasc Res 38:113–119
116. Sata M, Fukuda D, Tanaka K, Kaneda Y, Yashiro H, Shirakawa I (2005) The role of circulating precursors in vascular repair and lesion formation. J Cell Mol Med 9: 557–568
117. Simper D, Stalboerger PG, Panetta CJ, Wang S, Caplice NM (2002) Smooth muscle progenitor cells in human blood. Circulation 106:1199–1204
118. Sata M, Saiura A, Kunisato A, Tojo A, Okada S, Tokuhisa T, Hirai H, Makuuchi M, Hirata Y, Nagai R (2002) Hematopoietic stem cells differentiate into vascular cells that participate in the pathogenesis of atherosclerosis. Nat Med 8:403–409
119. Deb A, Skelding KA, Wang S, Reeder M, Simper D, Caplice NM (2004) Integrin profile and in vivo homing of human smooth muscle progenitor cells. Circulation 110:2673–2677
120. Saiura A, Sata M, Hirata Y, Nagai R, Makuuchi M (2001) Circulating smooth muscle progenitor cells contribute to atherosclerosis. Nat Med 7:382–383

121. Shimizu K, Sugiyama S, Aikawa M, Fukumoto Y, Rabkin E, Libby P, Mitchell RN (2001) Host bone-marrow cells are a source of donor intimal smooth- muscle-like cells in murine aortic transplant arteriopathy. Nat Med 7:738–741
122. Kobayashi N, Yasu T, Ueba H, Sata M, Hashimoto S, Kuroki M, Saito M, Kawakami M (2004) Mechanical stress promotes the expression of smooth muscle-like properties in marrow stromal cells. Exp Hematol 32:1238–1245
123. Minasi MG, Riminucci M, De Angelis L, Borello U, Berarducci B, Innocenzi A, Caprioli A, Sirabella D, Baiocchi M, De Maria R, Boratto R, Jaffredo T, Broccoli V, Bianco P, Cossu G (2002) The meso-angioblast: a multipotent, self-renewing cell that originates from the dorsal aorta and differentiates into most mesodermal tissues. Development 129: 2773–2783
124. Palumbo R, Sampaolesi M, De Marchis F, Tonlorenzi R, Colombetti S, Mondino A, Cossu G, Bianchi ME (2004) Extracellular HMGB1, a signal of tissue damage, induces mesoangioblast migration and proliferation. J Cell Biol 164:441–449
125. Cossu G, Bianco P (2003) Mesoangioblasts--vascular progenitors for extravascular mesodermal tissues. Curr Opin Genet Dev 13:537–542
126. Tagliafico E, Brunelli S, Bergamaschi A, De Angelis L, Scardigli R, Galli D, Battini R, Bianco P, Ferrari S, Cossu G (2004) TGFbeta/BMP activate the smooth muscle/bone differentiation programs in mesoangioblasts. J Cell Sci 117:4377–4388
127. Brunelli S, Cossu G (2005) A role for MSX2 and necdin in smooth muscle differentiation of mesoangioblasts and other mesoderm progenitor cells. Trends Cardiovasc Med 15: 96–100
128. Brunelli S, Tagliafico E, De Angelis FG, Tonlorenzi R, Baesso S, Ferrari S, Niinobe M, Yoshikawa K, Schwartz RJ, Bozzoni I, Cossu G (2004) Msx2 and necdin combined activities are required for smooth muscle differentiation in mesoangioblast stem cells. Circ Res 94:1571–1578
129. Galli D, Innocenzi A, Staszewsky L, Zanetta L, Sampaolesi M, Bai A, Martinoli E, Carlo E, Balconi G, Fiordaliso F, Chimenti S, Cusella G, Dejana E, Cossu G, Latini R (2005) Mesoangioblasts, vessel-associated multipotent stem cells, repair the infarcted heart by multiple cellular mechanisms: a comparison with bone marrow progenitors, fibroblasts, and endothelial cells. Arterioscler Thromb Vasc Biol 25:692–697
130. Sampaolesi M, Torrente Y, Innocenzi A, Tonlorenzi R, D'Antona G, Pellegrino MA, Barresi R, Bresolin N, De Angelis MG, Campbell KP, Bottinelli R, Cossu G (2003) Cell therapy of alpha-sarcoglycan null dystrophic mice through intra-arterial delivery of mesoangioblasts. Science 301:487–492
131. Bailey AS, Fleming WH (2003) Converging roads: evidence for an adult hemangioblast. Exp Hematol 31:987–993
132. Gehling UM, Ergun S, Schumacher U, Wagener C, Pantel K, Otte M, Schuch G, Schafhausen P, Mende T, Kilic N, Kluge K, Schafer B, Hossfeld DK, Fiedler W (2000) In vitro differentiation of endothelial cells from AC133-positive progenitor cells. Blood 95:3106–3112
133. Rehman J, Li J, Orschell CM, March KL (2003) Peripheral blood "endothelial progenitor cells" are derived from monocyte/macrophages and secrete angiogenic growth factors. Circulation 107:1164–1169
134. Harraz M, Jiao C, Hanlon HD, Hartley RS, Schatteman GC (2001) CD34- blood-derived human endothelial cell progenitors. Stem Cells 19:304–312
135. Kamihata H, Matsubara H, Nishiue T, Fujiyama S, Tsutsumi Y, Ozono R, Masaki H, Mori Y, Iba O, Tateishi E, Kosaki A, Shintani S, Murohara T, Imaizumi T, Iwasaka T (2001) Implantation of bone marrow mononuclear cells into ischemic myocardium enhances collateral perfusion and regional function via side supply of angioblasts, angiogenic ligands, and cytokines. Circulation 104:1046–1052
136. Iba O, Matsubara H, Nozawa Y, Fujiyama S, Amano K, Mori Y, Kojima H, Iwasaka T (2002) Angiogenesis by implantation of peripheral blood mononuclear cells and platelets into ischemic limbs. Circulation 106:2019–2025

137. Fujiyama S, Amano K, Uehira K, Yoshida M, Nishiwaki Y, Nozawa Y, Jin D, Takai S, Miyazaki M, Egashira K, Imada T, Iwasaka T, Matsubara H (2003) Bone marrow monocyte lineage cells adhere on injured endothelium in a monocyte chemoattractant protein-1-dependent manner and accelerate reendothelialization as endothelial progenitor cells. Circ Res 93:980–989
138. Ohtani K, Egashira K, Hiasa K, Zhao Q, Kitamoto S, Ishibashi M, Usui M, Inoue S, Yonemitsu Y, Sueishi K, Sata M, Shibuya M, Sunagawa K (2004) Blockade of vascular endothelial growth factor suppresses experimental restenosis after intraluminal injury by inhibiting recruitment of monocyte lineage cells. Circulation 110:2444–2452

Part V
Cell Cycle and Differentiation in the Muscle

Chapter 11
Cell Cycle Regulation in Myogenesis

Cristina Giacinti and Antonio Giordano

Abstract Skeletal myoblasts differentiation begins with irreversible withdrawal of myoblasts from the cell cycle. This growth arrest is mediated and maintained by the Rb protein, together with p21 and other inhibitors of cell cycle progression. The subsequent activation of muscle-specific promoters at the onset of differentiation is regulated by myogenic bHLH transcription factors, such as MyoD. MyoD protein binds DNA and recruits coactivators and corepressors of transcription to specific promoters and orchestrates the early differentiation events that lead to the fusion of myoblasts into myotubes and the formation of skeletal muscle.

Keywords Myoblasts · Muscle cell differentiation · Retinoblastoma gene family · bHLH transcription factors · Cyclin kinase inhibitors

1 Introduction

When the preconditions for dividing cells are not met, cells can cease progressing the division cycle and withdraw into a quiescent state G0; when the circumstances change so as to favor cell division, cells resume their progress through the cycle. In contrast skeletal myoblasts differentiation begins with irreversible withdrawal of myoblasts from the cell cycle. This permanent cell cycle withdrawal is characterized by a loss in their ability to reenter the cell cycle in the presence of growth factor stimulation [1]. This growth arrest is mediated and maintained by the Rb protein, together with p21 and other inhibitors of cell cycle progression. The subsequent activation of muscle-specific promoters at the onset of differentiation is regulated by myogenic bHLH transcription factors, such as MyoD. MyoD protein binds DNA and recruits coactivators and corepressors of transcription to specific

C. Giacinti (✉)
Department of Histology and Medical Embryology, University "La Sapienza", Rome, Italy
e-mail: cristina.giacinti@uniroma1.it

A. Giordano, U. Galderisi (eds.), *Cell Cycle Regulation and Differentiation in Cardiovascular and Neural Systems*, DOI 10.1007/978-1-60327-153-0_11,

promoters [2] and orchestrates the early differentiation events that lead to the fusion of myoblasts into myotubes and the formation of skeletal muscle. The MyoD family of bHLH proteins are transcription factors that form heterodimers with ubiquitously expressed members of the E-protein family, E12/47, E2-2, E2-5, and HEB, and bind to a DNA consensus CANNTG, known as an E-box, found in the promoters of several muscle-specific genes [3, 4]. Even though MyoD and myf-5 are expressed in dividing myoblasts [5], their function is kept in check and only when the myoblast has withdrawn from the cell cycle upon mitogens reduction, muscle-specific genes are activated. This implies that there are regulatory mechanisms that control myogenic factor function by sensing the cell cycle status of myoblast as it responds to growth factors and external signals.

2 Cell Cycle Overview

2.1 Cell Cycle

The cell cycle is a process through which cells duplicate themselves. The cell cycle is divided into four phases: G1, S, G2, M. In eukaryotic cells, cell cycle progression involves a series of tightly regulated and coordinated events.

Two of the most important protein involved in the cell cycle machinery mechanisms are cyclin-dependent kinase (cdks) and cyclins. A variety of cyclin:cdk complexes are in fact able to guide the cdks to appropriate substrates and activate their catalytic activity. Cyclin:cdk complexes are formed during distinct phases of the cell cycle and are specifically involved in the phosphorylation of a distinct set of target proteins. Mammalian G1 cyclins D and E mediate progression through G1/S phases. Three D-type cyclins exist (cyclin D1, D2, and D3), which are expressed differently in various cell lineages, with most cells expressing cyclin D3 and either D1 or D2 [6, 7]. Two types of cyclin E (E1 and E2) exist, which show overlapping expression patterns in mouse tissues and can be co-overexpressed in human tumors [8]. Mitotic cyclins A and B mediate progression through the S/G2/M phases. Cyclin A1 is expressed in meiosis and in early embryogenesis, whereas cyclin A2 is found in proliferating somatic cells [9]. Cyclin B2 probably plays a role in Golgi remodeling during mitosis [9], while cyclin B1 controls other functions of this cyclin type. Until now at least 16 cyclins have been discovered, for many of which binding partners and functions have yet to be identified.

D-type cyclins are short-lived proteins whose synthesis and assembly with Cdk4 or Cdk6 in G1 are dependent on mitogenic signaling [10]. Cyclin D/Cdk activity goes on through the first and subsequent cycles as long as mitogenic stimulation continues. Cyclin E-protein levels peak at the G1/S progression, followed by an increase in cyclin A levels in the S phase. Both cyclin E and A interact with and activate Cdk2, while cyclin A can also bind Cdk1 [11]. At the G2/M boundary, cyclin B levels increase, resulting in activation of its partner, Cdk1. This fluctuation in cyclin expression and resultant oscillation in Cdk activity form the basis of a coordinated cell cycle progression.

The cell responds to mitogenic stimuli and advances through the various phases of the cell cycle only for a limited phase of its cycle. In fact they need stimulation only during the first two-thirds of their G1 phase where the cell may decide to continue its advance and complete its cell cycle. This point is termed "restriction point" (R point); it is a central event in normal cellular proliferation control [12]. It has been demonstrated that pRb is the molecular device that serves as the R point switch. pRb results to be hypo-phosphorylated in resting G0 cells, to be increasingly phosphorylated during progression through G1, and to be maintained in a hyperphosphorylated state until late mitosis [13]. pRb phosphorylation seems to be related to mitogenic signals which converge on the cell cycle machinery, represented by the cyclinD/cdk4 [6] complex in the early and mid-G1 and composed of cyclinE/cdk2 in late G1.

The Rb gene family includes three members, pRb105, p107, and p130, collectively referred to as "pocket proteins" [14]. The term "pocket proteins" is derived from the conserved binding pocket region through which pRb, p107, and p130 bind viral oncoproteins and cellular factors such as the E2F family of transcription factors. pRB/105 and p107 are the best substrate for cyclinD1–cdk4/6 complex activity, while p130 phosphorylation depends on the cyclinD3–cdk4 complex activity. All of the three pocket proteins are phosphorylated by cyclinE/cdk2 complex, instead both p130 and p107 are phosphorylated by cyclin A–cdk2 but not pRb/105. When in its actively growth-suppressing hyperphosphorylated state, pRb physically associates with E2F factors and blocks their ability to activate expression of genes that encode products necessary for S phase progression. In addition pRb recruits chromatin-remodeling factors such as histone deacetylase 1 (HDAC) [15], SWI/SNF factors (Harbour et al., 2000), Polycomb group proteins [16], or methyl transferase [17] that act on the nearby surrounding nucleosome structure.

The Rb proteins repress gene transcription, required for transition from G1 to S phase, by directly binding to the transactivation domain of E2F and by binding to the promoter of these genes as a complex with E2F [18]. Pocket protein regulates G1–S transition also through E2F-independent mechanisms: (1) p107 and p130 bind and inhibit the cyclinE/cdk2 and cyclinA/cdk2 kinases, (2) pRb inhibits cdk activity and G1–S progression by increasing the expression of p27 and stabilizing the p27 protein by binding the Skp2–p27 complex, avoiding p27 ubiquitination. Progression of the cells through G1 and S phase requires inactivation of Rb protein phosphorylation. The phosphorylation status of these proteins is regulated by CKI binding to the cyclin/cdk complexes [19]. The CKI inhibitor p15, p16, and p17 specifically act on cyclin D kinase activity, p21, p27, and p57 on all other cyclin/cdk complexes. It has been reported that both cyclinD1 overexpression and related CKI inhibitors alterations produce persistent hyperphosphorylation of pRb, resulting in cell cycle arrest.

Extracellular physiological signals induce the phosphorylation of Rb protein, affecting the decision to transit the R point. Mitogens directly induce rapid expression of cyclin D which begins to inactivate pRb through its ability to associate with its previously cited cdk4 [6] partners [20]. Each type of cyclin D could receive different upstream signals that converge on it, so that we can say that the control of cell proliferation in various cell type, known to be exercised by different mitogens, is modulated via expression of a distinct D-type cyclin gene, which is under the control of a distinct transcriptional promoter. The common point is all these different signals converge to a common target which is pRb. There is no evidence that mitogenic stimuli modulate directly the levels of cyclin E or cycline/cdk2 complex. Extracellular signals can affect the cyclin/cdk machinery negatively as well. Serum starvation leads to a collapse of cyclin D levels and activity and to an increase of specific cdk inhibitor (CKI) [21].

3 Cell Cycle and Myogenesis

3.1 Regulation of MyoD Function in Dividing Myoblast

The bHLH protein MyoD is functionally regulated during myoblast proliferation; different models explain how MyoD activity might be suppressed in the dividing myoblasts.

During myoblasts proliferation MyoD activity is regulated by interactions with the Id family of proteins [22]. The Id family proteins are negative regulators that contain HLH domains, but lack the basic DNA-binding domain, and consequently fail to bind DNA. Therefore, the HLH domain of Id proteins allows heterodimerization with other HLH proteins, such as the MyoD and E-proteins, thus sequestering MyoD and/or E-proteins into complexes that fail to bind DNA [23]. However at the onset of differentiation, Id gene expression is downregulated, resulting in MyoD and E-protein dimerization, which leads to the activation of muscle-specific genes [22].

In addition phosphorylation on serine 200 by cdk2 and cdk1 has been shown to inhibit MyoD activity, thereby triggering its degradation. Ser 200 phosphorylation of MyoD may also reduce the interaction of MyoD with its associated factors, such as Rb [1], muscle enhancer factor MEF-2, and the co-activator p300 [24].

Olson and colleagues observed that a conserved threonine residue in the basic region of all the myogenic factors was essential for activity. This threonine was also a consensus site for protein kinase C (PKC) phosphorylation in vitro, a kinase activated by β-FGF and TGF-β. These factors inhibit the activity of the myogenic factors through the direct PKC-dependent phosphorylation [25].

Finally, cyclin D1 overexpression promotes the nuclear accumulation of cdk4, which in turn binds the C-terminus of MyoD and prevents its binding to the DNA. Indeed, increasing amounts of cdk4 could specifically disrupt DNA binding of both MyoD homodimers and MyoD/E12 heterodimers [25, 26] . Disruption of MyoD DNA binding did not require the active kinase since cdk4 alone could inhibit the interaction with DNA. Since cdk4 is expressed at similar levels in both myoblasts and myotubes [27], this result suggested that cdk4 would likely not be present in myotube nuclei to potentially disrupt MyoD function. In dividing myoblasts both cyclin D1 and cdk4 are nuclear, but in well-formed myotubes cyclin D1 is absent and cdk4 is in the cytoplasmic compartment of newly differentiated muscle, as postulated. Cyclin D1 is the mitogen sensor and the limiting factor in the assembly of active cdk4/cyclin D1 complexes, and it is absent in differentiated myotubes [28]. The specific MyoD–cdk4 interaction in dividing myoblasts, coupled with the cyclin D1-dependent nuclear targeting of cdk4, suggested a mitogen-sensitive mechanism whereby cyclin D1 can regulate MyoD function and the onset of myogenesis by controlling the cellular location of cdk4 rather than the phosphorylation status of MyoD [26].

Genes that respond rapidly to growth stimuli, the immediate early genes (myc, jun, and fos) also repress myogenic factor function and expression [23, 29]. In addition, peptide growth factors and phorbol esters induce PKC, an activator of cJun and cFos [30]. Initial studies suggested that the leucine zipper of the bZip protein cJun interacted directly with the bHLH domain of MyoD to suppress myogenic activity (Bengal, 1992 1 /id).

3.2 Regulation of MyoD During Terminal Growth Arrest and Differentiation of Myoblasts

3.2.1 MyoD and Rb Functional Interaction

The functional interaction between MyoD and Rb during muscle differentiation suggests that they act in the same pathway governing cell cycle arrest and muscle differentiation. Rb is drastically induced during terminal differentiation of skeletal muscle cells [31] suggesting the accumulation of Rb protein is likely to be required for growth arrest during differentiation of mitogenic cells.

One of the early events that occurs during differentiation is the accumulation of Rb in the hypo-phosphorylated form. Hypo-phosphorylated Rb is correlated with lack of cell growth, repression of genes involved in DNA replication, and the differentiated state of a variety of cell types. Indeed, the hypo-phosphorylated form of Rb in association with E2F ensures myoblast G1 arrest. Functional MyoD, liberated from cdk inhibition upon the reduction in cyclin D1 levels in response to decreased mitogen signaling, activates myogenesis as well as genes that maintain cell cycle exit, such as p21 and Rb.

MyoD induces Rb expression as early as 48 h after induction of muscle differentiation [32], reinforcing the Rb-mediating inhibition of the G1–S phase transition. The fact that hypo-phosphorylated pRb accumulates upon muscle differentiation, exceeding the levels of E2F, suggests that pRb might have other targets during the myogenic program. It has been reported that MyoD directly binds hypo-phosphorylated Rb both in vitro and in vivo [33]. This result suggested that a direct MyoD/Rb interaction blocked Rb phosphorylation by cdks during the early stages of muscle differentiation and led to the suppression of myoblast cell growth, cell cycle exit, and differentiation.

Thus, in differentiating myocytes, MyoD-mediated expression of the cdk inhibitor p21 inhibits the residual cyclin/cdk activity [34] and prevents the formation of E2F complexes containing the kinase cyclin E/cdk2 and cyclin A/cdk2

[35], which are possibly involved in the initiation of DNA synthesis. In addition, induction of the cdk inhibitors p18 and p21 has been observed to couple cell cycle arrest to myogenic differentiation [36]. The positive feedback loop is then accomplished by the hypo-phosphorylated form of pRb that co-activates MyoD resulting in further upregulation of expression of p21 and of genes necessary for continued differentiation.

The bHLH MyoD protein can also directly inhibit cyclin kinase activity, which would in turn lead to an increase in the hypo-phosphorylated form of Rb and terminal cell cycle arrest of myoblasts. Indeed, MyoD interacts with the kinase cdk4 through a conserved 15 amino acid domain in the C-terminus of MyoD [26]. Surprisingly, full MyoD, the C-terminus beyond the bHLH region, or just the 15 amino acid cdk4-binding domain inhibit cdk4 phosphorylation of an Rb target in vitro, whereas myogenin does not. Expression of only the MyoD 15 amino acid cdk4-binding domain in vivo fused either to GFP or GST with an NLS, blocks Rb phosphorylation, inhibits cell growth and promotes differentiation of C2C12 cells in growth medium.

Unlike other cyclin/cdk complexes, cyclin T2a/cdk9 levels and activity are not downregulated during muscle differentiation [37]. Furthermore, MyoD-mediated recruitment of cyclin T2/cdk9 on muscle-specific regulatory regions activates transcription, whereas inhibition of cdk9 kinase activity prevents the activation of the myogenic program [38, 39]. Intriguingly, cdk9 is able to phosphorylate pRb in vitro, and this kinase activity peaks at 96 h after the induction of the differentiation program in C2C12 cells [40–42]. Phosphopeptide analysis of p56/pRb after phosphorylation by cdk9, compared to that mediated by cdk2 and cdk1, indicates that, at least in vitro, the three cdks share several target phosphosites, but cdk9 affects only serine residues [41]. By studying the composition of cyclin T/cdk9 complexes, we identified pRb as a cdk9-interacting protein in murine C2C12 myoblasts, in murine NIH3T3 fibroblasts, and in two human tumor cell lines, HeLa and Jurkat. Cyclin T2/cdk9 binds to pRb, involving residues 129–195 of cdk9, the first 642 amino acids of cycT2 and the C-terminal region of the RB protein [835–928] and phosphorylates the pRb region spanning amino acids 793–834. This region contains at least three proline-directed serines (sp), S795, S807, and S811, which have been reported to be phosphorylated in vivo and which could be targeted by the cdk9 complex [39].

3.2.2 MyoD and Chromatin Remodeling Factors

Transcription is achieved by the MyoD binding to its cognate sequences (E-box, CANNTG), and consequent recruitment of HATs, SWI/SNF chromatin-remodeling complexes, and polymerase II-activating kinases [38, 43]. When differentiation is induced, cAMP responsive element binding protein (CREB) transcription factor is upregulated and phosphorylated on serine 133, a critical residue for CREB-CBP/p300 association and transcription activation. After that step, CREB is able to recruit a multiprotein complex containing MyoD and the HATs p300 and PCAF on RB promoter to induce gene expression [44]. The engagement of HATs draws attention to the balance between acetylation and deacetylation of histone and non-histone proteins that controls gene expression in a variety of cellular processes, with transcription being activated by HATs and silenced by HDAC. Furthermore it has been demonstrated that MyoD is acetylated by p300/CBP or pCAF on two lysines located at the boundary of the MyoD DNA-binding domain, and this leads to an increase in MyoD transactivation activity. However, it now appears that p300 acts as a scaffold for HAT, P/CAF, and it is the HAT activity of P/CAF that is important for the myogenic activity of MyoD [45, 46]. Injection of P/CAF antibodies into C2C12 muscle cells also inhibited differentiation. Only deletions in P/CAF that remove HAT activity blocked myogenesis but HAT deletions in p300 had no effect. P/CAF also acetylates MyoD in vitro and replacement of the acetylated lysine residues with arginines reduced reporter activation substantially but this had little effect on DNA binding in vitro with or without E12 [46]. The disparity between DNA binding and reporter activation suggests the acetylated domain adjacent to the MyoD basic region may interact with additional transcriptional activators yet to be identified. These studies indicate that p300–pCAF can facilitate MyoD activity through direct acetylation.

It is still not known if all Rbs in differentiated myotubes are inactive or in association with the histone deacetylase transcriptional repressors, such as HDAC1 [47]. The physical interaction between the hypo-phosphorylated form of pRb and HDAC1 in growth-arrested cells has suggested an additional mechanism of cooperation between pRb and MyoD in the absence of a direct interaction. In fact, once hypo-phosphorylated in response to cdk inactivation by serum withdrawal, pRb disassembles HDAC1 from MyoD, thereby removing its inhibitory effect on myogenic transcription [24]. As a consequence, the HDAC1–pRb complex in myotubes can both allow accumulation of hyperacetylated MyoD and block the E2F-dependent transcription of genes, which antagonizes myogenesis. Thus, the hypo-phosphorylated form of pRb can switch HDAC1 from being a direct repressor to an indirect activator of muscle-specific transcription.

3.3 Myogenesis and the CKIs

Progression through the G1 phase of the cell cycle is closely modulated by a family of cdks, cdk4, and cdk6, whose activities are in turn constrained by a group of proteins known as the CKIs (see review, [48]).

The CKIs are divided into two groups based upon their structure and cdk targets. The first group is known as the INK4 proteins since they bind specifically to and inhibit the catalytic subunit of cdk4 and cdk6 and include p15, p16, p18, and p19.

The second group, the cip/kip family, binds to the cyclin D-, A- and E-dependent kinases via both the cyclin and the catalytic subunits and include p21, p27, and p57. It has been shown in transfection studies that MyoD can upregulate the promoter for the CKI p21, implying similar regulation during cell cycle exit in myogenesis [49].

However, mice lacking both MyoD and myogenin genes have normal expression patterns of p21 and mice lacking the p21 gene develop normally, suggesting that the role of p21 in myogenesis is either redundant or minor {Deng, 1995 2 /id}. Interestingly, mice lacking both p21 and p57 fail to form myotubes and myoblasts show increased proliferation and apoptosis, suggesting that p21 and p57 redundantly play a role in skeletal muscle terminal cell cycle withdrawal [50]. p57 is predominantly expressed in differentiated tissues, and the 7 kb mRNA for this inhibitor is only detectable in skeletal muscle and heart, so its importance in cdk regulation in muscle may be dominant compared with p21 [50, 51]. Unexpectedly, the p21 and p27 CKIs are also essential activators of cyclin D-dependent kinases in mouse fibroblast. p21/p27 remain associated with cyclin D–cdk4 in an active kinase complex and are liberated later in the cell cycle to inhibit cyclin E/cdk2. Both p21 and p27 are required for the assembly and nuclear import of cyclin D1/cdk4 and their participation in this process is thought to provide the nuclear import signal that is lacking in the kinase. The calcium-binding protein calmodulin is also a key regulator of the cell cycle [52]. Recent studies have shown that calmodulin is essential for cdk4 activity and the nuclear accumulation of cyclin D1–cdk4 complexes during G1 [52, 53] and that this accumulation likely involves Hsp90.

3.4 Differentiated Muscle Cells

Myotubes cannot be induced to reenter S phase in response to growth factor stimulation. Although the mechanisms preventing myotubes from reinitiating DNA synthesis are still unclear, there are several studies that indicate that Rb, p21, and cyclinD3 are involved. However, a large body of evidence suggests that Rb and p21 also function to prohibit the replication of DNA in differentiated muscle cells. For example, myocytes lacking p21 can synthesize DNA indicating that p21 is also involved in preventing DNA synthesis in differentiated muscle cells [52, 54].

The expression of cyclins in muscle cells is downregulated at the onset of terminal differentiation, as cells arrest in the G0/G1 phase of the cell cycle [55], with the exception of cyclin D3, whose expression is induced during terminal differentiation . By using a hormone-activated MyoD cell line, cyclin D3 was induced as well as stabilized following MyoD expression. Moreover, cyclin D3 is found in inactive complexes with cdk4, cdk2, and PCNA [32]. Therefore, the mechanism by which cyclin D3 prevents differentiated muscle cells from reinitiating DNA synthesis is likely due to the ability of cyclin D3 to trap proliferating cell nuclear antigen (PCNA), a positive regulator for DNA synthesis, into inactive complexes with cdk4, cdk2, and Rb.

3.5 Differentiation Checkpoint

Recently it has been demonstrated that the differentiation checkpoint mechanism to be independent of the cell cycle and indeed to become activated in cells that have already entered into the early differentiation program [56].

For many cell types, arrest of proliferation induces the differentiation program in an irreversible manner. Therefore, growth arrest of precursor cells due to DNA damage could conceivably cause inappropriate differentiation prior to DNA repair, leading to the formation of abnormal, non-functional tissues. This differentiation of damaged precursor cells would be avoided if the checkpoint-induced growth arrest occurred at points within the cell cycle that are incompatible with the differentiation process. For example, for many cells types, differentiation can only occur when the cells are in early G1/G0 phase and will not be induced by growth arrest in G2/M. However, some of the cell cycle checkpoint may coincide with the exit into differentiation. In this case, multicellular organisms will need an additional filtering system, a "differentiation checkpoint," which will not allow cells with massive mutations to differentiate and form tissues.

They demonstrate that while various genotoxic agents arrest myoblasts at distinct points of cell cycle progression, they are not sufficient to promote differentiation. On the contrary, treatment with genotoxic agents significantly decreases the efficiency of myogenic differentiation. This observation clearly indicates the existence of two distinct checkpoints that regulate the fate of myogenic cells: a first checkpoint that blocks cell cycle progression in response to DNA damage and a second checkpoint that blocks terminal differentiation.

The differentiation checkpoint requires phosphorylation of MyoD at Tyr30, a modification that may change the transcriptional activity of this key myogenic factor. The kinase that mediates the activation of the differentiation checkpoint and phosphorylates MyoD protein was identified as the nuclear tyrosine kinase c-Abl. It should be noted that c-Abl is a known mediator of DNA damage checkpoints and promotes apoptosis in proliferating cells in response to DNA damage [57]. It appears that the differentiation checkpoint is MyoD dependent and is either activated by the cell cycle checkpoint or is independently induced by DNA damage.

However, the growth arrest of myoblasts, caused by genotoxic agents, induces expression of p53, whereas the differentiation check point itself is p53 independent. Moreover, the c-Abl kinase, which is typically inactive in the early G1/G0 cells, in the presence of active Rb [57] appears to be the key regulator of the myogenic differentiation checkpoint.

4 Cell Cycle Deregulation in Muscle Cells

Permanent withdrawal from the cell cycle is a crucial event during terminal differentiation. Dysfunction of either cell cycle control or differentiation machinery is responsible for deregulated growth and transformed phenotype [58]. Control of G1/S transition is regulated by a set of specific CDK and cyclin complexes, sequentially expressed, activated, and degraded to ensure both entry and progress in the cell cycle [58] (see above). In large part, the cyclin/CDK complexes are needed to phosphorylate pRb, which in turn releases E2F and leads to the transcription of growth regulating genes such as cyclin A {Reed, 1997 3 /id}. p21WAF1, a cyclin-dependent kinase inhibitor (CKI), which inhibits all cyclin/CDK complexes, particularly those in the G1 phase, has been found to be associated with the growth arrest of both normal and malignant cells [48]. Enhanced p21WAF1 mRNA expression occurs through both p53-dependent and -independent mechanisms [59] and as a result of mRNA and protein stabilization induced in a number of different cell lines and signal transduction mechanisms [60].

Rhabdomyosarcoma (RMS), the most common soft-tissue sarcoma arising from undifferentiated mesenchymal cells bearing developing skeletal muscle features consists of several subtypes, with ERMS, the embryonal subtype, and ARMS, the alveolar subtype, being among the most frequent tumors in children [61]. RMS presents a number of genetic alterations which define the embryonal [61] and the alveolar subtype [62]. These different subtypes also share molecular changes, including disruption of the p53 pathway through mutation or MDM2 amplification and deregulation of imprinted genes at the chromosome region 11p15.5 [63].

The established RD cell line, originating from the ERMS tumor, is one of the most representative models of pathological myogenesis. RD cells fail to control cell cycle mechanisms [64] and differentiation progress in spite of the expression of the myogenic-specific transcription factors MyoD and myogenin, which are transcriptionally inactive despite apparently being able to bind DNA [61]. MyoD and myogenin, when ectopically expressed in RD cells, do not induce muscle differentiation, even in the presence of cyclin-dependent kinase inhibitors (CKIs) or myogenic co-factors[65], while ectopic expression of MRF4, which is undetectable in RD, induces exit from the cell cycle and myogenic differentiation, both of which are enhanced in the presence of CKIs.

The MEK/ERK pathways control the growth and survival of a broad spectrum of human tumors [66] and have also been involved in differentiation [67]. Indeed, a role of the MEK/ERK pathway in growth inhibition has been reported to be dependent upon whether activation is acute or chronic [68]. Although ERKs are constitutively activated in tumor growth and are involved in the induction of proliferation, a high p38 level is believed to be a negative regulator [6]. Furthermore, the ERK and p38 pathways have recently been reported to cooperate to cause sustained G1 cell cycle arrest requiring p21WAF1 expression [6, 69].

Recently it has been demonstrated that the mechanism of ERK mediated and ERK-independent growth arrest and myogenic differentiation in RD cells, particularly with regard to the expression of proteins involved in cell cycle control, such as p21WAF1. p21WAF1 expression is post-transcriptionally regulated by TPA-mediated MEK/ERK activation, but transcriptionally induced by MEK/ERK inhibition and p38 activation.

In this study they highlight the importance of targeting the MEK/ERK pathway as a means of restoring the expression of the tumor suppressor p21WAF1 as well as the growth arrest mechanism. The results of this study suggest that the targeting of ERKs to rescue p21WAF1 expression and myogenic transcription factor functions leads to the reversal of the Rhabdomyosarcoma phenotype [70].

References

1. Gu W, Schneider JW, Condorelli G, Kaushal S, Mahdavi V, Nadalginard B (1993) Interaction of myogenic factors and the retinoblastoma protein mediates muscle-cell commitment and differentiation. Cell 72(3):309–324
2. Sabourin LA, Rudnicki MA (2000) The molecular regulation of myogenesis. Clin Genet 57(1):16–25
3. Lassar AB, Davis RL, Wright WE, Kadesch T, Murre C, Voronova A et al (1991) Functional-activity of myogenic Hlh proteins requires hetero-oligomerization with E12/E47-like proteins invivo. Cell 66(2):305–315
4. Shirakata M, Friedman FK, Wei Q, Paterson BM (1993) Dimerization Specificity of myogenic helix-loop-helix DNA-binding factors directed by nonconserved hydrophilic residues. Genes Development 7(12A):2456–2470
5. Lindon C, Montarras D, Pinset C (1998) Cell cycle-regulated expression of the muscle determination factor Myf5 in proliferating myoblasts. J Cell Biol 140(1):111–118
6. Aguirre-Ghiso JA, Estrada Y, Liu D, Ossowski L (2003) ERKMAPK activity as a determinant of tumor growth and dormancy; regulation by p38(SAPK) Cancer Res 63(7):1684–1695
7. Evan GI, Wyllie AH, Gilbert CS et al (1993) Hot papers - biochemistry - induction of apoptosis in fibroblasts by C-Myc protein. Scientist 7(14):15
8. Geng Y, Yu QY, Whoriskey W, Dick F, Tsai KY, Ford HL et al (2001) Expression of cyclins E1 and E2 during mouse development and in neoplasia. Proc Natl Acad Sci U S Am 98(23):13138–13143
9. Jackman M, Firth M, Pines J (1995) Human cyclins B1 and B2 are localized to strikingly different structures - B1 to microtubules, B2 primarily to the Golgi-apparatus. Embo J 14(8):1646–1654
10. Stacey DW, Gu Y, Yang K, Hitomi M (2004) Image analysis reveals cyclin D1 role in cell cycle control. Cytometry Part A 59A(1):70
11. Lees E, Tsai LH, Faha BF, Riabowol K, Harlow E (1993) Cyclin-e displays properties appropriate for A G1-cyclin. J Cell Biochem 1993:294
12. Lundberg AS, Weinberg RA (1999) Control of the cell cycle and apoptosis. Eur J Cancer 35(4):531–539
13. Weinberg RA (1995) The molecular-basis of oncogenes and tumor-suppressor genes. Ann NY Acad Sci 758:331–338
14. Giacinti C, Giordano A (2006) RB and cell cycle progression. Oncogene 25(38):5220–5227
15. Luo RX, Postigo AA, Dean DC (1998) Rb interacts with histone deacetylase to repress transcription. Cell 92(4):463–473
16. Dahiya A, Wong S, Gonzalo S, Gavin M, Dean DC (2001) Linking the Rb and polycomb pathways. Mol Cell 8(3):557–568
17. Nielsen SJ, Schneider R, Bauer UM, Bannister AJ, Morrison A, O'Carroll D et al (2001) Rb targets histone H3 methylation and HP1 to promoters. Nature 412(6846):561–565
18. Adams PD, Kaelin WG (1995) Transcriptional control by E2F. Sem Cancer Biol 6(2):99–108
19. Sherr CJ, Roberts JM (1995) Inhibitors of mammalian G(1) cyclin-dependent kinases. Genes Dev 9(10):1149–1163
20. Matsushime H, Roussel MF, Ashmun RA, Sherr CJ (1991) Colony-stimulating factor-I regulates novel cyclins during the G1 phase of the cell-cycle. Cell 65(4):701–713
21. Resnitzky D, Gossen M, Bujard H, Reed SI (1994) Acceleration of the G(1)/S phase-transition by expression of cyclin-D1 and cyclin-e with an inducible system. Mol Cell Biol 14(3): 1669–1679
22. Neuhold LA, Wold B (1999) Myod tethered to Itf-1 activates a muscle-specific enhancer in undifferentiated myoblast. J Cell Biochem 1993:203
23. Benezra R, Davis RL, Lassar A, Tapscott S, Thayer M, Lockshon D et al (1990) Id - a negative regulator of helix-loop-helix dna-binding proteins - control of terminal myogenic differentiation. Ann NY Acad Sci 599:1–11

24. Puri PL, Sartorelli V, Yang XJ, Hamamori Y, Ogryzko VV, Howard BH et al (1997) Differential roles of p300 and PCAF acetyltransferases in muscle differentiation. Mol Cell 1(1):35–45
25. Li L, James G, Hellerharrison R, Czech MP, Olson EN (1992) Fgf inactivates myogenic helix-loop-helix proteins through phosphorylation of a conserved protein-kinase-C site in their dna-binding domains. Cell 71(7):1181–1194
26. Zhang JM, Chen LS, Krause M, Fire A, Paterson BM (1999) Evolutionary conservation of MyoD function and differential utilization of E proteins. Dev Biol 208(2):465–472
27. Skapek SX, Rhee J, Spicer DB, Lassar AB (1994) Inhibition of myogenic differentiation in proliferating myoblasts by cyclin D1-dependent kinase. Science 267(5200):1022–1024
28. Diehl JA, Sherr CJ (1997) A dominant-negative cyclin D1 mutant prevents nuclear import of cyclin-dependent kinase 4 (CDK4) and its phosphorylation by CDK-activating kinase. Molecular and Cellular Biology 17(12):7362–7374
29. Miner JH, Wold BJ (1991) C-Myc inhibition of myod and myogenin-initiated myogenic differentiation. Mol Cell Biol 11(5):2842–2851
30. Herschman HR (1991) Primary response genes induced by growth-factors and tumor promoters. Ann Rev Biochem 60:281–319
31. Endo T, Goto S (1992) Retinoblastoma gene-product Rb accumulates during myogenic differentiation and is deinduced by the expression of Sv40 large T-antigen. J Biochem 112(4):427–430
32. Cenciarelli C, De Santa F, Puri PL, Mattei E, Ricci L, Bucci F et al (1999) Critical role played by cyclin D3 in the MyoD-mediated arrest of cell cycle during myoblast differentiation. Mol Cell Biol 19(7):5203–5217
33. Gu W, Schneider J, Mahdavi V, Nadalginard B (1993) Down-regulation of tumor-suppressor pocket proteins reverses the terminally differentiated state of muscle-cells. Circulation 88(4):190
34. Walsh K, Perlman H (1997) Cell cycle exit upon myogenic differentiation. Cur Opin Gen Dev 7(5):597–602
35. Puri PL, Balsano C, Burgio VL, Chirillo P, Natoli G, Ricci L et al (1997) MyoD prevents cyclin A cdk2 containing E2F complexes formation in terminally differentiated myocytes. Oncogene 14(10):1171–1184
36. Franklin DS, Xiong Y (1996) Induction of p18(INK4c) and its predominant association with CDK4 and CDK6 during myogenic differentiation. Mol Biol Cell 7(10):1587–1599
37. Simone C, Giordano A (2001) New insight in Cdk9 function: from Tat to MyoD. Front Biosci 6:D1074–D1082
38. Giacinti C, Bagella L, Puri PL, Giordano A, Simone C (2006) MyoD recruits the cdk9/cyclin T2 complex on myogenic-genes regulatory regions. J Cell Physiol 206(3):807–813
39. Simone C, Stiegler P, Bagella L, Pucci B, Bellan C, De Falco G et al (2002) Activation of MyoD-dependent transcription by cdk9/cyclin T2. Oncogene 21(26):4137–4148
40. Bagella L, MacLachlan TK, Buono RJ, Pisano MM, Giordano A, De Luca A (1998) Cloning of murine CDK9/PITALRE and its tissue-specific expression in development. J Cell Physiol 177(2):206–213
41. De Luca A, Tosolini A, Russo P, Severino A, Baldi A, de Luca L et al (2001) Cyclin T2A gene maps on human chromosome 2q21. J Histochem Cytochem 49(6):693–697
42. Grana X, Deluca A, Sang N, Fu Y, Claudio PP, Rosenblatt J et al (1994) Pitalre, a nuclear Cdc2-related protein-kinase that phosphorylates the retinoblastoma protein in-vitro. Proc Natl Acad Sci U S A 91(9):3834–3838
43. Simone C, Forcales SV, Hill DA, Imbalzano AN, Latella L, Puri PL (2004) p38 pathway targets SWI-SNF chromatin-remodeling complex to muscle-specific loci. Nat Gen 36(7): 738–743
44. Magenta A, Cenciarelli C, De Santa F, Fuschi P, Martelli F, Caruso M et al (2003) MyoD stimulates RB promoter activity via the CREB/p300 nuclear transduction pathway. Mol Cell Biol 23(8):2893–2906

45. Puri PL, Avantaggiati ML, Balsano C, Sang NL, Graessmann A, Giordano A et al (1997) P300 is required for MyoD-dependent cell cycle arrest and muscle-specific gene transcription. Embo J16(2):369–383
46. Sartorelli V, Puri PL, Hamamori Y, Ogryzko V, Chung G, Nakatani Y et al (1999) Acetylation of MyoD directed by PCAF is necessary for the execution of the muscle program. Mol Cell 4(5):725–734
47. Brehm A, Miska EA, McCance DJ, Reid JL, Bannister AJ, Kouzarides T (1998) Retinoblastoma protein recruits histone deacetylase to repress transcription. Nature 391(6667):597–601
48. Sherr CJ, Roberts JM (1999) CDK inhibitors: positive and negative regulators of G(1)-phase progression. Genes Dev 13(12):1501–1512
49. Halevy O, Novitch BG, Spicer DB, Skapek SX, Rhee J, Hannon GJ et al (1995) Correlation of terminal cell-cycle arrest of skeletal-muscle with induction of P21 by Myod. Science 267(5200):1018–1021
50. Zhang PM, Wong C, Liu D, Finegold M, Harper JW, Elledge SJ (1999) p21(CIP1) and p57(KIP2) control muscle differentiation at the myogenin step. Genes Dev 13(2):213–224
51. Lee MH, Reynisdottir I, Massague J (1995) Cloning Or P57(Kip2), a cyclin-dependent kinase inhibitor with unique domain-structure and tissue distribution. Genes Dev 9(6):639–649
52. Lu KP, Means AR (1993) Regulation of the cell-cycle by calcium and calmodulin. Endocr Rev 14(1):40–58
53. Taules M, Rius E, Talaya D, Lopez-Girona A, Bachs O, Agell N (1998) Calmodulin is essential for cyclin-dependent kinase 4 (Cdk4) activity and nuclear accumulation of cyclin D1-Cdk4 during G(1) J Biol Chem 273(50):33279–33286
54. Andres V, Walsh K (1996) Myogenin expression, cell cycle withdrawal, and phenotypic differentiation are temporally separable events that precede cell fusion upon myogenesis. J Cell Biol 132(4):657–666
55. Kiess M, Gill RM, Hamel PA (1995) Expression and activity of the retinoblastoma protein (Prb)-family proteins, P107 and P130, during L(6) myoblast differentiation. Cell Growth Differ 6(10):1287–1298
56. Puri PL, Bhakta K, Wood LD, Costanzo A, Zhu JY, Wang JYJ (2002) A myogenic differentiation checkpoint activated by genotoxic stress. Nat Genet 32(4):585–593
57. Liu ZG, Baskaran R, LeaChou ET, Wood LD, Chen Y, Karin M et al (1996) Three distinct signalling responses by murine fibroblasts to genotoxic stress. Nature 384(6606):273–276
58. Deshpande A, Sicinski P, Hinds PW (2005) Cyclins and cdks in development and cancer: a perspective. Oncogene 24(17):2909–2915
59. Gartel AL, Goufman E, Tevosian SG, Shih H, Yee AS, Tyner AL (1998) Activation and repression of p21(WAF1/CIP1) transcription by RB binding proteins. Oncogene 17(26):3463–3469
60. Ostrovsky O, Bengal E (2003) The mitogen-activated protein kinase cascade promotes myoblast cell survival by stabilizing the cyclin-dependent kinase inhibitor, p21(WAF1) protein. J Biol Chem 278(23):21221–21231
61. Tapscott SJ, Thayer MJ, Weintraub H (1993) Deficiency in rhabdomyosarcomas of a factor required for myod activity and myogenesis. Science 259(5100):1450–1453
62. Barr FG (2001) Gene fusions involving PAX and FOX family members in alveolar rhabdomyosarcoma. Oncogene 20(40):5736–5746
63. Keleti J, Quezado MM, Abaza MM, Raffeld M, Tsokos M (1996) The MDM2 oncoprotein is overexpressed in rhabdomyosarcoma cell lines and stabilizes wild-type p53 protein. Am J Pathol 149(1):143–151
64. Knudsen ES, Pazzagli C, Born TL, Bertolaet BL, Knudsen KE, Arden KC et al (1998) Elevated cyclins and cyclin-dependent kinase activity in the rhabdomyosarcoma cell line RD1. Cancer Res 58(9):2042–2049
65. Otten AD, Firpo EJ, Gerber AN, Brody LL, Roberts JM, Tapscott SJ (1997) Inactivation of MyoD-mediated expression of p21 in tumor cell lines. Cell Growth Differ 8(11):1151–1160

66. Hoshino R, Chatani Y, Yamori T, Tsuruo T, Oka H, Yoshida O et al (1999) Constitutive activation of the 41-/43-kDa mitogen-activated protein kinase signaling pathway in human tumors. Oncogene 18(3):813–822
67. Bennett AM, Tonks NK (1997) Regulation of distinct stages of skeletal muscle differentiation by mitogen-activated protein kinases. Science 278(5341):1288–1291
68. Tombes RM, Auer KL, Mikkelsen R, Valerie K, Wymann MP, Marshall CJ et al (1998) The mitogen-activated protein (MAP) kinase cascade can either stimulate or inhibit DNA synthesis in primary cultures of rat hepatocytes depending upon whether its activation is acute/phasic or chronic. Biochem J 330:1451–1460
69. Todd DE, Densham RM, Molton SA, Balmanno K, Newson C, Weston CR et al (2004) ERK1/2 and p38 cooperate to induce a p21(CIP1)-dependent G(1) cell cycle arrest. Oncogene 23(19):3284–3295
70. Ciccarelli C, Marampon F, Scoglio A, Mauro A, Giacinti C, De Cesaris P et al (2005) p21(WAF1) Expression induced by MEK/ERK pathway activation or inhibition correlates with growth arrest, myogenic differentiation and onco-phenotype reversal in rhabdomyosarcoma cells. Mol Cancer 4

Subject Index

A. Giordano, U. Galderisi (eds.), *Cell Cycle Regulation and Differentiation in Cardiovascular and Neural Systems,* DOI 10.1007/978-1-60327-153-0,

D

E

MIX
Papier aus verantwortungsvollen Quellen
Paper from responsible sources
FSC® C105338

If you have any concerns about our products,
you can contact us on
ProductSafety@springernature.com

In case Publisher is established outside the EU,
the EU authorized representative is:
Springer Nature Customer Service Center GmbH
Europaplatz 3, 69115 Heidelberg, Germany

Printed by Libri Plureos GmbH
in Hamburg, Germany